Vibration of Mechanical Systems

C. Nataraj

Villanova University

CENGAGE
Learning·

Andover • Melbourne • Mexico City • Stamford, CT • Toronto • Hong Kong • New Delhi • Seoul • Singapore • Tokyo

CENGAGE
Learning·

**Vibration of Mechanical Systems,
1st Edition**

Author: C. Nataraj

For product information and technology assistance,
contact **emea.info@cengage.com**.

For permission to use material from this text or product,
and for permission queries,
email **emea.permissions@cengage.com**.

British Library Cataloguing-in-Publication Data
A catalogue record for this book is available from the British Library.

ISBN: 9781408072653

Cengage Learning EMEA
Cheriton House, North Way, Andover, Hampshire, SP10 5BE
United Kingdom

Cengage Learning products are represented in Canada by Nelson Education Ltd.

For your lifelong learning solutions, visit
www.cengage.co.uk

Purchase your next print book, e-book or e-chapter at
www.cengagebrain.com

Printed in the UK by Lightning Source
1 2 3 4 5 6 7 8 9 10 – 13 12 11

This book is dedicated to

My Dear Father, A. S. Chandrasekhar

who taught me the value of life-long learning
and is still an inspiration to me at 92

Contents

Dear Student: Please Read This

Have no fear of perfection. You'll never reach it[1].

—Salvador Felipe Jacinto Dalí

I have tried to use my light lecturing style in this book, and it is hence somewhat different from a traditional serious-minded tome. Hopefully, you will not interpret this light approach to a heavy subject to mean that the subject is not important. In fact, nothing could be further from the truth: from tiny molecular vibrations to huge oscillating binary stars, vibrations comprise a fundamental phenomenon that governs the universe. Of course, in this book, most of the time we won't discuss all that; we will just stick to vibrations in the macroscopic world and on Mother Earth. But it is still good to know that the principles you learn here come in very useful if you decide to pursue, say, quantum mechanics or astronomy or something in between.

Textbooks are of course used by instructors as well as students; and *you* know that professors are quite different from you! It is quite difficult to make a book appealing to both the students and the teachers; given this choice, I decided to come down on the side of the students in this book. In other words, I have tried to think more of the students' approach to learning than of the instructor's approach to teaching. To emphasize, *this book is student-focused.* That is because I want you to read it, reread it and revel in this wonderful topic. The world of vibrations, which I was introduced to about 30 years ago, has fascinated me ever since, and I am writing this book because I want you to share that fascination.

So, what is the subject of vibrations all about? It is the study of mechanical systems that usually have oscillatory motion. Hence, it is essentially a branch of dynamics. You will also quickly discover that dynamics leads to differential equations. Some of these differential equations can be solved analytically and some cannot be. The latter would have to be dealt with using numerical techniques. So, we are interested in looking at mechanical systems that vibrate, we want to model them using principles of dynamics, and finally we would like to solve the differential equations to predict how these systems might behave in practice.

The above paragraph then tells us that we need the following subjects and concepts to get the most from this course.

- **Dynamics.** You probably still have nightmares from this course (professors probably used the word "challenging" and "fascinating" to describe dynamics—that is one of those euphemisms for

---

[1] Just kidding!

"really tough!"). Don't shudder yet! You may have a pleasant surprise at the end of Chapter 2, which discusses pretty much all you need to study vibrations.

- **Mechanics of Materials.** Vibration usually happens because of elasticity in structures and systems, and elastic deformation is pretty much what you studied in your mechanics of materials books. Chances are—this book was not as much of a nightmare as dynamics, but, you do need to remember a few things from there. A good text might help you quickly refresh yourself on this topic.

- **Differential Equations.** Unfortunately, there is no getting away from this, since they will turn up in every problem of every chapter in this book. The sooner you learn (or relearn) this stuff, the better you are going to be able to understand all the good things in this course.

- **Numerical Methods.** We are going to use MATLAB^{TM} to carry out all numerical studies. However, you are free to use many programming tools that exist today, all of which can do the job quite well. That makes the whole process less burdensome, although you will have to acquaint yourself with the computational tools. Once you learn it however, you will find that they will serve you well in other courses as well as in a working engineering environment.

If you are not a fan of mathematics and do not believe in doing it just for its sake (not too many of those kinds of people around!), you should probably try to master the fairly elementary mathematics quickly so that you can get to the fun stuff as soon as possible. If, on the other hand, you keep avoiding the mathematics, you will never be able to get the best value and enjoyment out of this material, and concepts that start out being murky will stay murky.

The point is this: even though we are going to talk equations *ad nauseum*, the objective in this course is *not* mathematics; it is the engineering principles of analyzing and designing vibrating systems. And *that* is what is going to be most interesting to a budding Mechanical Engineer like yourself.

I would like to encourage you strongly to make an effort to go beyond the basic material. To help in that process, I have what I call "think boxes" interspersed throughout the book; they are marked by a special symbol.[2] These are additional questions that should make you think. Many of them do not have simple closed-form answers, and a single person may not be able to answer them; in fact, if they spark a discussion (in the class room or outside), they would have served their purpose. In addition, at many places, and especially in problems, I have given links to papers and articles that will serve as jump-off points, and will help open up the practical and the research world to you. If you have access to a good library (many of them are also accessible on the World Wide Web), do follow the links and see where that takes you.

Among the problems at the end of the chapter you will occasionally notice symbols like ⚿ . One of these symbols (signifying a curvy road) means the problem is somewhat challenging, and you should watch out. Two of these symbols means the problem is going to give your grey cells a real workout (like the cases Sherlock Holmes would describe as a "three-pipe" problem!).

You should probably ask me what gives me the right to write this book and to foist it on you. The answer is that, in addition to pursuing research in the areas of rotor dynamics and vibrations for 30 years, I have taught vibrations and rotor dynamics to undergraduates and graduate students for

---

2

24 years. It has truly been a passion for me as the thousands of students who have gone through my classes can testify. (It is perhaps beside the point that many of them would compare the experience to a medieval torture chamber!) This book is somewhat of a culmination of that long teaching process.

Finally, I have listed information on getting some feedback to me on page xxi. Use it and let me know what I could have done better; I would love to have you catch me out on mistakes!

To the Instructor

गुरू ब्रह्मा गुरू विष्णु गुरू देवो महेश्वरा
गुरू साक्षात परम ब्रह्मा तस्मयी श्री गुरवे नमः [1]

If you have read my statement to the students on the previous pages, you are probably wondering if you can teach from a student-centric textbook like this! While it is indeed true that I tried to make this book enjoyable and stimulating so that the students would be motivated to read it, the material is no less serious in its intent and purpose than any standard textbook.

I started to write this book around 1995 (that is fifteen years ago as of this edition) not because I wanted to add another one with my name to the existing panoply of textbooks in vibration. There are clearly some classic books in vibration including an early one by den Hartog, and later ones by Timoshenko and Meriovitch, which excited and influenced me in my own journey in this fascinating topic. However, there is no question that in the fast-changing engineering world, both systems and computational tools change, and the books have to change constantly to reflect that. Hence, although I love to curl up with den Hartog's book once in a while for his lucid explanation of the mysteries of dynamics and vibrations, it is doubtful that it could serve as a good contemporary textbook for modern times.

I also discovered that many contemporary textbooks (which are excellent in their own right) did not always connect vibration to dynamics in a fundamental way, as well as explore the design of vibrating systems in a way that suited my approach to engineering. Finally, and most importantly, I found that most students (even the ones who loved to come to my classes) hated to read the textbook with a passion! So, I set out to write something that the students would like (or at least, not *hate!*), and here is my version of what I believe we should be teaching budding Mechanical Engineers, and I hope you agree with me.

The intended level of this course is introductory either at the undergraduate level or the graduate level (for those who have not had any exposure to vibration theory). It starts with the basics of dynamics and explores single-degree-of-freedom (SDOF) and multi-degree-of-freedom (MDOF) systems. The coverage of continuous system modeling is admittedly perfunctory, and is meant to be more an introductory peek than a thorough analysis, which might be expected in a second and advanced course on vibrations. Some of the material not covered intentionally include random

---

[1] The teacher, "Guru" is the creator Brahma, Guru is the preserver Vishnu, Guru is the destroyer Shiva. Guru is directly the supreme spirit—I offer my salutations to this Guru. —Adi Shankara.

vibration, shock, practical vibration isolation materials and techniques, and dynamics of rotating machinery with the belief that it would be impossible to do justice to so many topics in a typical one-semester introductory course.

The book emphasizes the fact that even simple dynamic systems result in nonlinear equations, and there is continuous reference throughout the text to nonlinearities. This is a strong conviction of mine that we should not be sheltering the students from the fact that practically every engineering system is nonlinear. The evolution of cutting-edge industrial products now requires an engineer to understand the importance of nonlinearities in order to analyze and design systems that are optimal in performance, safety, and cost. Consider that an engineering student taking this course now would be potentially practicing engineering 30 years from now (2040!) when the engineering world could be far more sophisticated. In my opinion, preparing the students to remain relevant for that world would mean that we have to improve our students' modeling sophistication including the knowledge of nonlinearities. That said, even though the presence of nonlinearities is discussed, I decided that a discussion of nonlinear analysis techniques would be beyond the scope of an introductory course, and hence it is not included in the book. The typical approach in this book is therefore to linearize, or possibly to, use numerical simulation of nonlinear models.

We are going to use MATLAB^{TM} as a convenient illustration to carry out all numerical studies. However, given that it is a proprietary software, I recognize that it may not be suitable for all environments. The emphasis is really on the algorithms, and almost any programming tool that exists today would do the job quite well. Whether you use MATLAB or not, I would encourage using some computational environment; numerical analysis, in addition to making things more relevant to practice, also has a way of enhancing understanding with parametric analyses and graphical portrayal of results.

I have tried to integrate some design (or at least parametric analysis) in many problems. This is an effort to get the students to stretch their minds beyond simple "boxed answer" thinking that is common among undergraduate students. The ideas about parametric uncertainties and realistic constraints are critical to engineers, and should be imparted early in their studies. The fact that there is a fuzziness about real systems that we analyze and design is a very important one; not to be lost in the shuffle is the important fact that in spite of all this, the engineered systems *work*. The few problems explored here concern the design of some mechanical systems from the point of view of vibratory response. As you look through these problems, several things should be kept in mind. First, most design problems do not have unique solutions; i.e., many possible design configurations could fit the design specifications. On the other hand, the specifications can be so stringent that no solution may exist. Second, the design procedures as developed in this text have a necessarily narrow view of using vibration as the principal criterion. Most practical mechanical systems, however, cannot be boxed into any single course material and will have to be analyzed using concepts from such varied areas as dynamics, stress analysis, fluid mechanics, and heat transfer. It is good to transmit this *caveat* to the students.

In general, I would encourage the students to make an effort to go beyond the basic material. To help in that process, I have what I call "think boxes" interspersed throughout the book; they are marked by a special symbol.[2] These are additional questions that should make them think. Many of them do not have simple closed-form answers, and a single student may not be able to answer them; in fact, if they spark a discussion (in the class room or outside), they would have served their

---

2

purpose. In addition, at many places, and especially in problems, I have given links to papers and articles that will serve as jump-off points, and will help open up the practical and the research world to the students. If they have access to a good library (many of them are also accessible on the World Wide Web), I would encourage them to follow the links and see where that takes them. That kind of activity might also be more suitable for project groups as a team activity.

Among the problems at the end of the chapter, you will occasionally notice symbols like ⚠. One of these symbols (signifying a curvy road) means the problem is somewhat challenging and should be carefully considered before assigning it as a routine homework. Two of these symbols means the problem is going to be a real humdinger, and the students may need some hints or help (like the cases Sherlock Holmes would describe as a "three-pipe" problem!).

In addition to pursuing research in the areas of rotor dynamics and vibrations for 30 years, I have taught vibrations and rotor dynamics to undergraduates and graduate students for 24 years. It has truly been a passion for me as the thousands of students who have gone through my classes can testify. For many of those years, I have used earlier versions of this book in my classes. I have tuned the material in this text based on their feedback, and on my own observations of which material was easy, and which was particulary tricky.

In general, in a typical undergraduate course, it would be quite a challenge to teach all the material covered in this book. Given limited time, my suggestion would be to favor depth over breadth. That said, a reasonable course would cover dynamics (Chapter 2), SDOF undamped (Chapter 3) and damped (Chapter 4) vibrations, SDOF harmonic and periodic vibrations (Chapter 5), MDOF-free (Chapter 7) and harmonic vibrations (Chapter 8), experimental techniques (Chapter 10), and some design examples (interspersed through the text). Perhaps, in an introductory graduate book, or with an exceptional undergraduate student group, SDOF transient vibrations and transform methods (Chapter 6), and continuous systems (Chapter 9) could be covered as well.

To sensitize the student to the importance and subtleties of modeling I have a couple of running examples and problems. One concerns the vibration of a string; the second is the automobile suspension problem. These examples run through the text starting with simple undamped SDOF models and progressing through MDOF to continuous system models. Considerable numerical experimentation is possible when working with these problems to enhance student understanding.

Finally, I have listed information on getting your valuable feedback to me on page xxi. Please use it and let me know what I could have done better. Your input would be most appreciated and acknowledged.

Following is a typical course description including minimum outcomes for the purpose of accreditation. The material listed here is appropriate for approximately 45 hours of instruction. This text would conform to these outcomes.

- **Brief Description.** Free vibration of simple vibrating systems, harmonic excitation, transient vibration, MDOF systems, normal modes, modal analysis, applications.

  Prerequisites: Engineering dynamics, differential equations and linear algebra, some computer programming.
- **Course Objectives.** The main objective of this book is to introduce the student to design problems and solution techniques associated with the free and forced vibrations of various mechanical components. The book will deal with mathematical modeling techniques and the formulation of differential equations of motion for single, double, and multiple lumped mass systems. The emphasis throughout will be on computer solution techniques to reflect current industrial practices.

Topics Covered. The following topics are covered:

- Review of dynamics, fundamentals of vibration analysis, oscillatory motion.
- Free vibration with and without damping.
- Harmonic excitation.
- Periodic excitation.
- Transient vibration of single-DOF systems.
- MDOF systems, modes, coordinate coupling, principal coordinates.
- Forced vibration, modal analysis.
- Overview of experimental techniques.

- **Minimum Course Outcomes.** After successfully completing this book, at the very least the student should be able to do the following:

  - Derive equations of motion for particles.
  - Derive equations of motion for rigid bodies in a plane.
  - Solve linear one-DOF-free vibration problems.
  - Solve one-DOF harmonic excitation problems.
  - Solve simple rotating unbalance problems.
  - Solve vibration transmissibility problems.
  - Solve two-DOF-free vibration problems.
  - Set up and solve eigenvalue problems using a computer programming environment.

  Extra topics that could be covered include the following:

  - Laplace Transform methods of solution.
  - Transient vibration of MDOF systems.
  - Continuous system modeling and analysis.

Of course, as in all engineering courses, the material is best taught along with a laboratory course with suitable experiments, which would reinforce the theoretical concepts learned in class and inspire confidence in the mathematical formulations.

Acknowledgements

A good textbook (assuming that this one fits the description!) takes enormous efforts over a sustained period of time to write and bring to fruition. I certainly could not have done this without the inspiration, cooperation, and support from so many people. I will attempt to name some of them here, although it is inevitable that I will forget some.

My introduction to the area of vibration was at Indian Institute of Technology (Madras) by Professor V. Ramamurti, an animated and enthusiastic teacher whose lectures snared me into this fascinating area. His mentoring during my senior project in the area of torsional vibrations was exemplary in more things than technical. Later at Arizona State University, my courses in vibrations and mechanics from Professors Harold Nelson, Bill Bickford and C. E. Wallace were perhaps as good as—if not better than—any advanced courses I could have taken anywhere. The advanced knowledge, the stress on fundamentals, and the spirit of curiosity I picked up there are perhaps the principal driving forces for me even today. It was a privilege working on my research in rotor dynamics with Professor Harold Nelson, who, in his seemingly light-hearted way, insisted on unstinted accuracy, proper research methods, and an intellectual honesty that has shaped my own research.

I got started in my career with Professor Paul Robert Trumpler, an early and important contributor to rotor dynamics and lubrication theory. Although my association with him was unfortunately very short, he had enormous influence on my thinking and again reinforced all the critical things that I have come to value.

Later, at Villanova University, I have been fortunate to interact with so many of my colleagues (such as Hashem Ashrafiuon, James Peyton Jones, Jerry Jones, and Alan Whitman) in an intellectual atmosphere that fosters learning as well as teaching. I am glad to have spent the last two decades at a university where quality undergraduate teaching is encouraged and valued. In particular, I have to acknowledge my debt to Dean Barry Johnson for his strong support and encouragement and policy changes at Villanova that created an atmosphere to make such textbook projects possible at all. My thanks also to the undergraduate students who were the guinea pigs for the early versions of this book, and the graduate students who enrich my life in ways that they may not realize. In particular, my appreciation to my PhD student Peiman Naseradin Mousavi for his dedicated work with the arduous task of developing the solution manual. My gratitude to Karthik Chikmagalur who examined the final proofs painstakingly; it is certain that this text would have had many more errors without his critical eye.

I am also thankful to a host of vibration researchers, many of whom I met at conferences or came to know through the Technical Committee on Vibration and Sound committee of ASME, which I have been a part of for the past 15 years or so. People on this committee as well as the vibration folks in the international arena that I have come to know in general are truly an interesting bunch of people whose camaraderie I have cherished. I thank the editorial staff of Cengage Learning India for their encouragement, patience, and enthusiasm, which were crucial in making this book a reality.

On a personal level, as I mentioned in the dedication, my father is a seminal influence in my life with his love of learning and emphasis on the higher things in life; I truly believe that he embodies the "simple living and high thinking" philosophy. I am grateful for the immense happiness in my life for which my wife, Latha, and my delightful children, Chiraag and Neha are directly responsible.

Finally, and most importantly, my wife has been a constant and strong supporter, subordinating her own career ambitions to support me and the family. *Without her, this book would simply not be.*

Feedback

This is the very perfection of man, to find out his own imperfections.

—Saint Augustine

This book has been written for *you*, the student and not for the instructor. OK—that was an exaggeration—the instructor also needs to be able to use this book! Still, the focus is on the student, for if the book is not interesting to the reader, no instructor can make the student use it, read it, and *revel* in it. It is hence hoped that you found it interesting, reading and did not need to use it as a cure for insomnia! I have indeed tried to make the text clear, interesting, and even funny (OK, perhaps the humor angle doesn't always work!). In any event, I do recognize that there are possibly many lapses in the form of errors, omissions, and lack of clarity.

Please e-mail suggestions, corrections, or things that were not clear or were confusing to you at Dr.C.Nataraj@gmail.com. This invitation applies equally to the students and the instructors. With reference to instructors, I would be particularly interested in your experience teaching from this book, and any material that you think should have been included. If possible, I will get back to you for clarification and, possibly, acknowledgement of your name in future editions.

THANKS!

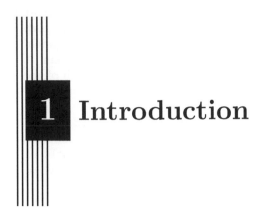

1 Introduction

To me there has never been a higher source of earthly honor or distinction than that connected with advances in science.

—Isaac Newton

As you doubtless know, dynamics is a study of objects and systems in motion. Being a subclass of dynamics, vibration is typically considered to be a study of systems in oscillatory motion. This, however, is a very approximate definition, as the theory and practice of vibration can and do involve systems that are not oscillating in a regular fashion. Hence, the line between vibration and the larger area of dynamics is fuzzy and not always clear. Also, with increasing emphasis on *controlling* the vibration in engineering systems, some aspects of control theory are becoming part of the vibration theory. Acoustics is another aspect of vibratory motion that has become its own discipline and is frequently not included in our concept of a vibrating system—this is of course somewhat artificial, since all sound waves arise from vibratory motion. Finally, nonlinear dynamics with its own unique phenomena and analysis techniques often interfaces with vibration, since most vibrating systems are nonlinear in practice. We will intentionally leave the fuzziness in the demarcations uncorrected so that we are not unduly restricted in our explorations. That said, most of the focus of vibration studies is on oscillatory motion, at least in an introductory text such as this one.

1.1 Historical Note

There have been several historical discussions in the literature, including an excellent book on the history of mechanics [1]; *Vibration and Impact* [2] also has a nice introductory chapter on the history of vibration. What follows here is a very brief note.[1] Clearly, early fascination with vibration has been with that particular euphonious vibration that we call *music*. Concepts of pitch and volume

---

[1]Since most of the readers of this text will be young people, it is important to observe that history seems to appeal only to older people (let us say, older than 50). As I get older, I am increasingly aware of this curious fact! Hence, I am somewhat skeptical of the amount of interest this chapter will evince from the typical young student.

were very well known, appreciated, and manipulated in most of the ancient cultures of the world including Hindu, Chinese, and Greek civilizations. This is certainly grounded in the fact that most human beings are blessed with an acoustic sensor that is extremely accurate, and has a very high frequency range and resolution. By "acoustic sensor" we are of course referring to the auditory system consisting of the human ear, the relevant nervous system, and the cognitive processing system. It turns out that the ear can distinguish pressure differences that are as minute as a billionth of the atmospheric pressure, which corresponds to air vibrations on the order of a tenth of an atomic diameter! It is not unusual for musically educated people to distinguish pitch differences as small as 3.5 Hz. Clearly, this extraordinary human capability has led to very finely *engineered* musical instruments with exquisite and demanding specifications. In a sense, these musical instruments were arguably the earliest sophisticated engineering products of the human civilization. We should also not forget that the very act of talking (or singing) is a consequence of the *vibration* of the vocal chords.

Stringed and pipe instruments were developed very early in the history of human civilization. The Hindu texts called *Atharvavedas*, which are undated, but are thought to go back to the Iron Age (tenth century BC), have explicit rules on musical notes, compositions, and rhythmic structures. It stands to reason that they had musical instruments capable of producing music that conformed to those rules. The flute, producing music from a standing wave pattern of longitudinal vibration of sound, is probably among the first instruments. An early invention was also drums of various kinds, producing a rhythmic pattern of sound from the vibration of membranes. *Sruti* was defined by the ancient Hindu scientist-philosopher Sarngdeva as a noticeable difference in pitch (or frequency); Bharata (with a text called *sarana catusthaya*) was another early experimenter (first century BC–third century AD).

Dimarogonas and Haddad [3] give an excellent exposition of the early Greek contribution to the science of music and make entertaining reading. In the western civilization, Pythagoras (582–507 BC) is widely considered to be among the first to study the vibration patterns of stringed instruments in a scientific manner. His remarkable achievements included establishment of the relationship of the frequency of vibration of a string to its length.

1.1.1   More Recent History

In the sixteenth century, Galileo stands out for many of his pioneering contributions to the understanding of dynamic systems. He established the relationship of the pendulum's oscillation frequency to its length, and, for the first time, related pitch in musical instruments to the frequency of vibration. Other contemporaries such as Hooke and Sauveur also discovered these results experimentally. Later, in the eighteenth century, Taylor (of the Taylor series fame) computed the frequency of a vibrating string and arrived at a result that agreed with Galileo's experiments.

The concept of modes is said to have originated by empirical observation by John Wallis and Joseph Sauveur in the seventeenth century. It can be said that the eighteenth century was probably the golden period for dynamics and vibrations when many advanced concepts were introduced by Daniel Bernoulli, Jean d'Alembert, Leonhard Euler, Joseph Lagrange, and J. B. J. Fourier. These concepts included Fourier analysis, Lagrange equations, calculus of variations, d'Alembert principle, and Euler equations of motion. In the mean time, the Hindu mathematicians continued to be focused on music; for instance, Ahobala (with a monograph called *Sangitaparijata* in the seventeenth century) independently established a correlation of the length of a string to its pitch.

The scientific and industrial revolution was bringing new problems to the fore that included vibration of structural systems. Euler and Bernoulli are said to have analyzed beam vibration for the first time. An energy principle for this purpose was enunciated by Lord Rayleigh; an early book in 1877 by Rayleigh [4] and another one on dynamics by Routh [5] are still relevant today! Extensions to plates and membranes followed a century later by G. R. Kirchoff and S. D. Poisson. Aurel Stodola is famous for a wide spectrum of advances in the theory of vibration that eventually led to the field of rotor dynamics and made turbine design possible. [You are urged to read *History of Rotor Dynamics* [6] for a fascinating history of this field.] More advances followed due to engineers such as C. G. P. de Laval in the nineteenth and twentieth centuries.

1.2 Relevance of Vibration

Vibration is a very important aspect to be considered during the design of all kinds of engineering systems. In general, there are two kinds of vibration. The first is where the vibration is undesirable, and should be either mitigated, or its effects on people or systems should be minimized. Examples include suspension design of automobiles and shock mounts on sensitive devices. The second is where vibration is introduced intentionally. Examples are musical instruments discussed above, massagers and similar devices, signaling devices such as the vibrators in your cell phones, etc. Of the two, the first aspect is much more important. It is also important to remember that oftentimes, vibration of a system can lead to damage or destruction, and hence may need to be reduced or eliminated even if we are not concerned with the immediate effect on surrounding systems or people.

Excessive vibration can cause great damage, sometimes in nonobvious ways. For instance, it is not unusual for a missile to become unstable and go way off target because of the vibration of its structure. The vibration in a surgical robot would cause errors with serious consequences for the person at the other end (namely, the patient!). Excessive vibration in a hand drill would make its operation dangerous at worst, and uncomfortable at the very least; in fact, it is not unusual to have what is called repetitive injury in workers who use tools regularly caused primarily by vibration. Anyone who has been in an automobile with bad shock absorbers can testify to the physical and mental disturbing effects on them.

The quantity of vibration that is tolerable depends on the system. In an earthquake, vibration of a few feet might be permissible depending on the flexibility of the structure. In a car, or a bridge, vibrations of many inches might be tolerated. However, finely tuned gas turbine rotors in aircraft engines, for example, can be destroyed by excessive vibration (in this case, excessive can mean as little as 50 μm). In micromachinery, sensors, and microactuators, vibrations have to be limited to micrometers for successful operation.

Hence, since vibration is so important to the successful operation of systems and to prevent damage, there are standards that specify the vibration levels that the system should survive in a laboratory setting before it can put in place. In the United States, NASA and the military tend to have more exacting demands [7] because of the nature of their businesses. All engineering designs are normally checked from the point of view of vibrations, and redesigned if they do not satisfy the specifications. For example, Fig. 1.1 and Fig. 1.2 show vibration testing of an automobile and an unmanned aircraft. Vibration-based design criteria are necessary even in things that would be termed electrical machinery such as motors and generators. Whenever there is a human interface, there are special and additional vibration require-

Figure 1.1: *Vibration testing of an automobile (courtesy of LMS International)*

Figure 1.2: *Vibration testing of the Predator (courtesy of LMS International)*

ments as you might guess.

1.2.1 Vibration as a Sensor

What is not very obvious is the usage of vibration as a "friend." Let us take the example of a very complicated machinery such as a gas turbine system. It is an expensive and *dangerous* system that we do not want to let fail. In other words, we would very much like an early predictive system that will tell us if the system has some defects or problems, that, if unchecked, could go on to destroy the machine in a catastrophic manner, resulting in tremendous loss of money and equipment, and possibly, even human lives. Well, this is where vibration comes in—as an early warning system. It turns out that, for most problems, vibration is often an early indicator, and is often easier to measure and analyze than something like pressure. Vibration signals also tend to have richer data with more information about the state of the machine than other quantities such as temperature and pressure. It is hence quite common to use vibration sensors constantly to monitor complex machinery, and to trigger a response if a predetermined pattern emerges in the vibration signals (often, the criterion can be something simple like vibration amplitude exceeding a certain value). As machinery become

Figure 1.3: *Vibration monitoring in a gas turbine engine (courtesy of Endevco)*

Figure 1.4: *Vibration monitoring in a helicopter (courtesy of Endevco)*

Figure 1.5: *Vibration monitoring of a boiler feedwater pump (courtesy of GE Energy)*

Figure 1.6: *Vibration monitoring of a wind turbine (courtesy of GE Energy)*

more autonomous with less human intervention or supervision, such warning systems are becoming almost indispensable. Examples of such machinery include gas turbine engines (Fig. 1.3), helicopters (Fig. 1.4), boiler feedwater pumps (Fig. 1.5), and wind turbines (Fig. 1.6).

What makes all of this complicated is the fact that the vibration occurs in practice due to a variety of reasons that are not always obvious. Analysis, on the other hand, has to focus on one phenomenon at a time, as you will see in this book. For example, for a screw compressor shown in Fig. 1.7, Fig. 1.8 shows the vibration signal captured on its housing [8]; notice how different it is from a textbook example of simple sinusoidal oscillation occurring in a mass-spring system. Unfortunately, real life is not divided into chapters like you see in this or any other text book! A positive perspective, however, is to say that this is what makes an engineer's job interesting and challenging (not to speak of providing job security!).

Figure 1.7: *A screw compressor (courtesy of Aerzen)*

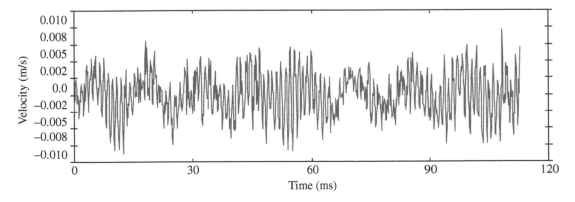

Figure 1.8: *Vibration signal measured on a reciprocating compressor*

1.3 Objectives

The intended level of this book is introductory either at the undergraduate level or at the graduate level (for those who have not had any exposure to vibration theory). It starts with the basics of dynamics and explores single- and multi-degree-freedom systems. There is also an introduction to continuous systems as well as an overview of some basic experimental techniques. There is an emphasis on the fact that engineered systems are invariably nonlinear. Also emphasized are parametric perspectives; in other words, many problems deal with a range of parameters rather than specific numerical values to emphasize a design outlook.

1.3.1 Needed Background

The background needed was already addressed in the note to the students on page xi (you did read that note, right?). To emphasize, knowledge of the following subjects is absolutely necessary for a comprehensive understanding of the concept of vibration.

- Elementary dynamics
- Differential equations
- Solid mechanics (tension, torsion, and bending)
- Some computer programming exposure

We will start the "real" book from the next chapter with a review of the first two topics mentioned above: dynamics and differential equations.

2 Dynamics: A Brief Review

Motion is the cause of all life.

—Leonardo da Vinci

This chapter outlines a general approach to solving simple problems in dynamics. It uses an example of a simple pendulum and a rigid body pendulum to illustrate the derivation of equations of motion and linearization. This short write-up is meant to be a review; it is not a substitute for a good text on dynamics (such as *Principles of Dynamics* [9]), which should be studied to supplement your understanding of the subject.

2.1 Dynamics of a Particle

Let us fast forward a few years; you are a successful engineer running your own company. Let us say, the local municipality approaches you seeking the engineering design of playground swings (Fig. 2.1) to be installed all across the city. They are obviously interested in minimizing the potential danger that would result in case of failure, but they also want to save money by selecting lighter and cheaper components. This is in fact the most common compromise in engineering design.

Clearly, the problem is dynamic (because the swing *moves*); it has uncertain parameters (because it should account for differing weights as well as different motion amplitudes); and, the principles of mechanics of materials and machine design are involved in designing the actual hardware such as the bolts, the hinges, the chains, etc. Without delving into all such details let us consider the basic dynamic problem so that we can predict

Figure 2.1: *Playground swing*

the forces that would result. These forces could subsequently be used for the mechanical design.

The simplest model (or *idealization*) for the swing is that of a simple pendulum (Fig. 2.2). The simple pendulum consists of a particle mass m, suspended from a rigid support by a massless inextensible string (or cable) of length ℓ in a gravity field. Please note that this is a *model*, and will hence attempt to predict the real system behavior, and will necessarily be approximate. The more complex the model, the more accurate its prediction (in general), but the more complicated the mathematics will be. We have hence chosen the simplest model we could think of; before we subject innocent children to the reality of our imperfect design, however, we may need to come up with better models. This would be important from the point of view of legality as well as morality!

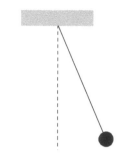

One of the principal assumptions here is that of a particle mass. Recall that a particle has mass, but is physically located at a point; in other words, it occupies no space and has no dimensions, from which it follows that it is a mythical quantity with infinite density! Clearly, it is a nonsensical

Figure 2.2: *Simple pendulum*

concept! Or is it that crazy? The truth is that it is not a bad approximation at all. If you go back to your textbook on dynamics you will find problems modeled with particle masses that has nothing do with the size. In fact, you can model as big an object as the earth as a particle and get decent answers on analyzing its motion around the sun. The assumption derives more from the type of motion that we need to model than the size of the mass. Because the particle has no dimension, it is not possible to model its rotational motion. Hence, in the case of the swing, if we wanted to model its wobbling motion, we would need something better than a particle.

This then is a problem in dynamics; given any problem in dynamics, the first step is always the identification of the system that you wish to analyze.[1] This could be achieved by drawing an imaginary boundary around the system of interest. Although this looks like a trivial step, you are highly encouraged to do this as an important element to guide the thinking process. This process results in two units: a *system* and its *environment*. The system is what you are interested in, and the environment is defined to be *everything* that is not in the system.

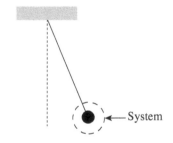

Following the above guidelines, we define the system to be the mass of the pendulum (Fig. 2.3). Then the environment is everything around it including the string, the support, a whale in the Arctic Ocean, an aircraft flying overhead, the moon, and a thousand other objects.

Figure 2.3: *Simple pendulum model*

Kinetics

The next step is to analyze the effect of the environment upon the system. Theoretically, everything in the environment affects the system. In dynamics, the way the environment affects the system is by the application of forces. In other words, the interaction of the system with the environment is by means of forces.

In general, there are forces of two kinds: forces of contact and forces at a distance. Examples of the first kind are friction, the tension in a string, normal reaction, etc. Forces at a distance act between non contacting bodies, and include those due to gravitational, magnetic, and electrostatic phenomena.

---

[1]This, or something similar (remember control volume?), is in fact a common step in diverse subjects such as fluid mechanics and thermodynamics.

A free body diagram (FBD) is essentially a depiction of the interaction of the system with its environment. Hence, *all* the forces acting on the system should be shown. This typically includes forces of both kinds. At this point, several approximations can be made and the effect of the environment upon the system can be simplified. This step is called *kinetics*.

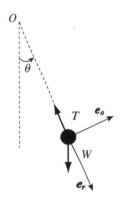

As an example, for the simple pendulum, although the moon was mentioned as having a possible effect upon its motion, it stands to reason that we can neglect that effect. Hence, in this case, we really need to include only the effect of the string (tension T) and the force due to the earth (gravity, W, force at a distance) upon the system (mass). This is shown in an FBD, Fig. 2.4,

Figure 2.4: *Simple pendulum FBD*

with the unit vectors, $\vec{e}_r$ and $\vec{e}_\theta$, chosen as shown in the figure. We can express the resultant of these forces as follows.

$$\vec{F} = (W\cos\theta - T)\vec{e}_r - (W\sin\theta)\vec{e}_\theta \tag{2.1}$$

Kinematics

Quite independently of the above process, we need to use the geometry of motion to come up with expressions for accelerations. This area of dynamics, called *kinematics*, does not make use of any of the information about the forces acting upon the system. Kinematics is derived from geometry and geometry alone.

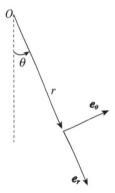

Returning to our example of the simple pendulum, we use geometry and certain assumptions (that the string is inextensible, and hence ℓ is constant) to derive expressions for velocities and accelerations. It is worth emphasizing that while doing this, one should *not bring forces* (such as gravity) into the picture.

Referring to Fig. 2.5, we find that the displacement, velocity, and acceleration of the mass are given below. Note that the the acceleration is computed with respect to the fixed point, O.

Figure 2.5: *Simple pendulum kinematics*

$$\vec{x} = \ell\vec{e}_r \tag{2.2}$$

$$\vec{v} = \frac{d\vec{x}}{dt} = \ell\dot{\theta}\vec{e}_\theta \tag{2.3}$$

$$\vec{a} = \frac{d\vec{v}}{dt} = \ell\ddot{\theta}\vec{e}_\theta - \ell\dot{\theta}^2\vec{e}_r \tag{2.4}$$

Newton's laws

The final step is to use all of the information obtained above by the application of Newton's laws. This is the linking step between kinematics (study of the motion itself) and kinetics (study of the forces that cause motion).

For the pendulum mass (treated as a particle), we have, from Newton's second law,

$$\underbrace{\vec{F}}_{\text{From kinetics}} = m \underbrace{\vec{a}}_{\text{From kinematics}} \tag{2.5}$$

Note that it is a vector equation. Also, recall that the acceleration to be used in second law should be the so-called absolute acceleration of the mass with respect to a point that is *fixed*.

An important concept to remember is that Newton's second law is essentially a "cause-effect" relationship: it relates the cause (forces) to the effect (motion).

Newton's second law requires a fixed point. This begs the question, "What *is* a fixed point?" Is the point of suspension of the simple pendulum a fixed point? For example, consider the case of the pendulum being fixed to the ceiling in a moving train. In that case it seems obvious that we would have to take the motion of the train into account. Even if it were in a "stationary" room consider the fact that the room is attached to the earth that is spinning about its axis and revolving around the sun, which is itself hurtling through space at an enormous speed. So, what **is** absolute acceleration and what **is** a fixed point? Solving Problem 2.25 on page 39 will get you thinking a little. Now, if you want to think a lot more, there is an excellent treatment of the dynamics of a falling body without ignoring the rotation of the earth in the book *Dynamics of a System of Rigid Bodies: Elementary Part* [5], where the eminent dynamicist Routh showed that the falling body does not fall straight down, but instead follows a curved path that depends on the latitude!

In our case of two dimensions, this gives rise to two scalar equations

$$F_r = ma_r \tag{2.6}$$
$$F_\theta = ma_\theta \tag{2.7}$$

Substituting for the forces and accelerations, we get

$$W \cos\theta - T = m(-\ell\dot{\theta}^2) \tag{2.8}$$
$$-W \sin\theta = m(\ell\ddot{\theta}) \tag{2.9}$$

Note that we have two equations to solve for the two unknowns, $\theta(t)$ and $T(t)$. Assuming a uniform gravitational field,

$$W = mg \tag{2.10}$$

Then, the second of the equations, Eq. (2.9) describes the dynamics of the pendulum and is called the *equation of motion*, often abbreviated by EOM.

$$\boxed{\ddot{\theta} + \frac{g}{\ell}\sin\theta = 0} \tag{2.11}$$

Note that it is a second-order nonlinear differential equation that needs to be solved to determine $\theta(t)$. Then, the other equation, Eq. (2.8), can be used to determine the unknown tension T.

$$T = m\ell\dot{\theta}^2 + mg\cos\theta \tag{2.12}$$

Summary of the modeling process

1. Identify the system and its environment by drawing an imaginary boundary.
2. Determine the complete effect of the environment upon the system, and draw an FBD showing all the forces acting on the system (kinetics). At this point, you may have to make some assumptions to determine which forces can be neglected.
3. From geometry derive a kinematic description for the system; this essentially results in expressions for accelerations. Again, some assumptions may have to be made.
4. Use Newton's laws to relate the forces in step 2 to the accelerations in step 3.
5. Do not quit yet! You are doing well; take a coffee break and start with step step 1 again for the next problem.

Since steps 2 and 3 are independent of each other, note that it is possible to carry out these two steps in the reverse order.

Here is a fairly simple—and old—puzzle on basic Newtonian physics. Consider a horse pulling a cart. The horse applies a force to the cart, and the cart applies an equal and opposite force on the horse by Newton's third law. Considering the two objects together, the forces add up and cancel. So, the net force is zero. How is that the two are still moving?

Here is another old puzzle on basic Newtonian physics. Consider a truck driver carrying a thousand pigeons in a *closed* container. He comes to a bridge that looks quite rickety and has a load limit that is exactly equal to the weight of his truck minus his cargo. Harking back to his high school physics, the driver divines a solution; he goes and bangs on the container, and gets the pigeons flying in the air, and figures that their weight is no longer added to the truck's weight. He gets back in the truck and drives. Does he survive the bridge? If you are stuck remember that, no matter what, FBD will get you moving.

2.1.1   Equilibrium Positions

The equilibrium positions of a dynamical system are defined mathematically to be the solutions of the governing equation with all the time derivatives set to zero. In order to determine the equilibrium positions of the pendulum, for example, we will set all the time derivatives to zero in Eq. (2.11). Then, if we denote the equilibrium positions by θ_0, we get

$$\sin \theta_0 = 0 \tag{2.13}$$

The solution of the above nonlinear algebraic equation yields infinite roots; however, only two are of physical significance:

$$\theta_{0,1} = 0 \tag{2.14}$$
$$\theta_{0,2} = \pi \tag{2.15}$$

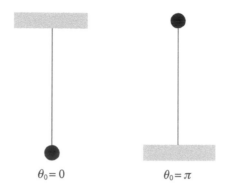

$\theta_0 = 0$ $\theta_0 = \pi$

Figure 2.6: *Simple pendulum equilibrium positions*

Clearly, the mathematical equilibrium positions determined above also correspond to positions of physical equilibrium; i.e., positions, where the pendulum can stay at rest indefinitely under the action of forces (tension T and gravity), which are in *static equilibrium*. Figure 2.6 demonstrates this fact for the two equilibrium positions, $\theta_{0,1}$ and $\theta_{0,2}$, that correspond to the hanging and the upright positions of the pendulum, respectively.

> Does the pendulum in the upright position strike you as an equilibrium position from intuition? If it does not, what is wrong with either your intuition or our solution process? Draw an FBD of the *static* system when the pendulum is in the upright position. What direction does the tension need to be if it is in static equilibrium?

2.1.2 Linearization

You may have noticed that Eq. (2.11) has the term $\sin\theta$ as one of its terms; i.e., one of the terms is a nonlinear function of the dependent variable (see A.2.1 for a discussion of linearity). Hence, the equation is *nonlinear*.

Nonlinear equations rarely have a closed-form solution, and hence, we need to come up with an approximation that will somehow make it easy to solve and, at the same time, be reasonably valid in the range of operation. The answer is linearization using Taylor series, Eq. (A.1.39), which ranks among the most important mathematical results used in engineering.

As a first step, we postulate something that is central to most areas of engineering: we assume that the system under consideration is usually in one of its equilibrium positions and that if it does get disturbed from this position, it still tends to stay in the neighborhood. In other words, we assume "small" deviations (or perturbations) from the equilibrium configuration. This will enable us to expand the nonlinear function about the equilibrium position using Taylor series and to truncate the series after the linear term. That is, we approximate a nonlinear function by

$$f(x) \approx f(x_0) + \left.\frac{df}{dx}\right|_{x_0} (x - x_0)$$

(2.16)

where x_0 would be, in the present case, an equilibrium position, and the above approximation would be reasonably valid provided the deviation $(x - x_0)$ is small.

How small is small? In the case of the simple pendulum, is 0.1 rad small? Or, can it be as large as 1/2 a radian? Clearly, there is no definite answer to that question. It would ultimately depend on the kind of accuracy that you, the analyst, would demand from your approximate model. What would be *your* criterion? See Problem A.8 on page 427 for a quantitative exploration of this concept.

In the case of the pendulum, the nonlinear function $\sin\theta$ is approximated by

$$\sin\theta \approx \sin\theta_0 + \cos\theta_0\,(\theta - \theta_0) \tag{2.17}$$

For the first of the two equilibrium positions determined above, this approximation becomes

$$\sin\theta \approx \phi \tag{2.18}$$

where ϕ has been defined as the deviation from the equilibrium position; i.e.,

$$\phi = \theta - \theta_0 \tag{2.19}$$

from which it follows that

$$\frac{d^2\phi}{dt^2} = \frac{d^2\theta}{dt^2} \tag{2.20}$$

Substitution of this approximation into the governing EOM, Eq. (2.11), yields a linear differential equation in ϕ.

$$\boxed{\frac{d^2\phi}{dt^2} + \frac{g}{\ell}\phi = 0} \tag{2.21}$$

For the second equilibrium position, $\theta_0 = \pi$, the approximation is

$$\sin\theta \approx -\phi \tag{2.22}$$

Substitution into the EOM yields a linear differential equation quite different from the previous case.

$$\boxed{\frac{d^2\phi}{dt^2} - \frac{g}{\ell}\phi = 0} \tag{2.23}$$

Comparison of the two linearized equations, Eqs. (2.21) and (2.23) shows that the same nonlinear equation can yield different equations depending upon the equilibrium position about which we carry out the linearization.

Note that the difference in the two cases is in the sign of the coefficient of ϕ; it is in fact this difference that makes the pendulum behave differently in the two cases. From physical experience, we know that the pendulum, when disturbed from the equilibrium position, $\theta_{0,1}$, tends to come back to that position; on the other hand, when disturbed from the second equilibrium position, $\theta_{0,2}$, it tends to move away. Hence, intuitively, the pendulum is *stable* about the $\theta_{0,1}$ equilibrium position, and *unstable* about the $\theta_{0,2}$ equilibrium position. We will see later how this can be shown by a simple solution of these equations. Suffice it to say for now that the linearized equations of a dynamic system do give us a lot of information about its dynamic behavior as evidenced by the example of the simple pendulum.

Stability

A system is said to be stable in an equilibrium position if, when disturbed from that position, it returns to that position. It is unstable if it continues to recede, and marginally (or neutrally) stable, if it does neither. An intuitive explanation can be provided using a small ball on three kinds of surfaces (Fig. 2.7). There are many general definitions of stability; for now, this intuitive definition is sufficient.

Figure 2.7: *Stable, unstable, and marginally stable configurations*

Consider the problem of balancing a stick on your palm. This is the so-called broom-balancing problem. How is that you can "stabilize" it, or turn an unstable situation into a stable one by giving it a base motion? Can you model the dynamics of it and determine the base motion you need to provide so that you can perhaps build a robot to do it instead of having to do it yourself?

2.1.3 What Not to Do with Gravity: A Weighty Problem

An error committed once too often during the analysis of a body in the gravity field is to include the acceleration due to gravity as an acceleration term in the kinematical description of the problem. It should be noted that gravity is a force acting on the body, which is proportional to the mass of the body provided the gravitational field is uniform. This is derived from Newton's law of gravitation and should not be confused with Newton's second law. It is this proportionality constant that is called the acceleration due to gravity (which is fairly constant all over the globe provided we are close enough to the earth's surface).

To illustrate this point, consider a freely falling body in vacuum and in the earth's gravitational field assumed uniform. The corresponding FBD is shown in (Fig. 2.8). Note that gravity (weight W) is the only force acting on the body.

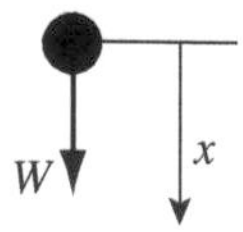

Figure 2.8: *Freely falling body FBD*

Hence, the application of Newton's second law now gives us

$$F = W = ma \tag{2.24}$$

Now, we can write, from Newton's law of gravitation,

$$W = mg \tag{2.25}$$

where g is the so-called acceleration due to gravity. From the above two equations, it follows quite trivially that the acceleration of a freely falling body in vacuum is equal to g. It should be realized however that if the body were not in vacuum (i.e., if we had to take air resistance into account), or if the gravitational field were not uniform (in which case you cannot write $W = mg$), then the acceleration of the body would not be equal to g. See Problem 2.24 at the end of this chapter that includes air resistance.

All this confusion is best avoided by the simple expedient of always showing the effect of gravity on the body as a term, say W, and perhaps substituting mg for it in the very last step. The important thing is to never bring gravity into the picture in the *kinematic* phase of the problem.

Consider the very interesting problem of digging a tunnel through the center of the earth; in other words, we have a tunnel extending the length of the diameter of the earth. Then, we drop a mass in it; can you analyze its motion? Note that the gravitational force is given by

$$F_g = G\frac{m_1 m_2}{r^2}$$

where G is the universal gravitational constant, m_1 here would be the mass of the earth, m_2 the mass of the particle, and r the distance to the center of the earth. This model is not necessarily very accurate, but it is OK for our thinking experiment. Of course, this is highly theoretical and could hardly be implemented (unless we needed to do this to mine energy for our power plants and automobiles in which case *we*, the engineers, would figure out some way!).

2.1.4   Summary

We have learnt from the example of a simple pendulum that even very elementary dynamic systems can give rise to nonlinear differential equations, which are often difficult to solve (if not impossible). Hence, typically we identify equilibrium positions at which the system can stay at rest. Then, we assume that the system can undergo small displacements, and use Taylor series to linearize the nonlinear equations about these equilibrium positions. In the next section, we will solve these linearized equations to obtain the displacement as a function of time.

2.1.5   Solution of the Linearized Equation of Motion

Motion about the hanging position

For small motions about the stable equilibrium position, $\theta_0 = 0$, the linearized EOM is

$$\ddot{\phi} + \frac{g}{\ell}\phi = 0 \tag{2.26}$$

Following the guidelines in A.2, we *seek*[2] a solution (based on experience) in the exponential form

$$\phi(t) = A e^{st} \tag{2.27}$$

where A and s are undetermined constants. Equation (2.27) is of course not the true solution unless it satisfies Eq. (2.26); recall that a solution to any equation is defined to be a function, which when substituted into the equation, satisfies it. Also,

$$\dot{\phi} = s A e^{st}, \quad \ddot{\phi} = s^2 A e^{st} \tag{2.28}$$

Next we substitute Eq. (2.27) into Eq. (2.26) to check if the assumed form is indeed the correct solution. That process yields

$$\left(s^2 + g\ell\right) A e^{st} = 0 \tag{2.29}$$

Ignoring the trivial solution ($A = 0 \Rightarrow \phi_h(t) \equiv 0$), we get an *algebraic* equation

$$\left(s^2 + \frac{g}{\ell}\right) = 0 \tag{2.30}$$

which *must* be satisfied for the assumed form of the solution to be the correct solution. Equation (2.30) is a quadratic equation in the constant s and is called the *characteristic equation* for this system. Solution yields two imaginary roots:

$$s_1 = +i\omega_n, \quad s_2 = -i\omega_n \tag{2.31}$$

where $i = \sqrt{-1}$, and

$$\boxed{\omega_n = \sqrt{\frac{g}{\ell}}} \tag{2.32}$$

ω_n is a very important quantity and is called the *natural frequency* of the system.

Again, drawing upon our encyclopedic (!) knowledge of differential equations, we note that we will need to collect *all* possible solutions of the linear differential equation and add them together to obtain the *complementary solution* to the equation. At this point, we also note that our solution process yielded no conditions on the arbitrary constant A, and also that this arbitrary constant could be different (in general) for the two families of solutions that correspond to s_1 and s_2. Hence, the total solution to the differential equation is

$$\phi(t) = A_1 e^{s_1 t} + A_2 e^{s_2 t} \tag{2.33}$$

where A_1 and A_2 are arbitrary constants, and s_1 and s_2 are given by Eq. (2.31).

Using Euler formula (Eq. A.1.7 on page 412), the solution becomes

$$\phi(t) = (A_1 + A_2) \cos \omega_n t + i (A_1 - A_2) \sin \omega_n t \tag{2.34}$$

With new constants, C_1 and C_2, which are both real,

$$\phi(t) = C_1 \cos \omega_n t + C_2 \sin \omega_n t \tag{2.35}$$

---

[2]This means that we do not know *yet* if this is indeed the true solution.

▬▬▬▬

Exercise 2.1

Prove that the constants C_1 and C_2 are real. (*Hint*: The solution of the equation, $\phi(t)$ represents a real angle and has to be real.)

▬▬▬▬

The above expression can also be written in the following form

$$\phi(t) = A \cos\left(\omega_n t - \psi\right) \tag{2.36}$$

where A and ψ are arbitrary constants.

Motion about the upright position

In the second case of the unstable equilibrium position, $\theta_0 = \pi$, when the linearized EOM is

$$\ddot{\phi} - \frac{g}{\ell}\phi = 0 \tag{2.37}$$

the solution procedure is identical. However, the characteristic equation in this case is

$$\left(s^2 - \frac{g}{\ell}\right) = 0 \tag{2.38}$$

The characteristic roots now become real

$$s_1 = \sqrt{\frac{g}{\ell}}, \quad s_2 = -\sqrt{\frac{g}{\ell}} \tag{2.39}$$

Hence, the solution is now exponential and non oscillatory.

$$\begin{aligned}
\phi(t) &= A_1 e^{s_1 t} + A_2 e^{s_2 t} \\
&= A_1 e^{\sqrt{\frac{g}{\ell}}t} + A_2 e^{-\sqrt{\frac{g}{\ell}}t}
\end{aligned} \tag{2.40}$$

A sketch of the two solutions, Fig. 2.9, confirms our physical intuition that the equilibrium position, $\theta_0 = 0$ is stable, and that the equilibrium position, $\theta_0 = \pi$ is unstable. Listing 2.1 shows the MATLAB program written to produce the plot.

Listing 2.1: *Local behavior about the hanging position*

```
% plots.m
% plotting in MATLAB

t=(0:.01:5*pi);
omega_n=2;
A=0.2;
psi=30*pi/180;
phi=A*cos(omega_n*t-psi);

plot(t, phi)
xlabel('Time');
ylabel('\phi');
grid on;
```

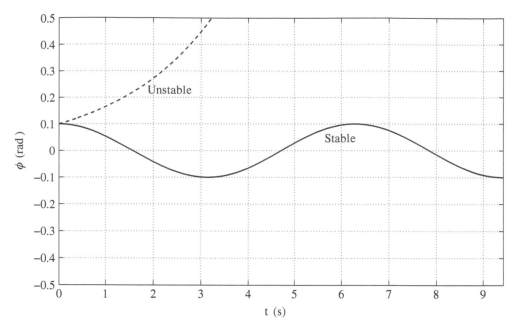

Figure 2.9: *Response of the linearized models*

Consider the prediction of the linearized equations in the unstable case. It seems to tell us that the response will increase forever. Does this make sense? Can the response increase forever? Where would the energy for all that motion come from? Also, we made a critical assumption when we linearized the equations (think Taylor series). Are we violating that assumption? If we wanted an accurate solution, do you think the *nonlinear* equation should be used to determine the motion when it is unstable? Why or why not?

The oscillatory solution

The stable solution that corresponds to an oscillatory solution is of most interest to us in our study of vibrating systems. The quantity ω_n that we defined by Eq. (2.32) is called the *natural frequency* of oscillation and is the frequency of oscillation when the pendulum is disturbed from its vertically hanging equilibrium position by a small amount. The period of natural oscillations is then found to be

$$\tau_n = \frac{2\pi}{\omega_n} \tag{2.41}$$

Exercise 2.2

The period of such an oscillating function is defined to be the interval between two closest instants of time when the function repeats itself. Using the solution given above, prove the above relation between the period and the frequency.

2.2   Dynamics of a Rigid Body

As opposed to a particle, a *rigid body* model has extended dimensions and can, therefore, rotate. Let us take an example of a pendulum again, but now consisting not of a point mass (as in a *simple pendulum*), but a uniform, slender rod, as shown in Fig. 2.10. The rod has mass m and length ℓ.

The forces acting on the rod are due to the support hinge and the gravitational field of the earth. R_x and R_y are the support reactions, or forces exerted on the rod by the support hinge. This is shown in an FBD, Fig. 2.11.

The forces can be expressed in terms of the Cartesian unit vectors as follows.

$$\vec{F} = R_x \vec{\imath} + (R_y - W)\vec{\jmath} \tag{2.42}$$

Figure 2.10: *Slender rod*

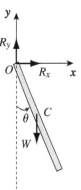

Figure 2.11: *Slender rod FBD*

Kinematics

Referring to Fig. 2.12, we find that the displacement, velocity, and acceleration of the *center of mass* are given as shown below. Since the rod is uniform, the center of mass is located at the center of the rod. In the case of the simple pendulum, we derived these using polar coordinates; here, just to be different, we are using Cartesian coordinates. Notice how the algebra gets more complicated; however, the unit vectors do not change direction like they did in polar coordinates and hence, the process becomes more straight forward. Of course, both the approaches are equally valid, and it is entirely your choice which coordinate system you should use. (The real pendulum couldn't care less which coordinate system you use!)

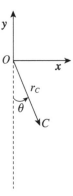

Figure 2.12: *Kinematics for the rod pendulum*

$$\vec{r}_C = \frac{\ell}{2} \left(\sin\theta\,\vec{\imath} - \cos\theta\,\vec{\jmath} \right) \tag{2.43}$$

$$\vec{v}_C = \frac{d\vec{r}_C}{dt} = \frac{\ell}{2}\dot{\theta} \left(\cos\theta\,\vec{\imath} + \sin\theta\,\vec{\jmath} \right) \tag{2.44}$$

$$\vec{a}_C = \frac{d\vec{v}_C}{dt} = \frac{\ell}{2}\ddot{\theta} \left(\cos\theta\,\vec{\imath} + \sin\theta\,\vec{\jmath} \right) + \frac{\ell}{2}\dot{\theta}^2 \left(-\sin\theta\,\vec{\imath} + \cos\theta\,\vec{\jmath} \right) \tag{2.45}$$

Since it is a rigid body, we should also determine its angular acceleration.

$$\alpha = \ddot{\theta} \tag{2.46}$$

Newton's laws

Since we now have a rigid body, Newton's second law is quite different from the case of the simple pendulum.[3]

$$\vec{F} = m\vec{a}_C \tag{2.47}$$

$$M_C = I_C\alpha \quad \text{or} \quad M_O = I_O\alpha \tag{2.48}$$

In general, different points on the rigid body have different accelerations; note that the acceleration in the above equation has to be that of the center of mass. Also, note that the moment equation is valid only about either a fixed point or the center of mass. Only three of the above four equations will turn out to be independent. That is of course as it should be, since we have only three unknown quantities, namely, θ, R_x, and R_y.

> A fixed point and the center of mass are not the only points about which the moment equation is valid. There is a third point that you would have discussed in your dynamics class. If you cannot recollect, this is the time to dig out your dynamics textbook (I hope you did not sell it!). When you are doing that, you may also want to see how the rigid body equations are derived from second law for the particle by considering the rigid body to be a constrained collection of infinite particles.

Looking at the the above equations we now realize that, if we are only interested in $\theta(t)$, and do not need to compute the support reactions, we can simply use the moment equation about O. Then,

$$-W\frac{\ell}{2}\sin\theta = \frac{m\ell^2}{3}\ddot{\theta} \tag{2.49}$$

where we have used $I_O = I_C + m(\frac{\ell}{2})^2 = \frac{m\ell^2}{3}$. Simplifying, we get the following EOM.

$$\boxed{\ddot{\theta} + \frac{3g}{2\ell}\sin\theta = 0} \tag{2.50}$$

Note that, since we did not use the force equations at all, we need not have computed the linear acceleration of the center of mass of the system and could have saved ourselves that much algebra. However, if we were interested in designing the hinge of the pendulum so that it could stand up to the dynamic forces, we *would* need to use the force equations to be able to compute R_x and R_y.

---

[3]These equations are valid for a rigid body whose motion is confined to a plane. The general case of a rigid body in three-dimensional space will not be dealt with in this book. Read the book *Principles of Dynamics* by Greenwood [9] for an authoritative treatment.

It is interesting to compare the above equation with Eq. (2.11) on page 10, the governing equation of the simple pendulum. Note that both the equations are nonlinear and only differ in terms of the constant coefficient in front of the second term. Clearly, any solution of the first equation would enable us to write the solution of the second one without having to solve it again. (Isn't mathematics marvelous?!) In fact, any oscillatory system, if modeled with simplifying assumptions, will result in the same (or similar) equation.

It should be clear that the equilibrium positions and linearizations will be the same as in the case of the simple pendulum; i.e., there is a stable equilibrium position, and an unstable one. For the stable case ($\theta_0 = 0$), the natural frequency of oscillation will now be

$$\omega_n = \sqrt{\frac{3g}{2\ell}} \tag{2.51}$$

The solution process is the same as in the previous treatment of the simple pendulum and will not be repeated here.

Now, suppose we *did* need the support forces. Then, we use the force equations, Eq. (2.47) and get the following expressions.

$$R_x = m\frac{\ell}{2}(\ddot{\theta}\cos\theta - \dot{\theta}^2\sin\theta) \tag{2.52}$$

$$R_y = m\frac{\ell}{2}(\ddot{\theta}\sin\theta - \dot{\theta}^2\cos\theta) + W \tag{2.53}$$

Note that the support reactions are functions of the motion variables of the pendulum. It is interesting to see what happens if they are linearized about the hanging position ($\theta_0 = 0$).

$$R_x = m\frac{\ell}{2}\ddot{\theta} = -\frac{3mg}{4}\theta \tag{2.54}$$

$$R_y = W = mg \tag{2.55}$$

See Problem 2.47 for more on computing the support reactions.

2.3 Energy of Vibration

We know from intuition that, for motion to occur, energy must be supplied. What then is the definition of energy and how do we relate it to the motion variables, stability and other issues we discussed above?

2.3.1 Work

First, we need to define *work*; consider a particle acted upon by a force, $\vec{F}$, whose position (in a Newtonian reference frame) is given by $\vec{r}(t)$, as shown schematically in Fig. 2.13.

Then, as the particle moves, the infinitesimal work performed is given by

$$dW = \vec{F} \cdot d\vec{r} = \vec{F} \cdot \vec{v}\, dt \tag{2.56}$$

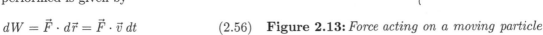

Figure 2.13: *Force acting on a moving particle*

where we have used the definition of velocity

$$\vec{v} = \frac{d\vec{r}}{dt} \tag{2.57}$$

Hence, as the particle moves over a time interval, (t_0, t_1), the work done by the force is given by

$$W = \int_{t_0}^{t_1} \vec{F} \cdot \vec{v} \, dt \tag{2.58}$$

It is important to note the following.

- Work is a scalar quantity; work done by many forces on the same particle would be simply additive.
- The velocity depends on the reference frame; hence, work done would also depend on the reference frame.

2.3.2 Law of Kinetic Energy

Consider a particle again with a force acting on it. Then, by Newton's second law,

$$\vec{F} = m\vec{a} = m\frac{d\vec{v}}{dt} \tag{2.59}$$

The infinitesimal work is

$$dW = \vec{F} \cdot d\vec{r}$$
$$= m\frac{d\vec{v}}{dt} \cdot d\vec{r}$$
$$= m\vec{v} \cdot d\vec{v}$$
$$= \frac{d}{dt}\left(\frac{1}{2}mv^2\right) dt \tag{2.60}$$

Integrating, the work done to go from point 1 to point 2 is given by

$$W_{12} = \frac{1}{2}mv_2^2 - \frac{1}{2}mv_1^2 = T_2 - T_1 = \Delta T \tag{2.61}$$

where T is the kinetic energy in the particle ($\frac{1}{2}mv^2$).

In other words (and generalizing to all situations),

The work done by all the forces on a mechanical system is equal to the increase in kinetic energy of the system.

The above statement is called the *law of kinetic energy*, or the *work energy principle*.

2.3.3 Conservative Forces

Suppose that the force acting on a particle satisfies the following two conditions:

- It is a function of position only.
- The line integral, $\int_{r_1}^{r_2} \vec{F} \cdot d\vec{r}$ is independent of the path taken from position 1 to position 2.

Then, it follows that, if we go from point 1 to point 2, and come back to point 1, the integral around such a closed path would be zero.

$$\oint \vec{F} \cdot d\vec{r} = 0 \tag{2.62}$$

Such a force is said to be a *conservative force*. In practice, this means that the force is not dissipative in nature. Friction, for example, would *not* be a conservative force.

2.3.4   Potential Energy

Consider a system on which only conservative forces act. Then, the work done depends only on the end points, and hence the integrand is an exact differential. Or

$$\vec{F} \cdot d\vec{r} = -dV \tag{2.63}$$

where V is called the potential energy, and the minus sign has been chosen (quite arbitrarily) because of convention. Thus,

$$W = \int_{r_1}^{r_2} \vec{F} \cdot d\vec{r} = -\int_{r_1}^{r_2} dV = V_1 - V_2 \tag{2.64}$$

The above equation states that the decrease in potential energy in moving a particle from point 1 to 2 is equal to the work done on the particle by the conservative field. If, for example, a particle is acted upon by gravity, the potential energy is given by $V = mgh$, where h represents the height above an arbitrary reference point.

2.3.5   Conservation of Energy

Putting this together with the law of kinetic energy Eq. (2.61), we have the following important result for conservative systems.

$$V_1 + T_1 = V_2 + T_2 \tag{2.65}$$

This is the well-known principle of conservation of energy; it states that, when a dynamic system is acted upon by conservative forces, all the mechanical energy is conserved.

2.3.6   Simple pendulum

Coming back to the simple pendulum we note that the kinetic energy is given very simply by

$$T = \frac{1}{2}mv^2 = \frac{1}{2}m\ell^2\dot{\theta}^2 \tag{2.66}$$

The potential energy is due to the gravity field and is given by

$$V = -mg\ell\cos\theta \tag{2.67}$$

using the suspension point as (an arbitrary) reference. Note that the tension in the string does not contribute to the potential energy. Hence, the total energy E is the sum of these two, and is conserved. Or,

$$E = T + V = \frac{1}{2}m\ell^2\dot{\theta}^2 - mg\ell\cos\theta = C = \text{constant} \tag{2.68}$$

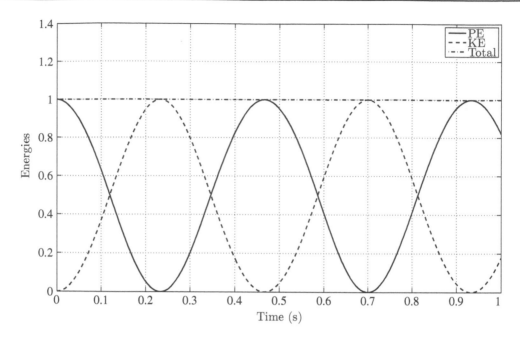

Figure 2.14: *Energies for the simple pendulum*

Note that this equation does not assume small displacements; hence, it is valid for the original nonlinear model. Figure 2.14 shows the variation of the two energies as the pendulum oscillates, starting with a non zero initial position. They were computed from a numerical solution of the original nonlinear equations (you will learn about the numerical methods at the end of this chapter). Clearly, the total energy remains constant.

There is yet another way to derive the energy conservation principle. Let us get back to the original differential EOM for the simple pendulum.

$$m\ell^2\ddot{\theta} + mg\ell\sin\theta = 0 \tag{2.69}$$

We have put back the constants we had divided out $(m\ell)$ to help understand the physics a little better. Next comes a trick where we rewrite $\ddot{\theta}$

$$\ddot{\theta} = \frac{d\dot{\theta}}{dt} = \frac{d\dot{\theta}}{d\theta}\frac{d\theta}{dt} = \dot{\theta}\frac{d\dot{\theta}}{d\theta} \tag{2.70}$$

Substituting this into the EOM, Eq. (2.69),

$$m\ell^2\dot{\theta}(d\dot{\theta}) + mg\ell\sin\theta(d\theta) = 0 \tag{2.71}$$

Integrating,

$$\frac{1}{2}m\ell^2\dot{\theta}^2 - mg\ell\cos\theta = C \tag{2.72}$$

where C is an arbitrary constant. This of course is the same energy conservation we wrote earlier, Eq. (2.68).

Since the total energy of the simple pendulum is conserved, its time derivative should be zero. Try enforcing this and see what you get. After you have done this, do you see the connection between energy conservation and Newton's laws? Would energy perhaps be an independent way to get the EOM? How about non-conservative systems when the total energy is *not* conserved?

When we derived the energy expression for the motion of the pendulum, we did not consider tension in the suspension rod at all. Looks like a terrible oversight; or is it? What would be the potential energy contribution made by the tension? If you are in doubt, go back to the *definition* of the potential energy and compute the work done from Eq. 2.58. Tension belongs to the category of the so-called *constraint forces*; it is the force that essentially ensures that the pendulum moves in a circular fashion (in this case). Do such constraint forces contribute to the energy? Can you think of more such constraint forces?

2.4   Numerical Solution

So far, we have solved the linearized equations by using standard procedures, or by the so-called *analytical* method. But, as already noted, the solutions we obtained were only valid for small perturbations from the equilibrium positions. So, what if we wanted to get a more accurate solution, or a solution valid for larger deviations from the equilibrium positions? In other words, what if we wanted to predict the *true* motion of the pendulum (or at least truer than the approximate linearized solutions)? Then, we have no recourse other than to go back to the original nonlinear equation we derived. Unfortunately, this is not quite a lollipop, and we are going to have to sweat a little.

Although there are analytical and asymptotic techniques available to solve the pendulum problem and a limited number of other nonlinear ordinary differential equations (ODEs), let us focus on one of the most common and popular techniques, the numerical method (also called numerical simulation, or direct integration), which has become the staple tool of the practicing engineer.

There is indeed a way to "solve" the pendulum equation analytically. We used quotes because it will result in a solution in terms of elliptical integrals that will still need a computer or tables to evaluate. More information can be found in a delightful book on nonlinear dynamics by Strogartz [10]. For a discussion and an overview that is more relevant to vibration read Chapter 4 in *Harris' Shock and Vibration Handbook* [11].

There are many methods available today for numerical simulation. Some are better suited for some specific problems, other routines for other problems. This section will not discuss the methods in general, but will describe two methods: an extremely simple Euler method, and a general-purpose popular routine called the Runge Kutta algorithm (which is more than 50 years old and predates modern computers!). General notions of accuracy and stability that are important concepts for the user of the algorithm will also be discussed. The assumption we will make is that you will have access to well-written programs that you can use such as the ones provided by MATLAB. However, it is still important to know how best to use these algorithms. It is fairly easy (and surprisingly common) to get garbage coming out of simulation codes. Knowing how to use them can help reduce some common errors.

There is another important cautionary statement that should be stated here. With the advent of powerful computers at every level, computer simulation including numerical integration has become pervasive and easy to implement and carry out. This is very tempting and the rush to get a computer solution can often trump basic understanding. Numerical integration should always be used in conjunction with the basic analytical calculations. This serves a couple of purposes. First, it gives a sense of what we can expect from the computer simulation before we start so that we will be better positioned to catch silly (and not so silly) errors that almost always happen in computer programs. Second, it often serves as a baseline for comparison of results with those from computer studies of more complex models. It should also be noted that there are indeed many analytical (and so-called semi analytical) techniques which can give better insights into the physics of nonlinear systems than a "brute-force" numerical integration. They can also help perform parametric analyses useful for design that is very painful and intractable with numerical simulation. Of course, numerical simulation is still very useful, especially for final verification of complex system designs. So, we will proceed to explore numerical integration keeping this cautionary note in the back of our mind.

2.4.1 Euler Method

Now, we are ready to solve a differential equation numerically. For simplicity, let us consider a single first-order equation, and solve it by the Euler method, which is a fairly simple and intuitive algorithm.

$$\dot{x} + 2x = 0.1 \sin 5t \tag{2.73}$$

with the initial condition, $x(0) = 0.25$. When we say "solve" we mean that we want to determine the future evolution of x; in other words, we want to know what happens to x as time evolves either in *explicit* analytical expressions, or in graphical form. It is important to define the word "solve" now, since sometimes (as you will see later), what we call a "solution" may still need extensive computation (such as elliptic integrals, or convolution integrals). This is also a good time to remind yourself that a differential equation is basically an *implicit* statement of the dependence of a variable (here, x, the dependent variable) on the independent variable (here, t).

The first step is to discretize time into many, many small intervals, each of length Δt (say equal to 0.01 s). We already know x at time 0, and we want to know x at Δt. This is actually quite easy as we can compute the slope of x at $t = 0$ from the equation we are trying to solve, Eq. (2.73).

$$
\begin{aligned}
\dot{x}(0) &= -2x + 0.1 \sin 5t \\
&= -2x(0) + 0.1 \sin (0) \\
&= 0.5
\end{aligned}
\tag{2.74}
$$

This means we can compute x at the first time step with the first-order Taylor series approximation (the same series we used when we linearized the simple pendulum equation) using

$$x(0.01) = x(0) + \Delta t\, \dot{x}(0)$$
$$= 0.25 + (0.01)(0.5) = 0.255 \tag{2.75}$$

Now, we have two points on the solution curve: $x(0)$ and $x(0.01)$. Then, we repeat this using $\dot{x}(0.01)$ and so on marching forward in time until we either run out of patience or we are satisfied that we have sufficient information to predict the response of the system.

The equation we are trying to solve is a simple linear differential equation and can be solved analytically as well. In fact, the solution is

$$x(t) = 0.2672e^{-2t} - 0.0172\cos 5t + 0.0069\sin 5t \tag{2.76}$$

So, we are in a position to compare the approximate solution to the exact solution and try to understand the extent of errors, and how to minimize them. Of course, typically, we do not have the exact solution to compare to; but there are still other ways of estimating the errors in all numerical methods, and should be used. This is *very* important; since we will always have errors, we need to know *how wrong* we are to have some confidence in our answers.

Figure 2.15 shows the solution obtained using the Euler method using various time steps (0.001, 0.05, and 0.1 s). The analytical solution is practically indistinguishable from the best numerical solution (with $\Delta t = 0.001$); it is hence easier to compare the errors, as shown in Fig. 2.16. Note how the errors increase as the integration time step is increased. Of course, the larger time step will result in a smaller computational time. Flush with gigabytes of RAM in some device, probably in your shirt pocket, you are probably wondering what the big deal is about the computation time. It is certainly not an issue for this problem; however, as the complexity increases with the order of

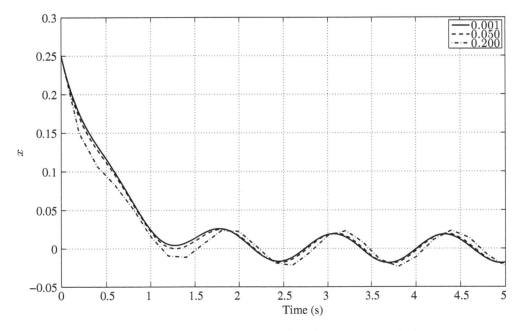

Figure 2.15: *Solution of Eq. (2.73) by Euler method*

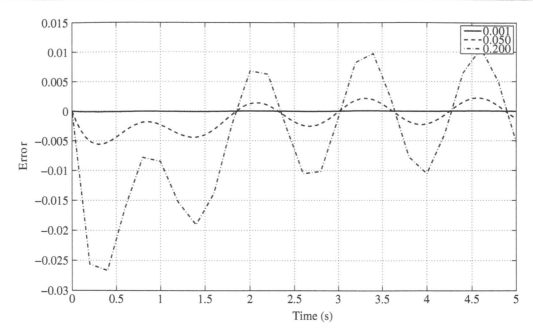

Figure 2.16: *Error in the solution of Eq. (2.73) by Euler method*

the system and nonlinearities as well, the computational time can add up quickly to the point that numerical simulation can become intractable if not impossible even on the most powerful computer hardware. Hence, it is always important to remember that accuracy and speed are often conflicting parameters, and one has to be willing to trade one for the other.

Listing 2.2 shows the MATLAB implementation of the Euler method; note that this program calls the function *numexfcn*, which supplies the ODE to be solved and is shown in Listing 2.3.

Listing 2.2: *Numerical solution by Euler method*

```
% solve ẋ + 2x = 0.1 sin 5t  using  the  Euler  method

    tfinal=5;
    delta_t=0.1;
    time=(0:delta_t:tfinal);
% initial  condition
    x(1)=0.6;

    for  i=1:length(time)−1,
        t=time(i);
%    compute xdot at ith time using  the ODE
        xdot(i)=numexfcn(t,x(i));
        x(i+1)=x(i)+delta_t*xdot(i);
    end;

    plot(time,x);
    grid on;
    xlabel('t');
    ylabel('x');
```

Listing 2.3: *Program supplying the ODE for the Euler method*

```
function xdot=numexfcn(t,x)
% supply the differential equation for the example

  xdot(1)=-2*x(1)+ 0.1*sin(5*t);
```

2.4.2   Runge Kutta Method

There is actually a family of Runge Kutta (RK) methods; what we describe here is a simple explicit fourth-order method and will work in many instances that you are likely to encounter. In this case, the incrementing procedure is more sophisticated than in the Euler method and is given as follows. Here, i denotes the integration step, and h is the integration time interval.

$$x_{i+1} = x_i + \frac{h}{6}(k_1 + 2k_2 + 2k_3 + k_4) \tag{2.77}$$

where

$$
\begin{aligned}
k_1 &= f(x_i, t_i) \\
k_2 &= f(x_i + hk_1/2, t_i + h/2) \\
k_3 &= f(x_i + hk_2/2, t_i + h/2) \\
k_4 &= f(x_i + hk_3, t_i + h)
\end{aligned}
\tag{2.78}
$$

For a rigorous discussion of the errors, see any good reference on numerical methods or ODEs (*Ordinary Differential Equations* [12], for example). An example of the RK implementation is shown in Listing 2.4; it uses the same ODE routine description that was shown in Listing 2.3.

Listing 2.4: *Numerical solution by the fourth-order RK method*

```
function [tout, yout] = oderk(FunFcn, t0, tfinal, y0, npts)
% Integrate a system of ordinary differential equations using
% fixed step 4th order Runge-Kutta method.
% [T,Y] = oderk('yprime', T0, Tfinal, Y0) integrates the system
% of ordinary differential equations described by the M-file
% YPRIME.M over the interval T0 to Tfinal and using initial
% conditions Y0.
% [T, Y] = oderk(F, T0, Tfinal, Y0, 125) uses 125 time pts)
% INPUT:
% F      - String containing name of user-supplied problem description.
%          Call: yprime = fun(t,y) where F = 'fun'.
%          t      - Time (scalar).
%          y      - Solution column-vector.
%          yprime - Returned derivative column-vector.
% t0     - Initial value of t.
% tfinal- Final value of t.
% y0     - Initial value column-vector.
% npts   - Total number of time points (Default: npts = 100).
%
% OUTPUT:
% T  - Returned integration time points (row-vector).
% Y  - Returned solution, one solution column-vector per tout-value.
% C. Nataraj

% integration time step
deltim = (tfinal - t0) / (npts-1);
```

```
time = t0 ;
tout (1 ,1) = time ;
y = y0 ;
yout (1 ,:) = y0 ;
yp = feval (FunFcn ,time ,y0 );
step = 0;

while (time<tfinal )
    step = step + 1;
    time = time + 0.5 * deltim ;
    ytem = y + 0.5 * deltim * yp;
    ak1 = feval (FunFcn ,time ,ytem );

    ytem = y + 0.5 * deltim * ak1;
    ak2 = feval (FunFcn ,time ,ytem );

    time = time + 0.5 * deltim ;
    ytem = y + deltim * ak2;
    ak3 = feval (FunFcn ,time ,ytem );

    yout (step +1 ,:) = y+deltim *(yp+2*(ak1+ak2)+ak3) / 6.0;
    y = yout (step +1 ,:);
    yp = feval (FunFcn ,time ,y );
    tout (step +1 ,1) = time ;
end ;      %  of  while  loop  on  time
```

Here, we will do an ad hoc comparison of the Euler method with the fourth-order RK method, just to provide a feel for numerical methods. Let us take the same simple problem for which we have an analytical solution and run it through both the methods. Figure 2.17 shows the comparison in the errors with a crude integration time step (0.1 s); note how much better the RK method is. This means that, in general, a better solution with less error can be obtained with the RK method; or,

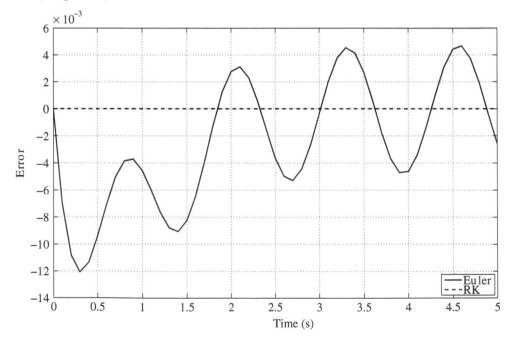

Figure 2.17: *Comparison of RK with Euler method*

a larger integration time step can be used (and therefore, faster execution time) for the same error. Either way, it is the error criterion that drives these decisions.

2.4.3 Multiple Differential Equations

In general, we do not have one first-order ODE to solve, but a higher-order ODE (like in the case of the simple pendulum), or a set of many ODEs. The extension of the above numerical methods to the general case is very straightforward, and is not described here. However, it is important to note that practically all computer systems expect the ODEs to be supplied in first-order form. In order to understand that, let us take a second look at the nonlinear simple pendulum EOM.

$$\ddot{\theta} + \frac{g}{\ell} \sin \theta = 0 \tag{2.79}$$

From basic mathematics it follows that any such second-order equation can be written in terms of two first-order equations without any loss of generality or accuracy. In fact, there are infinite ways of doing that, and hence, what we describe below is only one possible method. Let us define two new *state variables* as follows:

$$x_1 = \theta \tag{2.80}$$
$$x_2 = \dot{\theta} \tag{2.81}$$

Then, the EOM can be rewritten as the following two first-order equations.

$$\dot{x}_1 = x_2 \tag{2.82}$$
$$\dot{x}_2 = -\frac{g}{\ell} \sin x_1 \tag{2.83}$$

Or, in general matrix terms,

$$\dot{x} = f(t,x) \tag{2.84}$$

Note that we included t on the right-hand side, recognizing that an explicit function of time might get into the equation (most notably due to an applied force). This equation is now in the form suitable for numerical processing and can be coded appropriately.

To get our feet wet, let us solve the pendulum equation replete with the ugly nonlinearity—this will answer the question we raised about the accuracy of linearization. Using MATLAB code shown in Listing 2.5 and Listing 2.6 we get the responses shown in Fig. 2.18. Instead of the Euler code that we came up with we are now using a more sophisticated MATLAB code to solve the equations, but all the essential algorithmic concepts are still the same. The particular routine we are using here is called *ode45*, which is a variable step RK routine (not listed here). *ode45* calls the routine, *pendeqns*, which is where we supply the equations to be solved, namely, Eqs. (2.82) and (2.83). Figure 2.19 shows a flowchart that should give you an idea of how you will structure your program whether you are writing it in MATLAB or C or anything else. Later, in the problem section you will have plenty of opportunities (meaning, assignments!) to carry out this computation and to compare with the linearized response.

Listing 2.5: *Simple pendulum solution using MATLAB*

```
% simple pendulum nonlinear equation solution

% make the parameter values available in all functions
  global gravity ell

  gravity=9.81;
  ell=0.2;

  tinit = 0.0;
```

```
   tfinal = 3.0;
% generate a time vector automatically
   tspan=[tinit tfinal];

   inicon1 = [10*pi/180 0.0]'; %  ' denotes transpose
% this calls a MATLAB program, ode45
   [time1, soln1] = ode45('pendeqns',tspan,inicon1);

   inicon2 = [170*pi/180 0.0]'; %  ' denotes transpose
% this calls a MATLAB program, ode45
   [time2, soln2] = ode45('pendeqns',tspan,inicon2);

   plot(time1,soln1(:,1)*180/pi,time2,soln2(:,1)*180/pi);
   grid on;
   xlabel('Time_(s)');
   ylabel('\theta');
```

Listing 2.6: *Simple pendulum equations in a MATLAB function*

```
function xdot=pendeqns(time,x)
% supply the eom for the dynamics of a pendulum
% θ̈ + g/ℓ sin θ = 0
%
   global  gravity ell

   xdot(1,1) = x(2);
   xdot(2,1) = (−gravity*sin(x(1)))/ell;
```

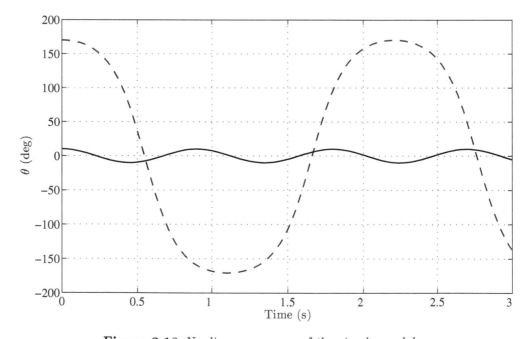

Figure 2.18: *Nonlinear response of the simple pendulum*

The two responses are shown for initial conditions close to the hanging and the upright positions (10° and 170°, respectively). Note that the nonlinear model predicts oscillations for both positions, which is different from the exponential instability, wrongly predicted by the linearized model.

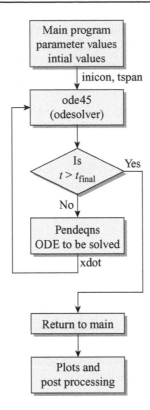

Figure 2.19: *Flowchart of the numerical simulation algorithm*

2.5   Review Questions

1. What is the need to even look at nonlinear equations?

2. What is the principal assumption used to derive linearized equations?

3. Define degrees of freedom of a dynamic system. Are the degrees of freedom unique?

4. How many degrees of freedom does a particle have if it is free to move on a two-dimensional surface?

5. How many degrees of freedom does a particle have if it is free to move in the three-dimensional space?

6. How many degrees of freedom does a system of two particles have if they are both free to move in the three-dimensional space?

7. How many degrees of freedom does a rigid body have if it is free to move on a two-dimensional surface?

8. How many degrees of freedom does a rigid body have if it is free to move in the three-dimensional space?

9. How many degrees of freedom does a system of two rigid bodies have if they are both free to move in the three-dimensional space?

10. How many degrees of freedom does a simple pendulum have if it is not constrained to move on a two-dimensional surface?

11. How many degrees of freedom does a Foucault pendulum have? (This is a good time to learn about the Foucault pendulum and how it can be used to prove rotation of the earth—go find a book or a web site!)

12. How is the number of differential equations obtained in the modeling process related to the number of degrees of freedom?

13. In Newton's second law, what is the important restriction on acceleration?

14. What makes the simple pendulum a nonlinear system?

15. In a simple pendulum, is the tension equal to the weight? If not, why not?

16. What is the role of gravity while working out the kinematics of a problem?

17. How would the kinematics of the simple pendulum change if its length were not fixed (for example, in the case when it is suspended by an elastic string)?

18. Does the natural period of a pendulum clock depend on its mass? Why or why not?

19. What is Euler's formula?

20. What is the form of the solution assumed when we solve linear homogeneous ODEs?

21. Why is the linear unstable solution not valid for large displacements?

22. Why does a man jumping out of a plane with a parachute reach a terminal velocity? What is that value typically?

23. When Newton's second law is applied to a rigid body, which point should be considered for linear acceleration?

24. In a rigid body, what are the special points about which the moment equation is valid?

25. In a conservative system, what is the work done on a closed path?

26. Is the law of kinetic energy only valid for conservative systems? Why or why not?

27. Which of the following are not conservative forces: gravity, sliding friction, spring force, magnetic pull, tension in a cable?

28. When energy is conserved in a mechanical motion, is heat generated? Why or why not?

29. Give an example of a practical system where energy is fully conserved.

30. When energy is conserved (in a simple pendulum, for example), what determines the amount of energy that is in the system [C in Eq. (2.68)]?

31. Which fundamental mathematical principle are the numerical methods of solving ODEs based on?

32. In numerical simulation, what is the fundamental trade-off?

33. In numerical simulation, how do you decide what the integration time step should be?

Problems

2.1 Linearize each of these nonlinear models about 0.

    (a) $\ddot{x} + \cos x = 0$

    (b) $\ddot{x} - \dot{x}^2 + x = 0$

    (c) $\ddot{x}^2 + 5\dot{x} + 3x = 0$

    (d) $\ddot{x} + e^{3x} = 0$

    (e) $e^x \ddot{x} + 2\dot{x} + \sin x = 0$

    (f) $\ddot{x} + (2x^3 - x^2 + 3x) = 0$

    (g) $\ddot{x} + (\sin t)\, x = 0$ (Be careful here!)

    (h) $\ddot{x} + \dot{x}^2 + x^3 = 0$ (Do you think you can solve the linearized equation here?)

(i) $\ddot{x} + \frac{x}{1-0.5x^2} = 0$

(j) $\ddot{x} + \sin 2x \cos x \tan x = 0$

2.2 Solve the following linear differential equations. Plot the response and look at them critically.

(a) $\dot{x} = \sin t$

(b) $\dot{x} + 2x = 0$

(c) $\dot{x} + 2x = \cos t$

(d) $\dot{x} + 3x = e^{-3t}$

(e) $\ddot{x} + 4x = 0$

(f) $\ddot{x} + 9x = \sin 2t$

(g) $\ddot{x} + 9x = \sin 3t$

(h) $\ddot{x} + 2\dot{x} + 100x = \cos 2t$

2.3 For each of the following characteristic equations for linear systems, determine if the response is periodic or exponential or neither; if periodic, determine the frequency and period of oscillation. Sketch each of them.

(a) $s^2 - 2 = 0$

(b) $s^2 + 9 = 0$

(c) $s^2 + 36 = 0$

(d) $s^2 + 100\pi = 0$

(e) $s + 3 = 0$

(f) $s^2 + s + 4 = 0$

(g) $s^2 + 10s + 4 = 0$

(h) $s^2 + 8s = 0$

(i) $s^2 = 0$

2.4 Express each of the following in trigonometric form, unless it is not possible to do so. If it *is* possible, determine the amplitude and frequency of oscillation. Sketch each of them. If you are using MATLAB, you can verify your answer as well, since MATLAB is aware of Euler's formula.

(a) $2e^{4t} - 5e^{-4t}$

(b) $6e^{8it} - 6e^{-8it}$

(c) $(2 + 3i)e^{3it} + (2 - 3i)e^{-3it}$

(d) $2.5e^{(3+2i)t} + 2.5e^{(3-2i)t}$

(e) $1.5e^{4it} + 1.5e^{-9it}$

(f) $2e^{(2-2i)t} - 2e^{(2+2i)t}$

(g) $e^{(2-6i)t} - e^{(3+6i)t}$

(h) $e^{(0)t}$

2.5 Determine whether or not each of the following is an oscillatory solution; if it is, find the frequency, period, amplitude, and phase for each of them. Verify by plotting them.

(a) $\theta(t) = 3\sin(2\pi t - 0.5)$

(b) $\theta(t) = 5\cos(5\pi t + \pi/2)$

(c) $\theta(t) = \sin(25t) - 2.2\cos(25t + \pi/4)$

(d) $\theta(t) = 10(\sin(3\pi t + \pi/8) + 2\cos(3\pi t - \pi/10))$

(e) $\theta(t) = 8\sin(50t) - 6\cos(55t)$

(f) $\theta(t) = 100\sin(9t) + 50\cos(9t - \pi/2)$

2.6 A harmonic motion has an amplitude x_0, and a period τ. Determine the maximum velocity and acceleration.

2.7 A harmonic motion has a frequency ω and its maximum velocity is v_0. Determine its amplitude, period, and maximum acceleration.

2.8 A harmonic motion has a frequency ω and its maximum acceleration is a_0. Determine its amplitude, period, and maximum velocity.

2.9 Determine the rotational rate of the earth about its axis in radians per second. Also determine the tangential linear velocity at its surface. (That is the speed with which you and I are moving—that sure beats a Ferrari, doesn't it?!)

2.10 A simple pendulum is set in oscillation from its rest position by giving it an angular velocity of 1 rad/s. It is found to oscillate with an amplitude of 0.5 rad. Find the natural frequency and length of the pendulum.

2.11 Assume a general form for displacement of a system in harmonic motion; show that its acceleration is always out of phase by 180° with the displacement.

2.12 We need to estimate the moment of inertia of an automobile connecting rod (Fig. 2.20) for dynamic studies. Hence, an experiment is set up by suspending it about one of its holes; the natural period of oscillation is found to be 2 s. The center of mass is located 0.3 m from the point of suspension, and the connecting rod has a mass of 2.5 kg. Find its moment of inertia about the center of mass.

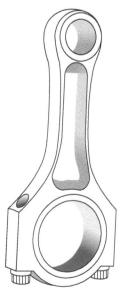

Figure 2.20: *An automobile connecting rod*

2.13 An bicycle wheel of mass 1.5 kg and radius 0.16 m is suspended about its rim. Its period of natural oscillation is found to be 0.5 s. Find the moment of inertia about its center of mass.

2.14 Consider the simple pendulum again. This time, derive the EOM using fixed Cartesian coordinates, x and y instead of the radial and tangential coordinates that we used.

2.15 Consider the simple pendulum. We identified the particle mass of the pendulum as a system of interest (Fig. 2.3), and applied Newton's equation for a particle. Now, instead of doing that, consider the entire pendulum as your system, as shown in Fig. 2.21; in this case, it is clearly not a particle, but is a rigid body. Derive the EOM (it better be the same equation that we obtained!).

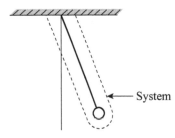

Figure 2.21: *Modeling the simple pendulum as a rigid body*

2.16 Consider the simple pendulum; from the modeling process we obtained *two* equations, but we carelessly cast aside one of them Eq. (2.12), which of course was the equation for tension. Now, linearize this equation also for both the equilibrium positions we found. Draw FBDs of the pendulum in static condition, and verify that you get the same answer for the tension.

2.17 For the simple pendulum, calculate the natural frequencies for each of the following pairs of cases and compare them. Explain the significance.

(a) $g = 9.81\,\mathrm{m/s^2}$, $\ell_1 = 2\,\mathrm{m}$, $\ell_2 = 8\,\mathrm{m}$.

(b) Identical pendulums; one on the earth, the second on saturn.

(c) $g = 9.81\,\mathrm{m/s^2}$, $\ell = 1.2\,\mathrm{m}$, $m_1 = 2\,\mathrm{kg}$, $m_2 = 5.5\,\mathrm{kg}$.

(d) Identical pendulums; one is in the lab, the second is on a rocket that is accelerating vertically upward at 3 g's.

2.18 For the rigid rod pendulum discussed in the text (on page 19), calculate and compare the natural frequencies for each pair of cases and discuss the significance.

(a) One with $I_{c1} = 0.5\,\mathrm{kgm^2}$ and the second with $I_{c2} = 0.8\,\mathrm{kgm^2}$.

(b) Identical lengths; $m_1 = 0.5\,\mathrm{kg}$, $m_2 = 0.75\,\mathrm{kg}$.

(c) Identical masses; $\ell_1 = 0.3\,\mathrm{m}$, $\ell_2 = 0.6\,\mathrm{m}$.

(d) Identical sizes; one made of steel; the second made of aluminum.

(e) Identical sizes; one made of steel; the second made of very light paper. (Is the mathematical model correct here, or do you need to go back to the FBD and modify it?)

2.19 Determine the response of the simple pendulum using the appropriate linearized equation to the following sets of initial conditions. Assume $g = 9.81\,\mathrm{m/s^2}$, $\ell = 0.8\,\mathrm{m}$. Plot them.

(a) $\theta(0) = 30°$, $\dot{\theta}(0) = 0°/s$

(b) $\theta(0) = 0°$, $\dot{\theta}(0) = 0°/s$

(c) $\theta(0) = 170°$, $\dot{\theta}(0) = 3°/s$

(d) $\theta(0) = 180°$, $\dot{\theta}(0) = 0°/s$

(e) $\theta(0) = 5°$, $\dot{\theta}(0) = -0.1°/s$

(f) $\theta(0) = 0°$, $\dot{\theta}(0) = 150°/s$

(g) $\theta(0) = -0.95\pi \, \text{rad}$ and $\dot{\theta}(0) = 2 \, \text{rad/s}$.

(h) $\theta(0) = 0.3 \, \text{rad}$ and $\dot{\theta}(0) = 1 \, \text{rad/s}$.

(i) $\theta(0) = 0$ and $\dot{\theta}(0) = -2 \, \text{rad/s}$.

2.20 The principle of operation of a ballistic pendulum used to determine the speed of bullets is as follows. A particle of mass m_0 traveling horizontally at a speed v_0 strikes the bob of a simple pendulum of mass m and length ℓ and sticks to it, as shown in Fig. 2.22. Determine the subsequent motion of the pendulum.

Figure 2.22: *Ballistic pendulum*

2.21 A pendulum has been suspended with the help of a rope, as shown in Fig. 2.23. Write the EOM and solve for the motion when the rope is cut. Assume $m = 1.5 \, \text{kg}$ and $\ell = 0.65 \, \text{m}$; the mass of the bar is negligible.

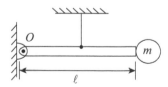

Figure 2.23: *A pendulum tied with a rope*

2.22 A cylinder of weight W and radius r rolls without slipping on a cylindrical surface of radius R, as shown in Fig. 2.24.

(a) Derive the nonlinear EOM.

(b) Linearize the equation about its vertical equilibrium position. Determine the natural frequency of oscillation of the linearized system.

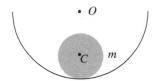

Figure 2.24: *Cylinder rolling inside another cylinder*

(c) Let $R \to \infty$, and determine the limiting value of the natural frequency. In other words, what happens as the cylindrical surface progressively becomes flat? (Think! What should happen physically?)

2.23 Consider an arbitrary-shaped body suspended about a point O, as shown in Fig. 2.25; it is often called the compound pendulum. Its center of mass is at C, and it has a mass moment of inertia, I_C about its center of mass. Derive the linearized EOM about the vertically hanging position. Determine its natural frequency of oscillation and plot it as a function of the distance OC. Explain the significance of this plot.

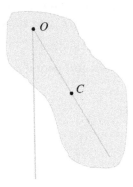

Figure 2.25: *Compound pendulum*

2.24 Consider a falling body in the atmosphere. As discussed earlier, if there is no other force acting on it, it would continue to accelerate until it reached the earth. Now, to be more realistic, let us add the buoyant force offered by the atmosphere. This force would act upwards and would be equal (by the principle of Archimedes) to the weight of the air displaced by the body. Note that the density of air is not constant; you may assume that it decreases linearly from the sea level as we go up in the atmosphere. Assume that the falling body is a sphere of radius R, but model it as a particle for writing Newton's second law.

(a) Derive the differential EOM.

(b) Obtain the homogeneous solution.

(c) Use your answer to explain what might happen if, jumping from a plane flying very high, a man opens a parachute right away.

2.25 Derive the EOM for a simple pendulum with a moving support, as shown in Fig. 2.26. Assume that the angle θ remains small, and that $x(t)$ is an arbitrary function of time.

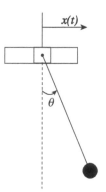

Figure 2.26: *Simple pendulum with moving support*

2.26 A thin uniform rod with mass m and length ℓ is suspended about a point distant a from the end ($AO = a$), as shown in Fig. 2.27. The mass moment of inertia of such a body is $\frac{m\ell^2}{12}$ about its center of mass; you may also need to use the parallel axis theorem: $I_P = I_C + md^2$, where d is the distance between an arbitrary point P and the center of mass C.

(a) Determine the natural frequency of oscillation.

(b) Find the value of $\frac{a}{\ell}$ for which the natural period is minimum. Explain the significance.

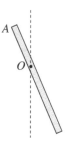

Figure 2.27: *A rod suspended about an arbitrary point*

2.27 Obtain the EOM for a planar compound pendulum shown in Fig. 2.28. Take $\ell_1 = 0.2\,\text{m}$, $\ell_2 = 0.25\,\text{m}$, $\ell_3 = 0.05\,\text{m}$, $\ell_4 = 0.03\,\text{m}$, $m = 2\,\text{kg}$, and I_c is a given known constant.

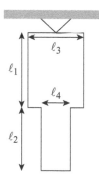

Figure 2.28: *A compound pendulum*

2.28 Archimedes, the Greek mathematician is supposed to have said: "Give me a place to stand and I can move the earth" [13] (Fig. 2.29). But since he predated Newton by almost two millennia, Archimedes did not consider the dynamic aspect of the problem! Suppose we do have the power to hold the earth and pivot it at one of its poles. The earth has a diameter of 6400 km and density, $\rho = 3000$ kg/m^3; assume that the earth is a prefect sphere. Derive the EOM and solve it.

Figure 2.29: *Archimedes and the really big lever (see Mechanics Magazine [14])*

2.29 For the compound pendulum shown in Fig. 2.30, a hook has been attached at its lower end. A man of mass 60 kg is sitting on the hook. Write the EOM and solve for the the initial conditions, $\theta = 0$ and $\dot{\theta} = 1.5$ rad/s. Model the man as a lumped mass and neglect the weight of the hook.

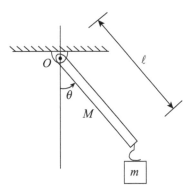

Figure 2.30: *A swinging man on a compound pendulum*

2.30 A hemispherical solid disk is pivoted as shown in Fig. 2.31.

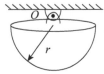

Figure 2.31: *A swinging hemispherical pendulum*

(a) Find the EOM.

(b) Solve for the motion for the initial values, $\theta = 0$ and $\dot{\theta} = 2\,\text{rad/s}$.

2.31 A uniform circular disc of radius a is pivoted at a point O, as shown in the figure. Find the natural frequency of the system. Also, find the value of b for which the natural frequency is a maximum. The mass moment of inertia of such a disc about its center of mass is $\frac{1}{2}ma^2$.

2.32 A swinging mechanism is shown in Fig. 2.32.

(a) Derive the EOM.

(b) Solve the obtained equation with the given initial conditions. The mass of the bar is negligible. Suppose $m = 10\,\text{kg}$, $g = 9.81\,\text{m/s}^2$, $\ell_1 = 1\,\text{m}$, $\ell_2 = 2.02$ m, $\theta(0) = 0$, and $\dot{\theta}(0) = 2\,\text{rad/s}$.

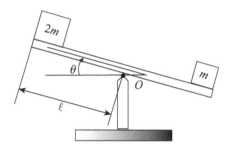

Figure 2.32: *A swinging mechanism*

2.33 A cylinder of mass m is welded to a bar of mass m pivoted around the point O. For the system shown in Fig. 2.33, do the following.

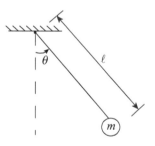

Figure 2.33: *A cylinder as a pendulum*

(a) Derive the EOM.

(b) Solve the equation for the given initial conditions. Suppose $m = 1\,\text{kg}$, $r = 0.05\,\text{m}$, $\ell = 0.5\,\text{m}$, $h = 0.25\,\text{m}$, $r = 0.05\,\text{m}$, $\theta = 30°$, and $\dot{\theta} = 0$. h is the height of the cylinder and r is its radius.

2.34 Find the EOM of an inverted pendulum shown in Fig. 2.34 including a bar of mass $2m$ and a welded mass m, at its end. Solve for its motion when the rope is cut. Use $m = 0.1\,\text{kg}$, $\ell = 0.6\,\text{m}$, and $\theta = 30°$.

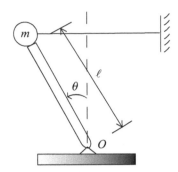

Figure 2.34: *An inverted pendulum*

2.35 Consider a block sliding on a curved surface that is circular with radius r, as shown in Fig. 2.35.

    (a) Find the angle at which the block would lose contact with the surface. In order to find this, you will need to carry out the necessary steps (including the FBD) *before* it leaves the surface.

    (b) Now, also write the equations of motion after it leaves the surface.

    (c)   Develop a numerical solution that includes both phases of the dynamics of the block (when it is sliding on the surface, and when it is flying through the air).

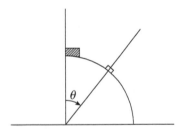

Figure 2.35: *Block sliding on a curved surface*

2.36 A uniform circular cylinder of mass m and radius a rolls without slipping on a plane inclined at an angle α with the horizontal, as shown in Fig. 2.36. Find its acceleration.

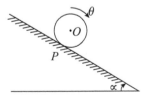

Figure 2.36: *Cylinder rolling on an inclined plane*

2.37 A rigid rod of mass m and length $4a$ is centered on a fixed cylinder of radius a, as shown in Fig. 2.37. Assume small rocking motions without slip and derive the EOM. Find the natural frequency of oscillation.

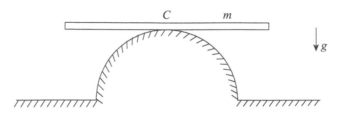

Figure 2.37: *Rigid rod rocking on a cylinder*

2.38 Solve the nonlinear equation governing the motion of the simple pendulum, Eq. (2.11) numerically. Let $\ell = 0.3$ m; and use initial conditions, (a) $\theta(0) = \pi/4$, and $\theta(0) = 1.1\pi$, with velocity equal to zero.

(a) Compare with the linearized solutions and comment on the extent of and reason for the difference between linearized and nonlinear solutions.

(b) Determine the hinge reactions as functions of time and compare it to the linearized solution. (Normalize the forces by dividing them by mg.)

2.39 For each of the cases in the Problem 2.19, find and plot the kinetic and potential energies of the nonlinear system.

2.40 For each of the cases in Problem 2.19, determine the nonlinear tension in the string, and plot it as a function of time.

2.41 For each of the differential equations in Problem 2.1, carry out a numerical integration of the nonlinear equations and compare with the analytical solution of the linearized equations. Use progressively larger initial conditions to simulate larger perturbations from the equilibrium positions and comment on the effect on the discrepancy between the linearized solution and the nonlinear solution (this comment needs to be *quantitative*).

2.42 The pendulum is constrained to remain in the plane of the sketch as the string is pulled through the hole at the uniform speed v_0, as shown in Fig. 2.38. This model represents for example the retrieval of a mass on a tether in a satellite. Somewhat unscientifically, it is also called the "sucking spaghetti" problem; imagine pulling a string of spaghetti (or noodle) into your

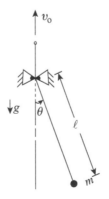

Figure 2.38: *Pendulum with changing length*

mouth at a constant speed. (*Hint*: do not do this at a formal dinner, and especially not if it is coated with tomato sauce!) Assume $\ell = \ell_0 - v_0 t$, where ℓ_0 is its initial length, and derive the EOM. Linearize the equation about the zero position. Then, solve it using a numerical method; experiment with different numerical values for the retrieval speed v_0. Also, carry out the solution when ℓ is increasing, or, $\ell = \ell_0 + v_0 t$.

2.43 A simplified non dimensional model of a magnetic levitation (MAGLEV) train leads to a nonlinear oscillator equation with displacement $x(t)$, with the magnetic force given by

$$F_s = F_1 - F_2 - F_3$$

where

$$F_1 = \frac{K}{(1-x)^2},$$

$$F_2 = \frac{K}{(1+x)^2},$$

and

$$F_3 = 4x.$$

The EOM is hence

$$m\ddot{x} = -F_s$$

$$\Rightarrow m\ddot{x} + \frac{4Kx}{(1-x^2)^2} - 4x = 0.$$

(a) Linearize the EOM with $x_0 = 0$ as the operating point.

(b) Assume $K > 1$ and solve the linearized EOM showing all the steps. What is the natural frequency of the system?

(c) What will happen to your solution if $K < 1$?

2.44 Pendulum vibration absorbers have been installed in high-rise buildings, bridges, and other structures to attenuate wind-excitation vibrations. Compared to a conventional vibration absorber made of a movable mass and a flexible member, a simple pendulum is more rugged, easily constructed, and suitable for heavy duty jobs. In a technical article [15], the author proposes a rotational pendulum absorber for neutralization of vertical disturbances. The characteristic frequency of the pendulum absorber can be dynamically tuned over a wide range by adjusting the rotational speed.

Figure 2.39 depicts a pendulum revolving about a vertical axis, in contrast to a simple pendulum swinging horizontally. For simplicity, the rotational pendulum is assumed to have a length ℓ and a lumped mass m at the tip.

(a) Derive the EOM of the pendulum.

(b) Linearize the equation about the equilibrium points.

(c) Investigate the natural frequency of the rotational pendulum when its base is turning at a constant speed, denoted by ω_0.

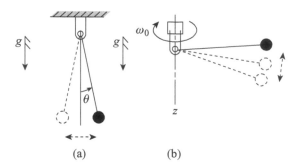

Figure 2.39: *Pendulum vibration absorber (see Wu [15])*

2.45 Each of these equations models a dynamic system (such as the simple pendulum). Find its equilibrium points, and linearize the equations about each of them. Compare the linearized response of the system with a numerical simulation of the nonlinear model. You may want to re-run the simulation for a variety of initial conditions to get a complete picture of the dynamic response.

(a) $\ddot{x} + 9x + 0.1x^3 = 0$ (This is a special equation called the Duffing oscillator; do you notice the similarity with the pendulum?)

(b) $\ddot{x} + 0.1(x^2 - 1)\dot{x} + 4x = 0$ (This is called the van der Pol oscillator; boy, are you going to be surprised by its nonlinear response!)

(c) $\ddot{x} + 3(x^2 - 1)\dot{x} + 4x = 0$ (This is the same van der Pol oscillator with a stronger nonlinearity; its response is said to constitute *relaxation oscillations*.)

(d) $\ddot{\theta} + (1 + 0.5\cos\theta)\dot{\theta} + \sin\theta = 0$

(e) $\ddot{x} + x^3\dot{x} + 2x + x^2 = 0$

(f) $\ddot{x} + 0.5\dot{x}\left(x^2 + \dot{x}^2 - 1\right) + x = 0$

(g) $\ddot{x} + 100(x^2 - 4)\dot{x} + x = 1$

(h) $\ddot{x} + (1 + 0.1 + 0.1\cos 2t)\sin x = 0$ (This is the model for a swing with a person on it who is pumping it periodically.)

(i) $\ddot{x} + (1 + 0.2\cos t)x = 0$ (This is called the Mathieu equation and represents various vibrating systems with the so-called *parametric excitation*.)

(j) $\ddot{x} + x + 0.15\dot{x} + 0.1x^3 = 0.1\sin t$ (This is the forced Duffing oscillator.)

2.46 Consider a hanging rigid bar that is rotating at a constant speed Ω about its axis. It will swing out because of the "centrifugal" effect. We are interested in the vibrational motion about that swung out position.

(a) Derive the nonlinear EOM, and determine the equilibrium positions.

(b) Investigate the stability of the equilibrium positions; clearly, it will be a function of the speed Ω.

(c) For small displacements about the equilibrium positions, determine the linearized response of the system; if oscillatory, determine the natural frequencies.

(d) Carry out a numerical simulation; you may use the following numerical values for this step. $m = 1$ kg, $\ell = 0.3$ m, $0 < \Omega < 500$ rad/s.

2.47 Consider the design of a playground swing (Fig. 2.40). Let us assume that it can be modeled as a simple pendulum. Assume the parameter values shown in Table 2.1.

Figure 2.40: *Playground swing*

Mass of the swing	5 kg
Mass of the "swingers"	150 kg
Length of the swing	3.7 m
Ultimate strength of steel	4×10^8 Pa

Table 2.1: *Parameter values*

(a) Solve the nonlinear EOM numerically; use three different sets of initial conditions (you pick them from your interpretation of reality!).

(b) Compare the solution to the linearized solution.

(c) Determine the tension in the cable using the numerically obtained solution and compare it to that obtained from the linearized solution.

(d) Assuming that the only design criterion is that the stress due to tension has to be less than the ultimate tensile strength, determine the minimum cross-sectional diameter of each cable (assume two cables).

3 SDOF Systems: Free Undamped Response

Results! Why, man, I have gotten a lot of results. I know several thousand things that won't work.

—Thomas A. Edison

In the previous chapter, we looked at the quintessential oscillatory system: the pendulum. We could spend the rest of the year studying the pendulum and we would be able to explore all kinds of fascinating phenomena that we are ever likely to observe in practical engineering systems. However, in order to keep us from falling asleep from ennui, we will look at other models of vibrating systems.

Before we do that, however, let us consider a practical example of a vibrating system that would probably come to your mind as soon as we mention the word, *vibration*: an automobile on suspension springs. An automobile is a very complicated dynamic system that is capable of motion in several forms. It can roll about its longitudinal axis, pitch about its midpoint, yaw about a vertical axis, and in addition, it can translate in all the three directions. In other words, if modeled as a single rigid body, it would have six degrees of freedom (DOF). Let us say, however, that we are not interested in a full-blown analysis of the car. We want to simplify it as much as

Figure 3.1: *Simplest model for a car*

possible. The simplest model for it then would be that of a particle mass on a spring and this is the model that is going to occupy our attention for much of this book (Fig. 3.1). Not only is this a good model for the basic bouncing motion that we would experience when the car goes over a bump, but it is also a good simplified single-degree-of-freedom (SDOF) model for most vibrating systems.

Notice how we have been saying "model" instead of "system." The physical system can be, and is often, very complicated. A model, on the other hand, is an idealization that suits our purpose. For example, we have taken a car with a large mass, uneven shapes, doors that open, rotating crankshafts, pulsing brakepads, and screaming babies in it, and modeled it as a simple particle that, by definition, has no length or area. Does this seem like an oversimplification likely to lead to gross errors? Actually, it is probably a sufficiently accurate model for preliminary studies of bouncing motion. Can you think of a model for the car that is slightly more accurate? In what way would your model complicate our mathematics?

3.1   Mass-Spring Model

Consider an idealized mass supported on a rigid support by means of a spring (Fig. 3.2). Considering the mass as our system of interest, if we draw a line around it to isolate it from its environment (Fig. 3.3), we note that the only contact force acting on it is that due to the spring. Among the noncontacting forces acting on it, we retain only gravitational attraction as being non-negligible. These forces are shown in the free body diagram (FBD) (Fig. 3.4).

Next, we denote by x, the possible vertical displacement of the mass. We note from experience that the force that needs to be applied to a spring is roughly proportional to the desired extension of the spring. In general, the force-extension for a spring would probably look like that in Fig. 3.5. However, as indicated in the figure, we will assume linearity (or proportionality) to keep matters simple. In other words, we use a linearized expression of the force about an operating point using the methods of Section 2.1.2 (on page 12), which amounts to assuming small displacements.

In the case of the present configuration, in view of the assumption of linearity, the mass needs to exert a force kx to the spring in order to stretch it by an amount x; k is the proportionality constant that we will call the spring constant. Hence, the force exerted by the spring on the mass F_s is, by Newton's third law, equal to kx and directed opposite to the displacement.

$$F_s = kx \tag{3.1}$$

A relation such as Eq. (3.1) that defines the nature of the forces applied to the mass by the spring is called the *constitutive relation* of the spring.

Figure 3.2: *Mass on a spring*

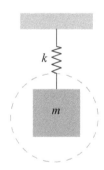

Figure 3.3: *Mass on a spring: system*

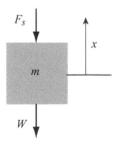

Figure 3.4: *FBD of mass-spring system*

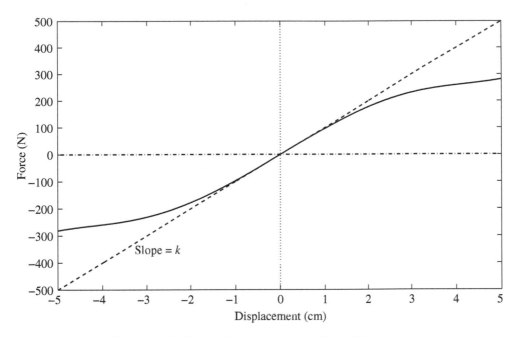

Figure 3.5: *Spring force characteristic and linearization*

The kinematics in this problem, which is derived from geometry, is very simple. Since the support to which the spring is attached is considered rigid and at rest, the absolute acceleration of the mass with respect to inertial space is given by

$$a = \ddot{x} \tag{3.2}$$

The final step is the assembly of the above pieces of information using Newton's second law,

$$\vec{F} = m\vec{a} \tag{3.3}$$

which reduces to a scalar equation

$$-F_s - F_w = m\ddot{x} \tag{3.4}$$

Here, F_w is the force due to the gravitational attraction of the earth on the body of interest, which we commonly refer to as the weight of the body. Hence,

$$F_w = mg \tag{3.5}$$

where g is the "acceleration due to gravity" assumed constant. Combination of the above equations results in

$$-kx - mg = m\ddot{x} \tag{3.6}$$

Reorganization yields a differential equation

$$m\ddot{x} + kx = -mg \tag{3.7}$$

Equation (3.7) is a second-order linear constant-coefficient inhomogeneous differential equation (see Section A.2 for a descriptive summary of the theory of differential equations). Note that the inhomogeneity comes from the action of gravity. It is, in particular, linear because of our assumptions of linearity in the spring force. It has constant coefficients because of the fact that the system parameters, the mass m, and the stiffness k, are not changing with respect to time.

> Can you think of a situation where one of the parameters in the equation of motion (EOM), say, the mass, could vary with time? Would that make the differential equation nonlinear? Think before you answer that question! You might want to check the definition of linearity in Section A.2.1 on page 416. Would the solution of these equations be any more complicated than the constant coefficient equations?

Equation (3.7) is a mathematical model for the physical system represented in Fig. 3.2, which we arrived at by making a series of approximations. Next, we want to determine the displacement of the mass x as a function of time (which is usually the end-objective in a typical problem). For example, if the model is an approximation for the vibration in a machine (such as a lathe) or a structure (such as a bridge), we want to know what the displacement in the machine (or structure) is, how severe it is likely to be, and how exactly it varies with time. In other words, we want to be able to predict the actual displacement in the structure at any future instant of time. This is accomplished mathematically by solving the equation that approximately represents the physical system; i.e., we need to solve Eq. (3.7), so that we can get an explicit expression for $x(t)$.

Recall that the solution of a linear inhomogeneous differential equation is found by adding together the solution of the homogeneous equation (the complementary solution) and the particular integral.

3.1.1 Homogeneous Solution

The homogeneous solution is the solution to the equation from which have been dropped all the terms that do not depend upon x or its derivatives. Or, we need to solve

$$m\ddot{x}_h + kx_h = 0 \tag{3.8}$$

As we did in the case of the simple pendulum [Eq. (2.27) on page 16], we *seek* a solution (based on experience) in the form

$$x_h = Ae^{st} \tag{3.9}$$

where A and s are undetermined constants. Equation (3.9) is of course not the true solution unless it satisfies Eq. (3.8); recall that a solution to any equation is defined to be a function, which when substituted into the equation, satisfies it. Hence, we substitute Eq. (3.9) into Eq. (3.8) to check if the assumed form is indeed the correct solution. That process yields

$$\left(s^2 m + k\right) Ae^{st} = 0 \tag{3.10}$$

Ignoring the trivial solution ($A = 0 \Rightarrow x_h(t) \equiv 0$), we get an *algebraic* equation

$$ms^2 + k = 0 \tag{3.11}$$

which *must* be satisfied for the assumed form of the solution to be the correct solution. Equation (3.11) is a quadratic equation in the constant s and is the *characteristic equation* for this system. Solution yields two imaginary roots

$$s_1 = +i\omega_n, \quad s_2 = -i\omega_n \tag{3.12}$$

where $i = \sqrt{-1}$, and

$$\boxed{\omega_n = \sqrt{\frac{k}{m}}} \tag{3.13}$$

We note that we will need to collect *all* possible solutions of the linear differential equation and add them together to obtain the complementary solution to the equation. At this point, we also note that our solution process yielded no conditions on the arbitrary constant A, and also that this arbitrary constant could be different (in general) for the two families of solutions that correspond to s_1 and s_2. Hence, the total solution to the homogeneous equation is

$$x_h(t) = A_1 e^{s_1 t} + A_2 e^{s_2 t} \tag{3.14}$$

where A_1 and A_2 are arbitrary constants, and s_1 and s_2 are given by Eq. (3.12).

Using Euler's formula [Eq. (A.7)], the solution becomes

$$x_h(t) = (A_1 + A_2) \cos \omega_n t + i (A_1 - A_2) \sin \omega_n t \tag{3.15}$$

With new constants, C_1 and C_2, which are both real,

$$x_h(t) = C_1 \cos \omega_n t + C_2 \sin \omega_n t \tag{3.16}$$

As in the case of the simple pendulum, the above expression can also be written in the following form

$$x_h(t) = A \cos (\omega_n t - \phi) \tag{3.17}$$

where A and ϕ are arbitrary constants.

3.1.2 Particular Solution

The particular solution to the differential equation is found by assuming a form that depends upon the form of the function on the right-hand side. Here, we need to find the particular solution of the following equation

$$m\ddot{x} + kx = -mg$$

The function on the right-hand side is a constant, and we guess a solution that is a constant

$$x_p(t) = C \tag{3.18}$$

Substitution of the assumed form into the differential equation yields

$$kC = -mg \Rightarrow x_p = C = -\frac{mg}{k} \tag{3.19}$$

Note that the particular solution does not involve any arbitrary constants. Also, note that the particular solution to the differential equation obtained above corresponds physically (in this case) to the static equilibrium position of the mass-spring system.

3.1.3 Discussion of Solution

The total solution to the differential equation is the sum of the homogeneous and particular solutions obtained above. Hence,

$$x(t) = A\cos\left(\omega_n t - \phi\right) - \frac{mg}{k} \tag{3.20}$$

The homogeneous solution for a vibrating system is called *free vibration* or *free response* and the particular solution is called the *forced response*. The free response is then what the mass would do, if you kicked it and walked away; the forced response is the response to an external force (that is independent of the displacement or velocity of the mass) that is acting on the system (such as gravity in this case).

In the linearized models, there is no interaction between the free and forced responses; for example, the natural frequency of the system would be independent of any external forces acting on the system. The system would vibrate naturally with the same frequency if we took it out into space for instance. This mathematical fact is very convenient, since we can undertake the analysis of free and forced vibration independently as long as we can keep in mind that the eventual response of the system is a sum of the free and forced responses. It is for this reason that we are going to concern ourselves exclusively with free vibration in this and the next chapter, and then look at forced vibration exclusively in the chapters after that.

The displacement expression is clearly an oscillating function with a frequency ω_n (the natural frequency of the mass-spring system), and such that its mean value is the static equilibrium position (Fig. 3.6). The period of oscillation is found to be

$$\tau = \frac{2\pi}{\omega_n} = 2\pi\sqrt{\frac{m}{k}} \tag{3.21}$$

The phase and the amplitude of the sine wave depend upon the arbitrary constants, which are evaluated from the initial conditions. In the present case, the magnitude of oscillation of the mass

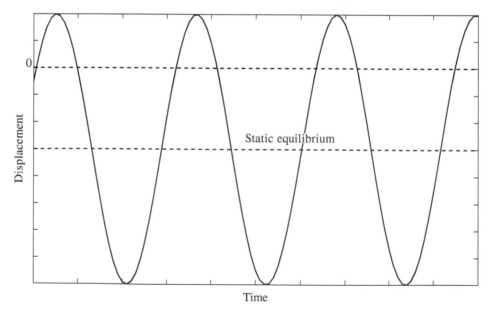

Figure 3.6: *Response of mass-spring system*

clearly depends upon the amount of initial displacement and velocity that are imposed upon it. As an example, suppose we depress the mass by an amount x_0 and let it go, the initial conditions would be

$$x(0) = x_0, \quad \dot{x}(0) = 0 \tag{3.22}$$

Then, the constants turn out to be

$$A = x_0 + \frac{mg}{k}, \quad \phi = 0 \tag{3.23}$$

The displacement of the mass for any subsequent instant of time is then found to be

$$x(t) = \left(x_0 + \frac{mg}{k}\right)\cos\left(\omega_n t\right) - \frac{mg}{k} \tag{3.24}$$

Exercise 3.1

Note that we obtained the homogeneous solution in three different forms, Eq. (3.14), Eq. (3.16), and Eq. (3.17); each form has two arbitrary constants that would need to be evaluated from the initial conditions. Given a set of general initial conditions, $x(0) = x_0$, and $\dot{x}(0) = v_0$, solve for each one of those constants.

— Example 3.1 Stabilized inverted pendulum

Consider a hinged massless rod of length ℓ with a particle mass m at the top of it, as shown in Fig. 3.7. The rod is attached to a wall with two identical springs of stiffness k. The springs are unstretched when the rod is vertical. Derive the nonlinear EOM. Linearize the equations for small excursions about the vertical equilibrium position. Identify the range of parameters when the configuration is stable.

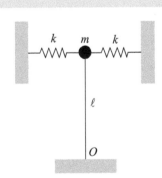

Figure 3.7: *Inverted pendulum*

Kinetics

The forces acting on the system are due to the support hinge, the springs, and the gravitational field of the earth. R_x and R_y are the support reactions, or forces exerted on the rod by the support hinge. This is shown in an FBD, Fig. 3.8.

The forces can be expressed in terms of the Cartesian unit vectors as follows.

$$\vec{F} = (R_x + 2F_s)\vec{\imath} + (R_y - W)\vec{\jmath} \tag{3.25}$$

Assuming the spring remains horizontal, the spring forces F_s are

$$F_s = k\ell \sin\theta \tag{3.26}$$

and the weight is

$$W = mg \tag{3.27}$$

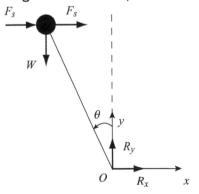

Figure 3.8: *Inverted pendulum FBD*

Kinematics

Since the rod is massless, the center of mass is located at the position of the bob. The displacement, velocity, and acceleration of the center of mass are given below.

$$\vec{r}_C = \ell\left(-\sin\theta\,\vec{i} + \cos\theta\,\vec{j}\right) \tag{3.28}$$

$$\vec{v}_C = \frac{d\vec{r}_C}{dt} = \ell\dot{\theta}\left(-\cos\theta\,\vec{i} - \sin\theta\,\vec{j}\right) \tag{3.29}$$

$$\vec{a}_C = \frac{d\vec{v}_C}{dt} = \ell\ddot{\theta}\left(-\cos\theta\,\vec{i} - \sin\theta\,\vec{j}\right) + \ell\dot{\theta}^2\left(\sin\theta\,\vec{i} - \cos\theta\,\vec{j}\right) \tag{3.30}$$

Also, since it is a rigid body, we can determine the angular acceleration.

$$\alpha = \ddot{\theta} \tag{3.31}$$

Newton's laws

For a rigid body, Newton's second law is as follows.

$$\vec{F} = m\vec{a}_C \tag{3.32}$$

$$M_C = I_C\alpha, \quad M_O = I_O\alpha \tag{3.33}$$

Only three of the above four equations will turn out to be independent. That is of course as it should be, since we have only three unknown quantities, namely, θ, R_x, and R_y.

Looking at the above equations we now realize that if we are only interested in $\theta(t)$, and do not need to compute the support reactions, we can simply use the moment equation about O. Then,

$$mg\ell\sin\theta - (2k\ell\sin\theta)(\ell\cos\theta) = m\ell^2\ddot{\theta} \tag{3.34}$$

where we have used $I_O = m\ell^2$. Simplifying, we get the following EOM.

$$\ddot{\theta} + \frac{2k}{m}\sin\theta\cos\theta - \frac{g}{\ell}\sin\theta = 0 \tag{3.35}$$

The equilibrium position is found from a solution of the following algebraic equation.

$$\left(\frac{2k}{m}\cos\theta - \frac{g}{\ell}\right)\sin\theta = 0 \tag{3.36}$$

Solving, we get the following equilibrium positions.

$$\theta_0 = 0, \pi, \cos^{-1}\left(\frac{mg}{2k\ell}\right) \tag{3.37}$$

Suppose, we consider small motions about the vertical position, $\theta_0 = 0$. Linearizing about that position, $\sin\theta \approx \theta$ and $\cos\theta \approx 1$. Then the EOM becomes

$$\ddot{\theta} + \left(\frac{2k}{m} - \frac{g}{\ell}\right)\theta = 0 \tag{3.38}$$

The system is stable and has an oscillatory solution when the coefficient of θ in the above equation is positive. Or, the system is stable about the position $\theta_0 = 0$ if

$$\boxed{\left(\frac{2k}{m} - \frac{g}{\ell}\right) > 0} \tag{3.39}$$

Then (and only then), the system would undergo a stable oscillation with a natural frequency,

$$\omega_n = \sqrt{\frac{2k}{m} - \frac{g}{\ell}} \tag{3.40}$$

How about that third equilibrium position we ignored: $\theta_0 = \cos^{-1}\left(\frac{mg}{2k\ell}\right)$? What does it physically correspond to? Give the parameters some numerical values and you can probably make better sense of this question. (For example, $m = 1\,\text{kg}$, $k = 100\,\text{N/m}$, etc.) Is the system stable about this equilibrium position?

3.2 What Is a Spring?

So far, we looked at systems with real springs. But clearly, you have experienced vibration when there was no real spring *per se*. For example, think about the vibration of an aircraft wing that you can observe from inside a flying plane; or the vibration you feel when you walk on a bridge, or even in buildings when a heavy truck goes by; or the vibration you can sense when you put your hand on a lathe, drill, or pretty much any kind of machinery. Referring back to our mass-spring model, we can clearly see mass in all these systems we mentioned here, but where is the spring?

The answer of course is that we do not need a physical spring for vibration to occur; we just need "springiness." In fact, anything you can think of acts like a spring; that is, *anything*. After all, what is *acting like a spring?* The most essential characteristic of a spring is that it deflects under a force. And, of course, almost everything in the world does so. So, now think about systems or components of systems in general, apply a metaphysical (or real) force, and you can imagine (or measure) the deflection it suffers. Then just divide the applied force by the deflection, and that would give one the linearized *stiffness* of that system. It is that easy to derive a mathematical model for stiffness.[1] Even if we do not always derive an expression for the stiffness, you need to internalize this concept of stiffness (into your brain's RAM!), and that will help you translate reality into a mathematical model much more easily.

Some practical aspects of stiffness are not that obvious to perceive; we need some scientific principles to establish their characteristics as they are more subtle, and the vibrations are very tiny (but damaging nevertheless). For example, it would be foolhardy to design a crankshaft in an automobile without considering its torsional vibration characteristics even though it is something that a lay person may never experience first hand. Rest assured that it *will* break if the engineers did not do their job, and just slapped it together. An aircraft engine shaft suffers transverse vibrations, minimization of which is critical to its operation even if it is something an air passenger may not touch and feel. And so on. Let us hence look at integrating concepts from your mechanics of materials to establish stiffness characteristics that we can use in vibration analysis. In fact, the entire solid mechanics course could be said to be an effort to compute k to supply a value for us to use in the vibrations course!

In mechanics of materials, you probably studied tension, torsion, and bending of engineering materials. In tension, for example, you would have derived the relationship between the applied load and the deflection of a prismatic rod (Fig. 3.9),

$$x = \frac{F\ell}{AE} \tag{3.41}$$

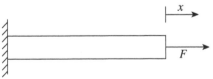

Figure 3.9: *Rod in tension*

---

[1] Actually, we are fudging a little here; it is not *that* easy! Especially if we wanted a good model. Rest assured however that we will discuss more complex models in the later chapters; for now, this definition will do.

where F is the applied force, ℓ is the length, A is the constant cross-sectional area, and E is Young's modulus. Now, we can rewrite this to get an effective stiffness

$$k = \frac{F}{x} = \frac{AE}{\ell} \tag{3.42}$$

So, if we had a situation of a rigid mass mounted on this rod, as shown in Fig. 3.10, we could model its axial (sometimes called longitudinal) vibration by simply modeling it as a mass-spring system with the above stiffness.

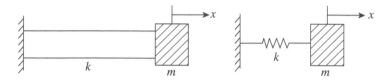

Figure 3.10: *Rigid mass on a flexible rod*

Of course, by doing this, we are essentially ignoring the mass of the rod (or possibly lumping a portion of it with the rigid mass). In other words, our notion of a spring is that of a *massless* device. Hence, all the vibration models we are considering here are essentially that of a rigid mass (or one with infinite stiffness) connected to a massless spring. Seems ridiculous, doesn't it? However, the truth is that this kind of model is not so bad at all. Many times, we can get decent predictions out of these simple models. Also, as you will learn in the rest of the book, there are better and more realistic models, and computerized techniques that will help us solve them.

It is important to remember that a spring is a *static* device, and that all the concepts from a mechanics of materials course deal with *statics*. But, everything we do in this book deals with *dynamics*. So, essentially, we derive a spring stiffness based on a static analysis (as in the tension example we just discussed), and then, turn around and use it in a dynamic setting. There is clearly an approximation here. Again, this approximation is a decent one for most situations; but even then, it is important to remember that this approximation is embedded in our models.

Now, let us think back to see where the tension equation came from. There were several steps involved including a kinematic assumption, a strain-displacement relationship, and a stress-strain model. You can read about the steps in your mechanics book. Let us just discuss the stress-strain relationship here briefly again focusing on the tension (and compression) problem. The relationship you most definitely used was

$$\sigma = E\epsilon \tag{3.43}$$

where σ is the normal stress, ϵ is the normal strain, and E is the Young's modulus for the material. It is called Hooke's law as you will no doubt remember. This is essentially a load-deflection relationship, and is linear. Since this is linear, and because of some other assumptions, we ended up with a linear spring in tension. Of course, the linearity has limitations of small strains, and is only applicable for some materials; fortunately, steel and aluminum, the most popular engineering materials, happen to be those materials. You probably established this relationship experimentally as well, by applying a load slowly to a thin rod in a laboratory. The key word is "slowly," because of what is called a "quasistatic" assumption. That is, we apply the load carefully and slowly so as not to bring dynamic effects into the picture. If you did apply a fast-varying dynamic load, the rod would also behave like a mass; by keeping it slow, we are essentially forcing it to behave as close to a pure spring as is practically possible. In any case, you need to remember that the stress-strain relationship is the basis of deriving stiffness characteristics for engineering systems. We will be revisiting this in the next chapter to examine its validity for practical situations.

Example 3.2 Torsional vibration

Consider a wheel suspended by a thin rod, as shown in Fig. 3.11. The length of the rod is ℓ, and its diameter is d. Derive a model for its oscillations.

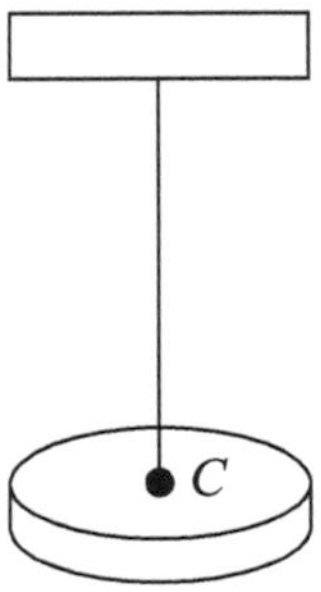

Kinetics

The only external force acting on the wheel is the restoring moment T_s from the rod and is shown in the FBD, Fig. 3.12.

The torsional moment due to the rod can be assumed to be linear if the displacement is small and within the elastic limit. Then,

$$T_s = k\theta \tag{3.44}$$

Figure 3.11: *Torsional pendulum*

Kinematics

The body undergoes pure rotation, and hence we have to figure out the angular acceleration only.

$$\alpha = \ddot{\theta} \tag{3.45}$$

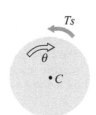

Newton's laws

The only relevant equation here is the moment equation.

Figure 3.12: *Torsional pendulum FBD*

$$M_C = I_C \alpha \tag{3.46}$$

Using the results from kinematics and kinetics,

$$-k\theta = I_C \ddot{\theta} \tag{3.47}$$

where I_C is the polar mass moment of inertia of the wheel and k is the torsional stiffness of the rod and can be determined using basic principles from mechanics of materials

$$k = \frac{GJ}{\ell} \tag{3.48}$$

where G is the shear modulus and J is the area polar moment of inertia, which is equal to $\frac{\pi d^4}{32}$ for a circular cross section.

Substituting and simplifying, the EOM becomes

$$\ddot{\theta} + \frac{GJ}{I_C \ell}\theta = 0 \tag{3.49}$$

Solving,

$$\theta(t) = C \sin(\omega_n + \phi) \tag{3.50}$$

where the natural frequency is given by

$$\omega_n = \sqrt{\frac{GJ}{I_C \ell}} \tag{3.51}$$

The period of oscillation is then

$$\tau_n = \frac{2\pi}{\omega_n}$$

$$= 2\pi\sqrt{\frac{I_c\ell}{GJ}}$$

$$= 2\pi\sqrt{\frac{32I_c\ell}{\pi d^4 G}} \tag{3.52}$$

— **Example 3.3 Torsional vibration of a rotating shaft** —

Consider a rotating shaft supported on bearings at one end with a propeller at the other end, as shown in Fig. 3.13. The propeller can be modeled as a rigid disk. Determine the natural frequency of oscillation.

Figure 3.13: *Propeller shaft*

Not withstanding gravity, this is the same problem as the one considered before! The bearing support can be considered to be a "fixed" end. Note that the shaft rotates and has torsional oscillations essentially about the same axis; for the torsional vibration problem however, it is acceptable to model it as if it is not rotating. In other words, free rotation does not interfere with the torsional analysis. Hence, the natural frequency would be the same as in the earlier example.

— —

3.3 Energy

Just like we did in the case of the simple pendulum, let us look at the variation of energy during the vibration of the mass-spring oscillator; this will give us a different insight into the physics of the system. Recall that the kinetic energy for a particle mass is

$$T = \frac{1}{2}mv^2 = \frac{1}{2}m\left[\dot{x}(t)\right]^2 \tag{3.53}$$

If a frictionless spring is acting on the particle, the energy is given by

$$V = \frac{1}{2}k\left[x(t)\right]^2 \tag{3.54}$$

where k is the spring constant, and x represents the displacement from the unstretched position of the spring. See Problem 3.16 for a derivation (which will get you hunting for your old physics text book that you thought you did not need any more!).

Consider now a set of initial conditions (which is after all how we "seed" the system with energy to start vibrating)

$$x(0) = x_0, \quad \dot{x}(0) = v_0 \tag{3.55}$$

Then, the system starts out with some potential and kinetic energies

$$V_0 = \frac{1}{2}kx_0^2, \quad T_0 = \frac{1}{2}mv_0^2 \tag{3.56}$$

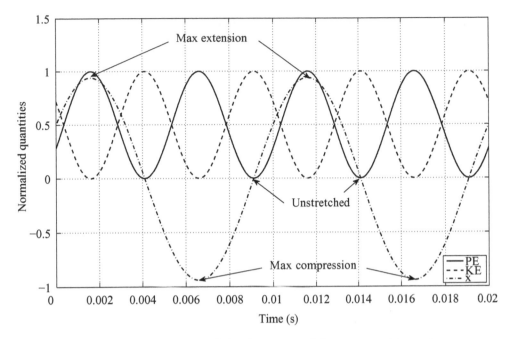

Figure 3.14: *Energies and displacement of the mass-spring system*

We substitute the solution for $x(t)$ and $\dot{x}(t)$ into the above energy expressions and plot them in Fig. 3.14. Note that the total energy remains constant (of course!). Also, the displacement is plotted on the same plot so that you can correlate the motion of the system with the energy. All the quantities are normalized; hence, you cannot deduce anything from the absolute magnitudes. Just look at the plot qualitatively. The potential energy disappears momentarily when the mass passes through the equilibrium position, and the kinetic energy disappears at the instant when it reaches the maximum compression or maximum extension. In fact, the two energies are exactly complementary.

There is another funny thing happening with the energy plot (Fig. 3.14). The energies are varying much faster than the displacement (and hence, the velocity). In fact, if you put on your critical glasses, you will see that it is exactly twice. In other words, since the displacement is varying as $\sin(\omega_n t)$, the energies are varying as $\sin(2\omega_n t)$. Why is that, and what does it imply?

3.4 Springs in Combination

Occasionally, we can model elastic members as springs in series or parallel. For example, consider an automobile on four suspension springs. If we are only considering the translational motion of the vehicle (up and down), then, the springs can be considered to be in parallel. For a more subtle example of springs in series, consider a motor on mounts that is fixed to a beam-like structure. The mounts act like springs, and the beam acts like one also.

3.4.1 Springs in Parallel

Consider two springs in a parallel combination, as shown in Fig. 3.15. Figure 3.16 shows the corresponding FBD. Kinematically, we restrict the displacements in the two springs to be the same. Then

$$x_1 = x_2 = x \tag{3.57}$$

From equilibrium,

$$F = F_1 + F_2 \tag{3.58}$$

From the individual constitutive equations,

$$F_1 = k_1 x_1 \quad F_2 = k_2 x_2 \tag{3.59}$$

A simpler equivalent system can be derived by replacing the original system of two springs by one spring with a spring constant k_{eq}, which offers the same displacement x, under the action of the same force F as the original system.

$$F = k_{eq} x \tag{3.60}$$

From the above equations, it follows simply that

$$k_{eq} = k_1 + k_2 \tag{3.61}$$

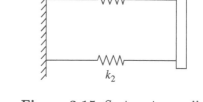

Figure 3.15: *Springs in parallel*

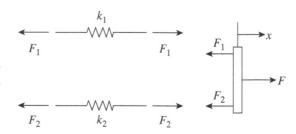

Figure 3.16: *FBD of springs in parallel*

Note that this is a similar relationship to that of electrical resistances in series.

3.4.2 Springs in Series

Consider two springs in series combination, as shown in Fig. 3.17. Figure 3.18 shows the corresponding FBD. Kinematically, the total displacement is simply the sum of the two displacements.

$$x = x_1 + x_2 \tag{3.62}$$

From equilibrium of the individual springs,

$$F = F_1 = F_2 \tag{3.63}$$

Hence, the equivalent system is now given by

$$k_1 x_1 = k_{eq}(x_1 + x_2) \tag{3.64}$$

It follows then that

$$\frac{1}{k_{eq}} = \frac{1}{k_1} + \frac{1}{k_2} \tag{3.65}$$

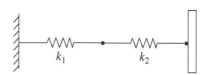

Figure 3.17: *Springs in series*

Note that this is a similar relationship to that of electrical resistances in parallel.

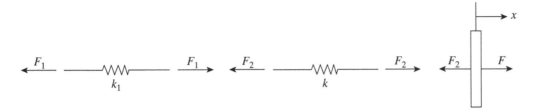

Figure 3.18: *FBD of springs in series*

> We referred to electrical resistances when we talked about springs. Although it was convenient, it was an entirely fallacious analogy. Resistances after all dissipate energy (which is why electric bulbs get hot), but springs emphatically do not. Springs are responsible for storing energy with the *potential to do work*. Hence the name, "potential energy." So, clearly, if we wanted an electrical analogy, we need to look for a device that stores energy that has the potential to do work. What kind of a device would that be? And, do you get similar expressions in that case? Sorry folks; if you do not even have a notion of what we are talking about here, you will need to dig up that physics book again.

3.5   Geared Shafts

Geared shafts are very common in the engineering world; they can be found in automobiles, all kinds of big and small machinery, aircraft, motor drives, etc. Gears are very diverse and complex and they experience many complex vibration mechanisms from tooth vibration to shaft vibration. Here, we will look at a simple gear train with a focus on the torsional vibration of the shaft. Consider a typical example of a prime mover (such as an engine or a motor) driving a load through a spur gear pair, as shown in Fig. 3.19.

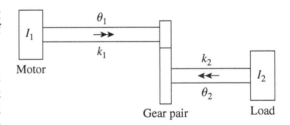

Figure 3.19: *Typical pair of geared shafts*

We will assume that the shafts have torsional stiffnesses k_1 and k_2, and that they are massless (and hence have zero moment of inertia). The gears are also assumed to have no mass. The motor has mass moment of inertia I_1, and the driven load has a moment of inertia I_2. The torsional displacements are θ_1 and θ_2 of the two shafts, respectively. Then, if there is no slip or backslash, the two velocities (and displacements) are related by the so-called gear ratio n. Note that we have assumed positive opposite directions for the displacements of the two shafts.

$$\theta_2 = n\theta_1, \quad \dot{\theta}_2 = n\dot{\theta}_1, \quad \ddot{\theta}_2 = n\ddot{\theta}_1 \tag{3.66}$$

This gear ratio is of course determined by the ratio of radii of the two gears. Or,

$$r_1 = nr_2 \tag{3.67}$$

We note at this point that, although there appear to be two rigid bodies, and hence two DOF in reality, we have an SDOF system because the two angular displacements are kinematically constrained.

As shown in Fig. 3.20, there is a force F that is applied by each gear on the other, which is equal and opposite by Newton's third law. This force results in a torque acting on each shaft. Figure 3.21 shows the FBD of the two rigid masses I_1 and I_2. The torques are given by

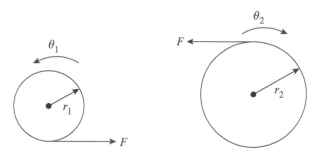

Figure 3.20: *Interaction force of the mating gears*

$$T_1 = Fr_1, \quad T_2 = Fr_2 \tag{3.68}$$

Hence,

$$T_1 = nT_2 \tag{3.69}$$

The kinematics is quite simple as we are only looking at the torsional motion: $\alpha_1 = \ddot{\theta}_1$ and $\alpha_2 = \ddot{\theta}_2$. Applying Newton's second law to each mass, we get two differential equations.

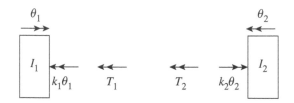

Figure 3.21: *FBDs of geared shafts*

$$I_1 \ddot{\theta}_1 = -k_1 \theta_1 + T_1 \tag{3.70}$$

$$I_2 \ddot{\theta}_2 = -k_2 \theta_2 - T_2 \tag{3.71}$$

Using Eq. (3.66) and Eq. (3.69), we can eliminate θ_2 and combine the above two equations.

$$\left(I_1 + n^2 I_2\right) \ddot{\theta}_1 + \left(k_1 + n^2 k_2\right) \theta_1 = 0 \tag{3.72}$$

Note how the moment of inertia and the stiffness of the second (load) shaft get multiplied by n^2 when they are *reflected* on the first (motor) shaft. Hence, the two-shaft system now behaves as an SDOF oscillator with a natural frequency given by

$$\omega_n = \frac{I_1 + n^2 I_2}{k_1 + n^2 k_2} \tag{3.73}$$

Alternatively, we could have written the equation in terms of θ_2 as follows.

$$\left(I_2 + \frac{I_1}{n^2}\right) \ddot{\theta}_2 + \left(k_2 + \frac{k_1}{n^2}\right) \theta_2 = 0 \tag{3.74}$$

This of course gives us the same natural frequency (did you expect something different?!).

RUNNING PROBLEM: STRING VIBRATION

SDOF model

Strings are useful models for a variety of problems including elevator cables, transmission cables, belts, guitar strings, etc. An example is shown in Fig. 3.22. In order to illustrate the value, validity, and limits of predictive capability of different models, we will use this as a continuing example throughout the text. Note that almost all the concepts explored here are not limited to strings, but are equally valid for all kinds of vibrating systems.

A string, shown in Fig. 3.23, vibrates because it has two essential properties: inertia (mass) and elasticity. Hence, in a sense, a string is just a mass-spring system in its simplest representation. The elasticity comes from the tension in the string, which we assume is uniform throughout the string; we also assume that the tension does not change with the displacement the string experiences. These assumptions are valid provided the tension is reasonable to prevent sag in the string, and if the displacements remain small (remember the linearity assumption from the chapter on dynamics?). We also ignore gravity. With these assumptions, we are ready to start modeling.

Clearly, the mass of the string is distributed throughout its length (to keep things simple, let us say it is uniform). Let ρ be the mass density, A the constant cross-sectional area, ℓ its length, and T the tension in the string. How do we convert the situation to a mass-spring system? One way is to just lump all the mass of the string at the center point of the string, as shown in Fig. 3.24. Being a very curious student you would probably want to challenge this. A critical question to ask is: Does it mean that the string vibrates as if its

Figure 3.22: *Cable cars*

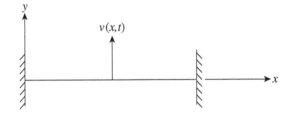

Figure 3.23: *String in transverse vibration*

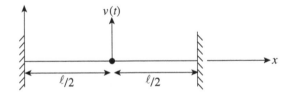

Figure 3.24: *String with mass lumped in the center*

entire mass were at the center? In other words, is the kinetic energy of vibration of the true string equal to the kinetic energy we are attributing to the entire mass by putting it at one point of the string. The answer is "Of course, no!" Clearly, every point on the string experiences a different vibrational amplitude, and by lumping all the mass at one point, we are denying that obvious fact, and, in the process, perhaps making a serious error. Or, are we? Let us examine this a little more.

The problem that we have raised here is a very fundamental one, and is an aspect of modeling of pretty much all kinds of physical systems (and is not limited to vibration). In addition to being a function of time, the displacement of the system is clearly a function of space (x) also. This is because, the inertial and elastic properties are distributed; hence, if we did include this fact, we would end up with a mathematical model of what we call a *distributed parameter system*. The problem of course is that we would end up with a partial differential equation in x and t, which is something we would rather not deal with at this stage of the game because of the added complexity (goodness knows we are still trying to get a grip on the ODEs!). On the contrary, if we lump the mass and elastic properties at one point (like we are proposing to do here), we would end up with a *lumped parameter system*; this will lead to an ODE, since we have artificially eliminated dependence on one of the independent variables, x, and only have to deal with t.

To answer the question we raised before,—yes, of course, we are making an error by ignoring the mass and elastic distribution. However, all modeling requires assumptions (just like we assumed that the simple pendulum can be modeled to be linear when it is really nonlinear). At this point, we cannot answer the question of how egregious this assumption is, or how bad the error will be in the answers we are going to get from the simplified model. However, as we go along in this text, we will continue to refine the model, and you will see how the answers get better and better. The bottom line is that, as modelers, we will have to draw the line somewhere between accuracy (and verisimilitude with reality), and effort and time involved with the solution. If we want very high accuracy, we will have to pay with more effort, and vice versa.

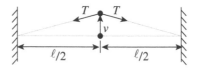

Figure 3.25: *FBD of string with mass lumped in the center*

So, with that abridged introduction to the beautiful complexities and dangerous pitfalls of the modeling process, let us get started with our simple model. Figure 3.25 shows the FBD with the assumed lumped mass at the center; we use $v(t)$ to designate the vertical (or *transverse*) displacement to distinguish it from x, the coordinate that notates the physical distance along the string. The lumped mass is the total mass, and is given by $m = \rho A \ell$.

The EOM follows easily from $F = ma$, with $a = \ddot{v}$

$$\rho A \ell \ddot{v} = -T \left(\frac{v}{\ell/2} + \frac{v}{\ell/2} \right) \tag{3.75}$$

Note that we have used linearization to determine the vertical components of the elastic restoring forces. Simplifying,

$$\ddot{v} + \frac{4T}{\rho A \ell^2} v = 0 \tag{3.76}$$

Note that the effective spring constant from each side of the string depends on the length. Here,

$$k_{\text{eff}} = \frac{T}{\ell/2} \tag{3.77}$$

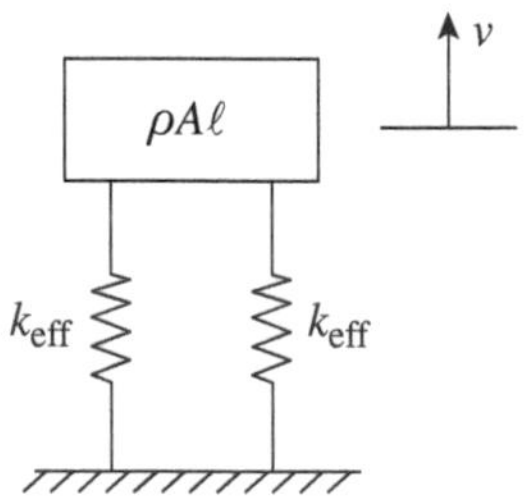

Figure 3.26 shows conceptually our SDOF model for the string.

The natural frequency is, by inspection,

$$\omega_n = \frac{2}{\ell} \sqrt{\frac{T}{\rho A}} \tag{3.78}$$

Figure 3.26: *Conceptual SDOF model for string vibration*

Remember this treatment because we will revisit this when we expand our modeling knowledge and can handle a little more complexity.

3.6   Stiffness Elements

Now that we have discussed several examples of spring-like elements, we are ready to summarize the effective stiffness properties of various components and systems that you are likely to encounter in practice. As discussed, they fall into three main categories: longitudinal (tension), torsional, and flexural (bending). The notation is as follows.

A—area of cross section
E—Young's modulus of elasticity
I—area moment of inertia about the neutral axis (not to be confused with the mass moment of inertia)
G—shear modulus
J—polar area moment of inertia

3.6.1   Longitudinal Stiffness

This applies to thin rods, cables, and such other elements.

$$k = \frac{AE}{\ell}$$

Coil spring:

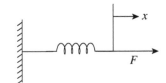

$$k = \frac{Gd^4}{64nR^3}$$

where n is the number of active coils, d is the coil diameter, and R is the mean helix radius.

3.6.2   Torsional Stiffness

$$k = \frac{GJ}{\ell}$$

3.6.3   Flexural Stiffness

Simply supported beam (pinned-pinned)

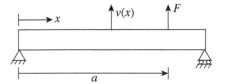

$$k = \frac{F}{v}\bigg|_{x=a}$$

where

$$
v = \begin{cases}
\dfrac{Fbx}{6EI\ell} \left(\ell^2 - x^2 - b^2\right) & x \le a \\[3mm]
\dfrac{Fbx}{6EI\ell} \left((\ell^2 - b^2)x - x^2 + \dfrac{1}{b}(x-a)^2\right) & x > a
\end{cases}
$$

Cantilever beam (fixed-free)

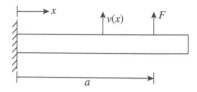

$$
k = \frac{F}{v}\bigg|_{x=a}
$$

where

$$
v = \begin{cases}
\dfrac{F}{6EI} \left(3ax^2 - x^3\right) & x \le a \\[3mm]
\dfrac{Fbx}{6EI} \left(3a^2x - a^3\right) & x > a
\end{cases}
$$

3.7   Review Questions

1. What is the constitutive relationship in a spring? Which two quantities does it relate?

2. Why did the ODE we derived for a mass-spring model turn out to be linear?

3. Practically speaking, what do the free and forced responses mean?

4. What is the mathematical equivalence of free response?

5. What is the mathematical equivalence of forced response?

6. In an SDOF system, if the spring stiffness is doubled, what happens to the natural frequency?

7. In an SDOF system, if the mass is doubled, what happens to the natural frequency?

8. Look around you in the room, and identify everything (or parts of it) that can be modeled as springs. How would you measure the stiffness for each of them?

9. Since a spring is rarely (!) massless (as we assumed), how would you include its mass in a more accurate analysis?

10. In the inverted pendulum, what is stabilizing, and what is destabilizing?

11. If we took the inverted pendulum out in space, when would it go unstable?

12. Would the inverted pendulum be stable or unstable in an elevator accelerating upward with an acceleration equal to g?

13. Which position of the spring would maximize the stability of the inverted pendulum?

14. What is a quasistatic assumption?

15. Which has higher axial stiffness—an aluminium or a steel rod (with identical dimensions)? Why?

16. Which has higher torsional stiffness—an aluminium or a steel rod (with identical dimensions)? Why?

17. Which has higher flexural stiffness—an aluminium or a steel rod (with identical dimensions)? Why?

18. When the mass-spring system passes through its equilibrium position, which energy is a maximum—potential or kinetic?

19. When the mass-spring system reaches an extremum point, which energy is a maximum—potential or kinetic?

20. When a mass-spring system is in free oscillation, what is the rate at which the potential or kinetic energy varies? Why?

21. If we wish to increase the stiffness in a system, should we put the springs in series or parallel?

22. For a system of two shafts coupled by gears, how many natural frequencies are there? Why?

23. What happens to the moment of inertia *reflected* on the mating shaft?

24. What happens to stiffness *reflected* on the mating shaft?

25. How does the natural frequency of a geared shaft pair depend on the gear ratio, n?

26. How does the natural frequency of a stretched string depend on its length?

27. How does the natural frequency of a stretched string depend on the tension?

28. How does the natural frequency of a stretched string depend on its area of cross section?

29. Does the effective flexural stiffness in a beam depend on the point of interest? Why or why not?

30. Where does the cantilever beam have the lowest stiffness?

31. Where does the cantilever beam have the highest stiffness?

32. Where does the simply supported beam have the lowest stiffness?

33. How does the flexural stiffness of a beam depend on its cross-sectional area?

34. How does the flexural stiffness of a beam depend on its length?

35. How does the natural frequency of a beam in flexural vibration depend on its length?

36. How does the natural frequency of a beam in flexural vibration depend on its Young's modulus?

37. How does the natural frequency of a rod in torsional vibration depend on its shear modulus?

38. Given identical beams, which configuration would give us a lower natural frequency—cantilever, or simply supported?

Problems

3.1 Determine the displacement, velocity, and acceleration of a spring-mass system with $\omega_n = 15\,\text{rad/s}$ for the initial conditions, $x_0 = 0.04\,\text{m}$, and $\dot{x}_0 = 1.5\,\text{m/s}$. Plot $x(t)$, $\dot{x}(t)$, and $\ddot{x}(t)$ for $t = 0$ to 15 s.

3.2 Determine the displacement, velocity, and acceleration of a spring-mass system with $k = 400\,\text{N/m}$, $m = 2.5\,\text{kg}$, and initial conditions, $x(0) = 0.09\,\text{m}$ and $\dot{x}(0) = 1.8\,\text{m/s}$.

3.3 A spring-mass system has a natural frequency of 9 Hz. When the spring is replaced by another spring whose stiffness is smaller by an amount of 800 N/m, the frequency is altered by 50%. Find the mass and spring constant of the original system.

3.4 The natural frequency of a spring-mass system is found to be 3.5 Hz. When an additional mass of 1 kg is added to the original mass m, the natural frequency is reduced to 3.2 Hz. Find the spring constant k and the mass m.

3.5 Consider the following pairs of mass-spring systems in a vertical configuration (i.e., gravity acts on them) shown in Fig. 3.27. For each pair, compare and contrast the natural frequencies and discuss your findings.

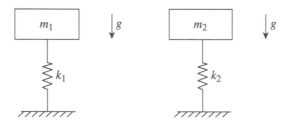

Figure 3.27: *Two mass-spring configurations*

(a) $k_1 = 4k_2$, $m_1 = m_2$.

(b) $k_1 = k_2$, $m_1 = 4m_2$.

(c) $k_1 = 2k_2$, $m_1 = 2m_2$.

(d) $k_1 = 100k_2$, $m_1 = m_2$.

(e) $k_1 = k_2$, $m_1 = m_2$. One pair is on the earth, the second is in a gravity-free environment.

(f) $k_1 = k_2$, $m_1 = m_2$. One pair is in the laboratory; the second is in a race car going around a curve at 250 km/h.

3.6 A refrigerator weighs 150 kg, and sits on four parallel springs, each with stiffness 1.2 kN/m. Determine its static deflection and the natural frequency of oscillation.

3.7 A DVD player is supported on four mounts, which can be modeled as springs. We wish to keep the static deflection less than 5 mm. The player weighs 8 kg. Determine the desired spring stiffness and the resulting natural frequency of oscillation.

3.8 A car weighs 1200 kg and is supported on four suspension springs. It is desired to have a natural frequency of 5 Hz. Determine the spring stiffness; also calculate the static deflection.

3.9 A truck-trailer weighs 4000 kg and is supported on eight suspension springs. It is desired to have a natural frequency of 3.5 Hz. Determine the spring stiffness; also calculate the static deflection.

3.10 A 500-gm mass hanging by a spring has a static deflection of 3 mm. Find its natural frequency.

3.11 A mass-spring system is found to have a natural frequency of 10 Hz. When a 2-kg mass is added to it, the natural frequency is found to be 8.5 Hz. Determine the original mass and the spring constant.

3.12 The natural frequency of a spring-mass system is found to be 2 Hz. When an additional mass of 1 kg is added to the original mass m, the natural frequency is reduced to 1 Hz. Find the spring constant k and the mass m.

3.13 For the two situations shown in Fig. 3.28, derive the equations of motion and determine the natural frequencies of oscillation. Which configuration results in a higher natural frequency?

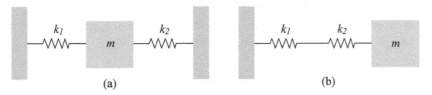

(a) (b)

Figure 3.28: *Mass-spring configurations*

3.14 Find the equivalent spring constant of the system shown in Fig. 3.29.

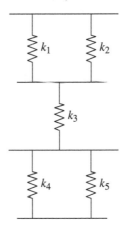

Figure 3.29: *Combination of springs*

3.15 Consider a hinged slender rod of mass m and length ℓ, as shown in Fig. 3.30. The rod is attached to a wall with a spring of stiffness k at a height a. ($OP = a$.) The spring is unstretched when the rod is vertical. Derive the nonlinear EOM. Linearize the equations for small excursions about the vertical equilibrium position. Identify the range of parameters when the configuration is stable. When it is stable, find the natural frequency of oscillation.

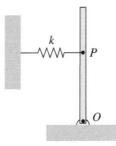

Figure 3.30: *Inverted pendulum*

3.16 Consider a force acting on a linear spring with spring constant k. Assume that there is no loss of energy; hence, the system is conservative. Recall the definition of the work done, and potential energy that we derived in Chapter 2.

$$W = \int_{r_1}^{r_2} \vec{F} \cdot d\vec{r} = -\int_{r_1}^{r_2} dV = V_1 - V_2$$

(a) Now, derive an expression for the potential energy of the spring with the usual convention that the reference position is that of the unstretched spring (when the potential energy is zero). Assume that the force is applied slowly so that there are no dynamic effects.

(b) ⚡ It is interesting to compare the work done by the force, in this case with a case of a *stretched* spring when a constant force acts on it. Why is this different?

3.17 ⚡ Consider the solid cylinder of mass m, and radius r, which is attached to a vertical spring of stiffness k, as shown in Fig. 3.31. Assume that the spring force is always directed along the length of the spring. The spring has a free length ℓ_0, and the center of the cylinder is at a vertical distance ℓ from the support to which the spring is attached. The cylinder is constrained to roll on the horizontal surface; i.e., it cannot lift off the surface. Assume also that the cylinder rolls without slipping. Derive the nonlinear EOM.

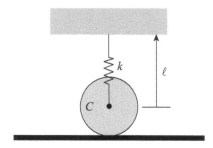

Figure 3.31: *Cylinder with a spring*

(a) Assume small motions about the vertical position and determine the frequency of oscillation. Assume that the spring force acts in the direction of the spring. Is there a condition for stable oscillations?

(b) ⚡ ⚡ Are there other equilibrium positions? If so, under what conditions are they stable?

(c) Solve the original equations numerically and verify the stability conditions by using different parametric values.

3.18 Find the natural frequency of the spring-mass system on an inclined plane, as shown in Fig. 3.32. How does the natural frequency depend upon the angle of inclination?

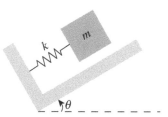

Figure 3.32: *Mass-spring on an inclined plane*

3.19 A cylinder of mass m and mass moment of inertia I_0 is free to roll without slipping, but is restrained by the spring of stiffness k, as shown in Fig. 3.33.

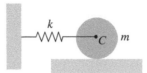

Figure 3.33: *Cylinder on a level surface*

(a) Model the cylinder as a particle mass and determine the natural frequency of oscillation.

(b) Now, model it as a rigid body and determine its natural frequency.

(c) Compare the two answers. What is your conclusion?

3.20 The model shown in Fig. 3.34 is used to predict the rolling oscillations in ships. The point M shown is the so-called metacenter, which represents the intersection of the line of action of the buoyant force and the center line of the ship. The distance between M and the center of mass C is h and can be assumed to be a known constant for small oscillations. The ship has a mass moment of inertia I_C about the roll axis, and has mass m. Determine the natural frequency of oscillation.

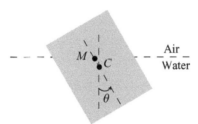

Figure 3.34: *Rolling oscillations in ships*

3.21 A pump is fixed to the end of a bracket and has a mass m (Fig. 3.35). The bracket's mass can be considered to be negligible; it has a length ℓ, modulus of elasticity E, area of cross section A, and area moment of inertia I. Determine the natural frequency of oscillation.

Figure 3.35: *Pump on a bracket*

3.22 Consider a wheel suspended by a thin rod, as shown in Fig. 3.36. The length of the rod is ℓ, and its diameter is d. When the wheel is given an angular displacement and released, it is found to oscillate with a period τ. Determine the polar mass moment of inertia of the wheel.

Figure 3.36: *Torsional pendulum*

3.23 An elevator with mass m is attached to a steel cable that is wrapped around a drum rotating with a constant angular velocity ω, as shown in Fig. 3.37. The radius of the drum is r; and the cable has a cross-sectional area A, and a modulus of elasticity E. Due to a malfunction in the motor drive system the drum stops suddenly when the elevator is moving down; at that instant, the length of the cable is ℓ. Ignore any damping and determine the amplitude of motion of the elevator car. Also, find the maximum stress in the cable. *After* deriving the equations, you may use the following numerical values: $m = 3600$ kg, $\omega = 3$ rad/s, $r = 30$ cm, $A = 6$ cm^2, $E = 8.3 \times 10^{10}$ Pa, $\ell = 15$ m [16].

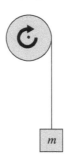

Figure 3.37: *Elevator*

3.24 A crane attempting to lift a weight may be modeled as an inverted simple pendulum of length ℓ, and mass m (Fig. 3.38). The spring of stiffness k models a support member and is fixed at a right angle to the massless rod at a distance a from the hinged end ($OA = a$). It is also unstretched at the position shown.

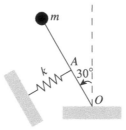

Figure 3.38: *Crane lifting a weight*

(a) Derive the nonlinear EOM.

(b) Linearize the EOM about the equilibrium position shown.

(c) Determine the range of parameters for which the system is stable.

3.25 Consider the spring-mass system shown in Fig. 3.39.

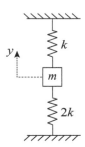

Figure 3.39: *Spring-mass system*

(a) Calculate the static deflection y_{st}, the system period, τ, and natural frequency ω_n.

(b) Obtain the complete solution of the EOM for the given initial conditions. $k = 144\,\mathrm{N/m}$, $m = 6\,\mathrm{kg}$, $y(0) = 0.1\,\mathrm{m}$, and $\dot{y}(0) = 0$.

3.26 A small particle of mass m is attached to two wires under tension, as shown in Fig. 3.40. Calculate the system natural frequency, ω_n, for small vertical oscillations if the tension T, in both wires is assumed to be constant. Use $m = 0.1\,\mathrm{kg}$ and $\ell = 0.4\,\mathrm{m}$.

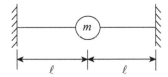

Figure 3.40: *A mass held by two wires*

3.27 An old car being moved by a magnetic crane pick up is dropped from a short distance above the ground (Fig. 3.41). Any damping effects of its worn-out shock absorbers are negligible. Determine the natural frequency, f_n, in cycles per second, (Hz), of the vertical vibration that occurs after impact with the ground. Use $k = 20\,\mathrm{kN/m}$ and $m = 1500\,\mathrm{kg}$.

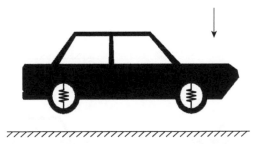

Figure 3.41: *A magnetic crane carrying a car*

3.28 Consider the system shown in Fig. 3.42.

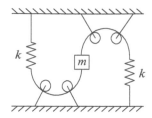

Figure 3.42: *A spring-mass configuration*

(a) Calculate the natural frequency, ω_n, of vertical oscillation of the spring-loaded cylinder when it is set into motion. Both springs are in tension at all times.

(b) Obtain the complete solution of the EOM and plot it. $m = 25\,\text{kg}$ and $k = 3000\,\text{N/m}$.

3.29 Consider the system shown in Fig. 3.43.

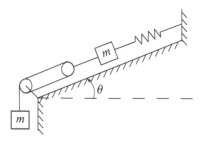

Figure 3.43: *A spring-mass system driven by a pulley*

(a) Determine the natural frequency, ω_n, of the system. The mass and friction of the pulleys are negligible.

(b) Obtain the complete solution of the EOM and plot it for the initial conditions. $x(0) = 0$, $\dot{x}(0) = -0.15\,\text{m/s}$, and $\theta = 30°$.

3.30 An automobile has been suspended in a mechanic's garage, as shown in Fig. 3.44. The mechanic has two choices to hang the car. Find the best choice if he wants to reduce the natural frequency of the system. $k_1 = k_2 = 40\,\text{kN/m}$, $m = 1200\,\text{kg}$.

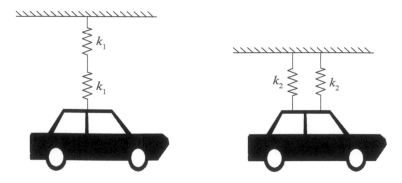

Figure 3.44: *A suspended automobile*

3.31 If the mechanisms given in the previous problem (Fig. 3.44) are equivalent regarding the natural frequency (i.e., they have same natural frequency), find the relationship between k_1 and k_2. Make your suggestion for an economical case, considering only the cost of the springs.

3.32 Figure 3.45 shows two mechanisms for hanging an automobile. The rods have lengths ℓ, areas of cross-section A, and are all made up of the same material. Then, can you tell which mechanism has the lower natural frequency? (*Hint*: First apply your intuition to think about the answer.)

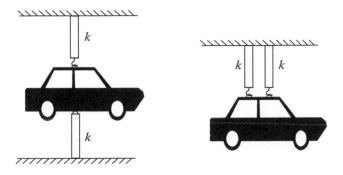

Figure 3.45: *Suspension mechanisms*

3.33 Consider the mechanism shown in Fig. 3.46.

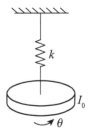

Figure 3.46: *A suspended tire*

(a) A tire has been suspended with the help of a solid steel rod. The material and geometric properties are given below. Find the natural frequency of the system. $E = 200\,\text{MPa}$, $G = 150\,\text{MPa}$, $\ell = 0.5\,\text{m}$, and $D = 0.1\,\text{m}$. (*Hint*: Consider torsional stiffness of the rod.)

(b) The solid rod is to be replaced with a hollow rod, with d and D being the inner and outer diameters, respectively. If both the systems have the same natural frequency (solid and hollow systems), design the hollow rod assuming $d = 0.75D$. Calculate and comment on the amount of material saved using a hollow rod.

3.34 A simple brake mechanism for an automobile has been shown in Fig. 3.47. Find the natural frequency of the system for $k = 100\,\text{N/m}$, $\ell = 0.45\,\text{m}$, and $m = 3\,\text{kg}$.

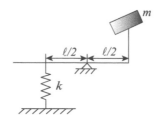

Figure 3.47: *A brake mechanism*

3.35 A slender rod of mass m is hinged at one end and is attached to a wall with the help of two springs with spring constants k_1 and k_2, as shown in Fig. 3.48. Derive the EOM of the system.

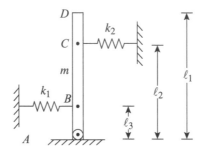

Figure 3.48: *Inverted pendulum supported by two springs*

3.36 A car moving with velocity v_0, as shown in Fig. 3.49, hits the wall and bounces back. Assume that the car is a rigid body and has a bumper that is a massless spring. Derive the EOM of the system, and derive its response after the crash.

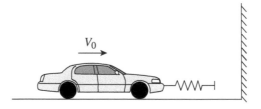

Figure 3.49: *Car hitting a wall*

3.37 Consider an elevator (Fig. 3.50) that has been stopped. It is supported on a system that can be approximated by springs of stiffness k. The mass of the elevator is 500 kg and the spring stiffness is 200 kN/m. Find the natural frequency of oscillation.

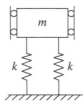

Figure 3.50: *Elevator*

3.38 Consider a traffic signal suspended from a flexible beam BC, as shown in Fig. 3.51. You may assume that AB is rigid. The beam BC is made of steel, has a length of 1.5 m, and a cross-sectional dimension of 15 × 10 cm. About half the mass of the beam may be lumped with the mass of the signal. The signal itself weighs 12 kg. Determine the natural frequency of oscillation.

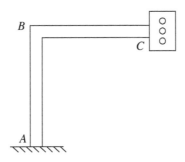

Figure 3.51: *Traffic signal*

3.39 Four springs and a mass are attached to a rigid, weightless bar AB that is hinged at one end, as shown in Fig. 3.52. Find the natural frequency of vibration of the system.

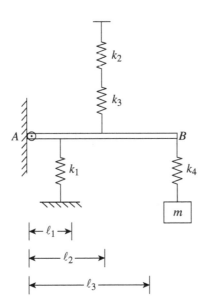

Figure 3.52: *Four springs and a mass attached to a rod*

3.40 A robot arm shown in Fig. 3.53 carries an object weighing 8 kg. Find the natural frequency of the robot arm in the axial direction for the following data.

$$\ell_1 = 15 \text{ in.}; \quad \ell_2 = 10 \text{ in.}; \quad D_1 = 2 \text{ in.}; \quad D_2 = 1.5 \text{ in.};$$

$$d_1 = 1.8 \text{ in.}; \quad d_2 = 1.2 \text{ in.}; \quad E_1 = E_2 = 3 \times 10^{10} \text{ Pa.}$$

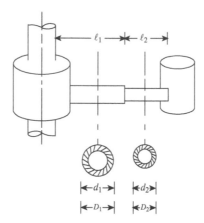

Figure 3.53: *Robot arm*

3.41 The atomic force microscope (AFM) is a very useful tool for characterizing nano-sized structures and biological materials and processes. It has become indispensable for researchers and engineers working with nanomechanics. The essential element of the AFM is the probe, whose vibrational displacement is used as a sensor to measure extremely small structures and forces, and is shown schematically in Fig. 3.54. It is essentially a cantilever beam and can be modeled as an SDOF system, as shown in Fig. 3.55. Derive the EOM assuming that AB is a rigid body, hinged at one end whose other end is coupled to springs. For a more involved discussion on modeling the AFM probe you might want to read the article by Mahdavi et al. [17].

Atomic force microscope

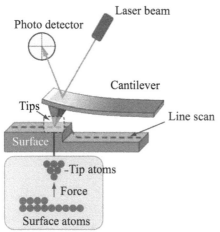

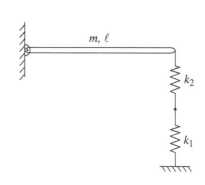

Figure 3.54: *Atomic force microscope sketch (courtesy of Agilent)*

Figure 3.55: *Atomic force microscope model*

3.42 A machine component can be modeled by a simple pendulum and a spring, as shown in Fig. 3.56. At the position shown (of 60° inclination from the horizontal), the system is in static equilibrium, and the spring is perpendicular to the rod. You may assume that the spring always remains perpendicular to the rod. Follow the steps enumerated below.

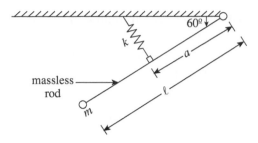

Figure 3.56: *Machine component*

(a) Draw an FBD and determine the nonlinear EOM.

(b) Assume small displacements and linearize the EOM.

(c) Find the condition for stability.

(d) Then, for the stable configuration, determine the natural frequency from the linearized equation.

3.43 A mass m_1 is supported on a spring of stiffness k and is in static equilibrium. A second mass m_2 drops through a height h and sticks to m_1 without rebound (Fig. 3.57). Determine the subsequent motion.

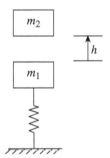

Figure 3.57: *Inelastic collision of two masses*

3.44 Derive the EOM for small vertical oscillations for the system shown in Fig. 3.58.

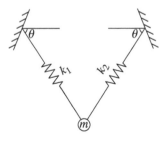

Figure 3.58: *System for Problem 3.44*

3.45 A cylinder contains gas initially under pressure p_0 and has a piston of area A and mass m. The enclosed gas obeys Boyle's law: $pV = p_0 V_0$, where V_0 is the initial volume of each half of the cylinder. The piston moves freely (without friction) and there is no leakage of gas between compartments. Assume that the pressures are uniformly distributed over the surfaces of the piston.

(a) Derive the nonlinear EOM for free oscillations of the piston.

(b) Linearize the EOM for small oscillations about the equilibrium position.

(c) Determine the natural frequency for small oscillations.

3.46 Consider a U-tube manometer filled with water (Fig. 3.59). It has a density ρ, acceleration due to gravity is g, and the length of the liquid column is ℓ. Assume laminar and inviscid motion and derive a mathematical model for the system. What is the natural frequency of oscillation? How would it change if it were filled with a heavier fluid?

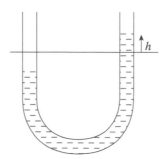

Figure 3.59: *U-tube manometer*

3.47 Find the equivalent stiffness and the natural frequency of the system with a rigid mass on a massless beam, as shown in Fig. 3.60. Assume $\ell = 2$ m and $m = 60$ kg.

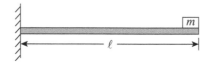

Figure 3.60: *System for Problem 3.47*

3.48 Find the equivalent stiffness and the natural frequency of the system with a rigid mass on a massless beam, as shown in Fig. 3.61. Assume $\ell = 2$ m and $m = 60$ kg.

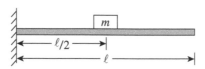

Figure 3.61: *System for Problem 3.48*

3.49 Find the equivalent stiffness and the natural frequency of the system with a rigid mass on a massless beam, as shown in Fig. 3.62. Assume $\ell = 2$ m and $m = 60$ kg.

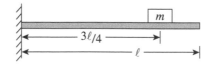

Figure 3.62: *System for Problem 3.49*

3.50 Find the equivalent stiffness and the natural frequency of the system with a rigid mass on a massless beam, as shown in Fig. 3.63. Assume $\ell = 2$ m and $m = 60$ kg.

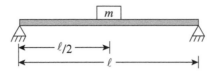

Figure 3.63: *System for Problem 3.50*

3.51 Find the equivalent stiffness and the natural frequency of the system with a rigid mass on a massless beam, as shown in Fig. 3.64. Assume $\ell = 2$ m and $m = 60$ kg.

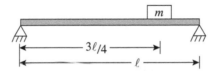

Figure 3.64: *System for Problem 3.51*

3.52 For the simple system shown in Fig. 3.65, derive the EOM.

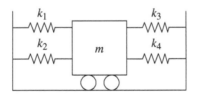

Figure 3.65: *System for Problem 3.52*

3.53 Determine the natural frequencies of the system shown in Fig. 3.66. The mass of the cylinder is m, radius is r, and the spring constants are k_1 and k_2. Assume that the rope passing over the cylinder does not slip. Assume that the mass of the cylinder is negligible.

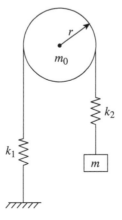

Figure 3.66: *System for Problem 3.53*

3.54 Determine the natural frequencies of the system in Fig. 3.67 if it is constrained to move vertically.

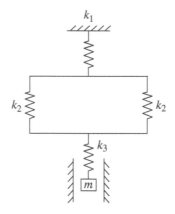

Figure 3.67: *System for Problem 3.54*

3.55 Derive the EOM for the crane modeled by a massless spring at the end of a rigid rod, as shown in Fig. 3.68. Determine its natural frequency.

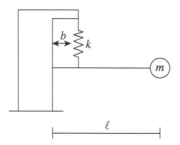

Figure 3.68: *A crane model*

3.56 A mass, m, can slide in a frictionless groove on a horizontal table that rotates at a constant angular speed Ω, as shown in Fig. 3.69. The mass is connected to the center of rotation by means of a spring of stiffness k, and unstretched length ℓ_0.

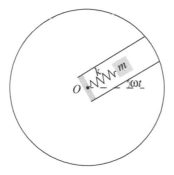

Figure 3.69: *Mass on a turntable*

(a) Assuming small displacements, derive the EOM of the mass.

(b) What are the conditions under which the system is stable?

(c) When it is stable, what is the natural frequency of oscillations?

3.57 An L-shaped massless rigid member is hinged at point O and has a mass m at the tip, as shown in Fig. 3.70. The member is supported by a spring of stiffness k, as shown.

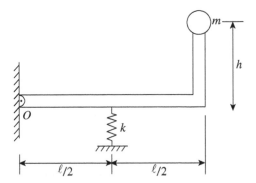

Figure 3.70: *L-shaped member*

(a) Determine the equilibrium position.

(b) Derive a linearized EOM for a small motion about the equilibrium position.

(c) Determine the natural frequency of small oscillations.

(d) Determine the height h for which the system becomes unstable.

3.58 A climbing robot has been invented for many applications. The robot can be modeled by a particle mass m climbing up a uniform massless aluminium beam of length ℓ, Young's modulus E, and radius r, as shown in Fig. 3.71.

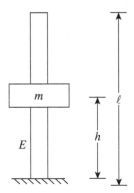

Figure 3.71: *Climbing robot*

(a) Determine the natural frequency of the system at the height shown ($h = \ell/2$).

(b) Consider the fact that the robot could be at any height h. Now, determine the natural frequency as a function of h, and plot it.

(c) Strictly speaking, if the robot is climbing fairly quickly, the effective stiffness of the beam is changing quickly as well. In that case, k would be a function of time. Let us say the robot is climbing up at a constant rate v. Use the following numerical values and carry out a numerical investigation to determine the vibrational response of the beam. $m = 20\,\text{kg}$, $\ell = 6\,\text{m}$, $r = 6\,\text{cm}$. For the climbing rate, investigate the response for varying values; some suggested numbers are 0.01, 0.1, and 0.5 m/s. Do you see a difference between what you get here, and what you obtained in (b)? The effect of a moving mass on an elastic structure is an important practical problem that leads to a dangerous instability; did you get any of that in your numerical simulation?

3.59 Robotic arms are being increasingly used in the industry for manufacturing as well as for demanding precise operations such as surgery. A robotic arm (especially long, flexible ones) can be modeled as a flexible beam. Figure 3.72 shows a robotic arm in a pharmaceutical facility. Assume that 2/3 of the robot's mass can be lumped at its free end along with the load it is picking up. A good assumption is that of a cantilever beam. Determine the natural frequency for the given parameter values. Robot mass $m_r = 3$ kg, length $\ell = 0.8$ m, rectangular cross section with width = 10 cm, height = 3 cm; made of steel. The load is variable. Perform the calculations for three values of the load: $m_\ell = 3$, 8, and 20 kg.

Figure 3.72: *A robotic arm*

3.60 The so-called zip-lines, shown in Fig. 3.73, are quite popular in jungle travels especially in Central America and attempt to give a Tarzan-like experience to the tourists to inject some excitement into the desk-strapped lives of the modern working people. They consist of long steel cables strung between tall trees often at altitudes of 150 ft (50 m) or more. Assume

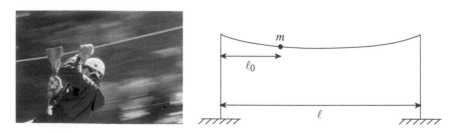

Figure 3.73: *The zip-line experience and a simple model*

that the tourist is a particle mass on a massless cable, and determine the natural frequency of the system as a function of his/her position ℓ_0. Ignore the fact that the person is actually a pendulum hanging from the cable, and ignore the dynamics of the forward motion of the person (that would make it very complicated!). Assume a uniform cable with tension T, and an overall length of ℓ. The mass of the tourist is m.

3.61 Consider the wing of an aircraft with an engine mounted on it, as shown in Fig. 3.74. The wing can be modeled as a uniform (OK, we are stretching the truth a little here!) cantilever beam of length ℓ, Young's modulus E, and area moment of inertia I. Assume that half its mass m_w can be lumped along with the mass of the engine m_e. Determine the effective natural frequency of the system.

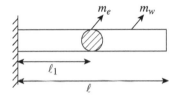

Figure 3.74: *An SDOF model of an aircraft wing*

3.62 Consider model of a grinding wheel as a solid steel cantilever shaft with a rigid mass at the end, as shown in Fig. 3.75. The shaft has length ℓ, Young's modulus E, and area moment of inertia I. Assume that its mass is negligible compared to that of the grinding wheel m. Determine the natural frequency of the system. Use the following numerical values: $\ell = 0.3\,\text{m}$, radius of the shaft $= 2.5$ cm, $m = 15\,\text{kg}$.

Figure 3.75: *Grinding wheel model*

3.63 The vibrational response of seated human subjects is an important problem. One of the simplest models for it is to consider the human being as one lumped mass, and the human-seat interface as a spring [18]. The mass is 56.8 ± 9.4 kg and the spring stiffness is 75.5 ± 28.3 kN/m. Determine the *range* of natural frequencies.

3.64 The upper torso of the human body is an excellent example of an inverted pendulum. The head mass sitting on top of the upper body is, in essence, unstable, but is actively stabilized by the neuromuscular system, as shown in Fig. 3.76.[2] Derive a mathematical model for the system. Use the following numerical values and determine the natural frequency of the system. $m_1 = 5.5\,\text{kg}$, $m_2 = 2.5\,\text{kg}$, $\ell = 0.8\,\text{m}$, $k = 2000\,\text{N/m}$. Assume that the spinal column modeled by the rigid rod is uniform, and that the "springs" are attached at the midpoint.

---

[2]This stabilizing neuromuscular system is inactive when we sleep—this is what leads to the "dropping heads" in people sleeping on trains and buses!

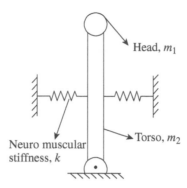

Figure 3.76: *Model for the upper part of the body*

3.65 The lower part of the human body can be modeled with an inverted pendulum, as shown in Fig. 3.77. In fact, most models of biped walking start with an inverted pendulum. (You might want to read the article by Kappaganthu and Nataraj [19] for an interesting dynamic analysis of a biped robot.) Assume that the legs are massless, and that there is a restoring torsional spring. Determine a mathematical model, and derive an expression for stable response and the natural frequency.

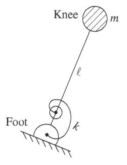

Figure 3.77: *Model for the lower part of the body*

3.66 Consider a crane lifting a heavy load using a steel cable, as shown in Fig. 3.78. We will assume that the point of suspension of the cable is stationary. The cable has Young's modulus E, a

Figure 3.78: *Crane lifting a heavy load*

circular cross section with diameter d, and length that varies between ℓ_1 and ℓ_2. (This means that you need to solve the problem for different values of ℓ; but do not treat ℓ as a function of time!) The load is W. Assume that the cable is massless, and model the system as a simple pendulum. Determine the maximum oscillation amplitude that the load can go through before the cable breaks. After setting up the equations, you may use the following numerical values: $W = 10{,}000$ N, $\ell_1 = 6$ m, $\ell_2 = 12$ m, $d = 10$ mm, $E = 200$ MPa, density $= 8{,}000$ kg/m^3, and ultimate tensile strength of 400 MPa. How is your answer affected if we include the mass of the cable (say, m_c) by "lumping" half of it with that of the load?

3.67 Consider a motor driving a load through a gear pair. The motor inertia is 0.025 kgm^2 and the load inertia is 0.32 kgm^2. The motor shaft and the load shaft are both 40 cm long. They each have an outer radius of 4.5 cm, and inner radius of 2.5 cm. They both are made of steel. Determine the natural frequency of torsional vibration for the system.

3.68 Consider a motor driving a load through a pair of gear trains, as shown in Fig. 3.79. Derive an EOM (there is only one DOF!). Ignore backlash and slip. Determine the natural frequency of oscillation.

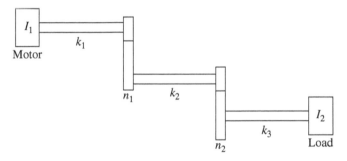

Figure 3.79: *Motor drive train*

3.69 The leading cause of turbine blade failure is high stresses and the resulting fatigue of the blade material. The stresses are due to fluctuating forces imposed by centrifugal and aerodynamic loading. One way to minimize the force fluctuations is by tuning the blade so that blade resonance cannot occur at the same frequency as the fluctuating excitation forces. In order to meet these requirements, the designer must have a clear understanding of the dynamics involved and an analysis procedure capable of accurately predicting both the natural frequencies and the alternating stresses that the blade must endure. The simplest model for it is that of a beam

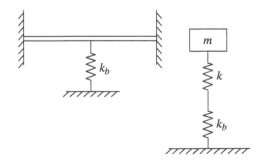

Figure 3.80: *Blade vibration: a simple model*

fixed at both ends with a spring attached to it at the center, as shown in Fig. 3.80. A slightly more complicated model is a beam with two springs attached at two points, as shown in Fig. 3.81. Elaborate models have been discussed in more detail in an article by Wagner and Griffin [20].

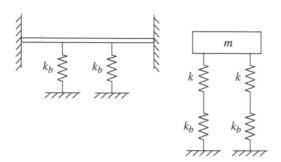

Figure 3.81: *Blade vibration: a better model*

(a) Find the EOM for both the models.

(b) Find the natural frequencies.

4 SDOF Systems: Free Damped Response

Scientists give us gobbledegook about friction and molecules. But they don't really know.
—Ray Bradbury (in Fahrenheit 451)

When we looked at a mass-spring system we noticed that the predicted response was a sinusoidal oscillation that lasted forever. This is of course patently untrue in that all real systems would stop oscillating after a while. This difference between reality and the predictions of our mathematical model is due to the absence of any kind of *dissipative* mechanism in our model.

Dissipation of energy can occur due to various friction (or damping) forms. There are three main types that we will briefly discuss here. The first is Coulomb damping (dry friction) that is caused whenever two solid bodies slide past each other. Without getting into a phenomenological description of why and how these forces come about, we restrict ourselves to an empirical description that the damping force would look as shown in Fig. 4.1. Note that the x-axis is velocity. The force is clearly a nonlinear function of velocity, and as such quite undesirable.

The second type of mechanism is structural (or internal) damping that comes about because of cyclic stresses in materials. If you take a paper clip and bend it back and forth a few times, you will notice that it gets warm; this increase in temperature is coming from the internal friction that causes dissipation of the energy that we put into the system by bending it. This is a common occurrence in all kinds of rotating machinery and needs to be included in more complicated models than we will encounter in this book. Also, some materials such as rubber have a high degree of internal friction and should always be modeled with it.

The third type of friction is the one we will be primarily concerned about and comes from interaction with fluids and arises due to viscous friction. This can be unintentional such as the frictional drag felt in an automobile (or aircraft, of course) as it moves through the medium of air. Or, it can be intentionally introduced such as by using lubricating oil to separate two solid objects that have relative motion (a door hinge is an obvious example). In this case, rather than eliminating friction, we are striving to replace Coulomb friction with fluid friction. Fluid-film bearings (including air bearings such as in computer hard drives) and automotive shock absorbers are some practical examples.

In the following detailed discussion, we will restrict ourselves mostly to fluid friction. Anytime we expect some amount of fluid friction, we will assume that it can be accurately modeled using a piston-cylinder mechanism or a dashpot. But, later in Chapter 5, we will also come back to the other types of damping and figure out how we can leverage the linear viscous thinking for those cases as well.

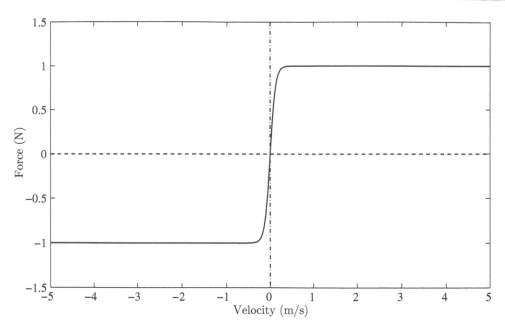

Figure 4.1: *Coulomb damping*

For linear viscous damping, the force-velocity characteristic is assumed to be linear, as shown in Fig. 4.2. Or,

$$F_d = c\dot{x} \tag{4.1}$$

where c is called the linear viscous damping coefficient.

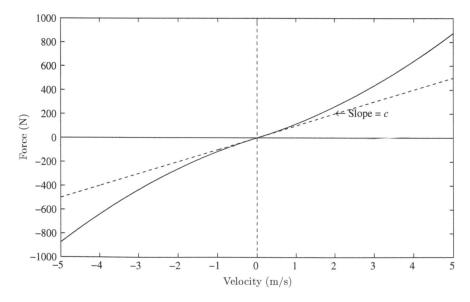

Figure 4.2: *Viscous damping*

4.1   Mass-Spring-Damper System

Suppose we consider the same mass-spring model as before with a damper added to it (Fig. 4.3). Figure 4.4 shows the corresponding free body diagram (FBD).

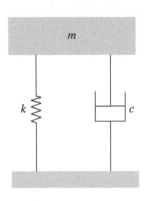

The kinematics in this problem is very simple. Since the support to which the spring and damper are attached is considered rigid and at rest, the absolute acceleration of the mass with respect to inertial space is equal to $\ddot{x}$.

The final step is the assembly of the above pieces of information using Newton's second law

Figure 4.3: *Mass with a spring and damper*

$$\vec{F} = m\vec{a} \tag{4.2}$$

which reduces to a scalar equation

$$-F_s - F_d = m\ddot{x} \tag{4.3}$$

The spring and damper forces are (with linear approximations),

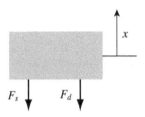

Figure 4.4: *FBD of mass-spring-damper system*

$$F_s = kx \tag{4.4}$$
$$F_d = c\dot{x} \tag{4.5}$$

The combination of the above equations results in

$$-kx - c\dot{x} = m\ddot{x} \tag{4.6}$$

Reorganization yields a second order differential equation

$$m\ddot{x} + c\dot{x} + kx = 0 \tag{4.7}$$

Equation (4.7) is a second-order linear constant-coefficient differential equation. Solution proceeds as in the case of the undamped system. We *seek* a solution (based on experience) in the form

$$x = Ae^{st} \tag{4.8}$$

where A and s are undetermined constants. Equation (4.8) is of course not the true solution unless it satisfies Eq. (4.7). We substitute Eq. (4.8) into Eq. (4.7) to check if the assumed form is indeed the correct solution. That process yields

$$\left(s^2 m + sc + k\right) Ae^{st} = 0 \tag{4.9}$$

Ignoring the trivial solution ($A = 0 \Rightarrow x_h(t) \equiv 0$), we get the characteristic equation

$$ms^2 + cs + k = 0 \tag{4.10}$$

which *must* be satisfied for the assumed form of the solution to be the correct solution. Solution of the characteristic equation yields

$$s_1,\ s_2 = -\frac{c}{2m} \pm \sqrt{\left(\frac{c}{2m}\right)^2 - \frac{k}{m}} \tag{4.11}$$

Clearly, whether the solution is real or complex depends upon the relative magnitudes of the parameters. In order to be able to discuss that with the least amount of algebraic complexity, we define some parameters. First, we define a value of the damping coefficient, called the critical damping coefficient, c_c, as that value when the term under the square root goes to zero; i.e.,

$$\left(\frac{c_c}{2m}\right)^2 - \frac{k}{m} = 0 \Rightarrow c_c = 2\sqrt{km} \tag{4.12}$$

Next, we relate the actual amount of damping in the system to this critical value using a nondimensional parameter, ζ, called the viscous damping ratio.

$$\zeta = \frac{c}{c_c} = \frac{c}{2\sqrt{km}} \tag{4.13}$$

With this definition, and using $\omega_n = \sqrt{k/m}$, we can rewrite the characteristic roots as

$$s_1,\ s_2 = \omega_n \left(-\zeta \pm \sqrt{\zeta^2 - 1}\right) \tag{4.14}$$

Now, it should be clear that the nature of the roots depends upon whether ζ is less than or greater than 1. We will consider each of these cases individually.

Exercise 4.1

Using SI units, verify that the viscous damping ratio is nondimensional.

4.1.1 Case 1: Underdamped system ($\zeta < 1$)

This is the case when the amount of damping is small such that the damping coefficient is less than the critical value. Such a system is said to be *underdamped*. Then, the characteristic roots become complex and are given by

$$s_1,\ s_2 = \omega_n \left(-\zeta \pm i\sqrt{1 - \zeta^2}\right) \tag{4.15}$$

We define

$$\omega_d = \omega_n \sqrt{1 - \zeta^2} \tag{4.16}$$

$$\text{Then, } s_1,\ s_2 = -\zeta\omega_n \pm i\omega_d \tag{4.17}$$

Substituting back into the assumed solution form and using the Euler formula,

$$\begin{aligned} x(t) &= A_1 e^{s_1 t} + A_2 e^{s_2 t} \\ &= C e^{-\zeta\omega_n t} \cos(\omega_d t + \phi) \end{aligned} \tag{4.18}$$

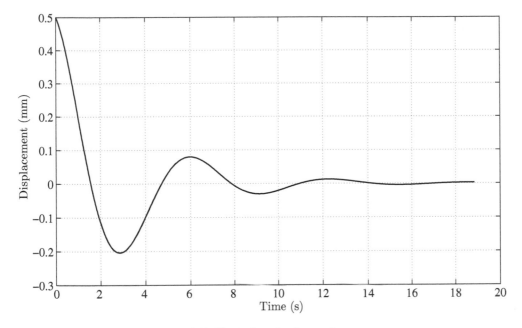

Figure 4.5: *Typical underdamped response*

The best way to look at this function is as a sine wave with a frequency of ω_d, and such that its amplitude (instead of being a constant C) gets *modulated* by the decreasing exponential. The response would hence be characterized as an exponentially damped oscillation. C and ϕ depend upon the initial conditions as usual. Figure 4.5 shows a typical response. ω_d is called the *damped natural frequency.*

4.1.2 Case 2: Overdamped system ($\zeta > 1$)

This is the case when the amount of damping is large enough that the damping coefficient is greater than the critical value. Such a system is said to be *overdamped*. Then, the characteristic roots are real and are given by

$$s_1, s_2 = \omega_n \left(-\zeta \pm \sqrt{\zeta^2 - 1}\right) \tag{4.19}$$

Substituting back into the assumed solution form,

$$x(t) = A_1 e^{s_1 t} + A_2 e^{s_2 t}$$
$$= A_1 e^{\omega_n \left(-\zeta + \sqrt{\zeta^2 - 1}\right) t} + A_2 e^{\omega_n \left(-\zeta - \sqrt{\zeta^2 - 1}\right) t} \tag{4.20}$$

This is just a sum of two decreasing exponentials and would typically look like that in Fig. 4.6. The important thing to note here is that there is no oscillation at all. In a sense, the addition of large amounts of damping deadens the system to such an extent that it stops oscillating.

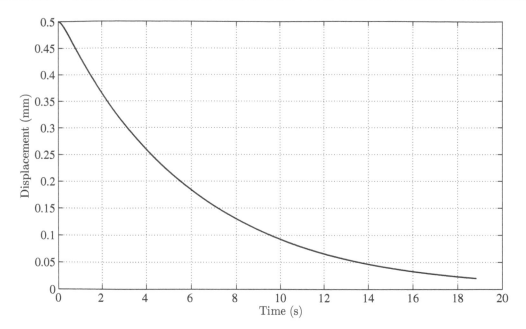

Figure 4.6: *Typical overdamped response*

4.1.3   Case 3: Critically damped system ($\zeta = 1$)

This is the case when the amount of damping is such that the damping ratio is exactly equal to 1; the system is then said to be *critically damped*. This could, of course, never happen in reality (although we can come close to it). Nevertheless, it is an interesting case that separates the two dramatically different types of underdamped and overdamped system responses.

Mathematically, this is a case of degeneracy, since the roots become equal and are both given by

$$s_1 = s_2 = -\omega_n \tag{4.21}$$

Then, substitution into the assumed solution form gives us only one independent solution,

$$x(t) = A_1 e^{-\omega_n t} \tag{4.22}$$

This breakdown in our standard solution procedure means that the other independent solution we are looking for is not in the standard form of a pure exponential. In other words, our original assumption was not right. Digging into differential equation theory, we find that the second solution in such *degenerate* cases can be sought in the form of te^{st}; simple substitution into the differential equation will in fact confirm that, and the solution will turn out to be

$$x(t) = A_1 e^{-\omega_n t} + A_2 te^{-\omega_n t} \tag{4.23}$$

Figure 4.7 shows a typical response.

Exercise 4.2

Prove that the limit of the above function Eq. (4.23) as time goes to infinity is indeed zero (as it should be from physical considerations). (*Hint*: You will need to use L'Hospital's rule.)

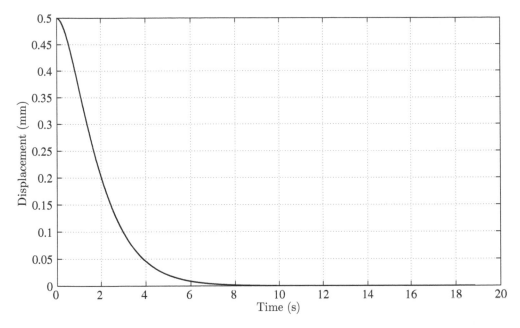

Figure 4.7: *Typical critically damped response*

So how is that, when the exponential assumption broke down, we assumed te^{st} as a possible solution? Why not t or $t\sin\omega t$ or any one of numerous other functions?

Finally, before you get caught up in a serious discussion on springs and dampers, look at Fig. 4.8 for a lay person's interpretation! (This is a true photo from a hiking trail in the Catskill Mountains in the state of New York in USA.)

Figure 4.8: *Springs and dampers on the trail*

4.1.4 Discussion

Figure 4.9 shows a comparison of responses for the three cases (with the same initial conditions). All of them will eventually get to zero (in infinite time theoretically!); however, we would normally be interested in the speed with which they get there. It is clear that the critically damped system responds the quickest of the three cases.

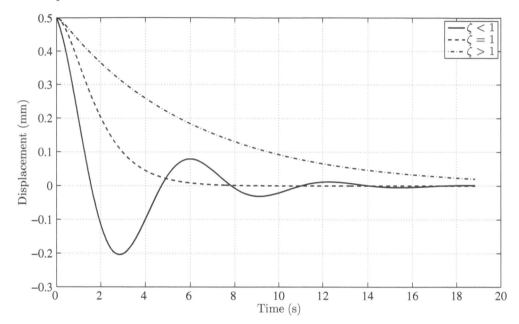

Figure 4.9: *Comparison of underdamped, critically damped, and overdamped responses*

Since we would be interested in the speed of decay of response, what quantitative criteria would you propose as a performance measure? Clearly, all stable systems will eventually reach zero, so the final value is not a good measure. How about the time to reach within 5% of the final value or 2%? Working from the displacement expressions, derive a relation for that time (it is called *settling time*).

The shock absorbers in automobiles are usually designed so that the system is underdamped with a ζ of 0.4–0.8. A door-closing mechanism should not oscillate under any circumstances and would have to be overdamped. If we wanted to add damping to a spring-mounted instrument, such as a compass, we would probably want it to be critically damped so that it settles on its final value as soon as possible.

Exercise 4.3

For a given system, plot damped natural frequency versus the damping ratio ($0 < \zeta < 2$).

4.1.5   Estimate of Damping From Experiment

If the vibrating system is underdamped ($\zeta < 1$), the system response will be an exponentially damped oscillation. Or,

$$x(t) = Ce^{-\zeta\omega_n t} \sin(\omega_d t + \phi) \tag{4.24}$$

where C and ϕ are determined from initial conditions. This response is shown in Fig. 4.10.

Suppose we use $x_1 = x(t_1)$ to denote one of the peaks in the response. Then,

$$x_1 = Ce^{-\zeta\omega_n t_1} \tag{4.25}$$

The next peak occurs after a period, or at $t_2 = t_1 + \tau_d = t_1 + \frac{2\pi}{\omega_d}$. Hence,

$$\begin{aligned}
x_2 &= x(t_2) \\
&= Ce^{-\zeta\omega_n t_2} \\
&= Ce^{-\zeta\omega_n \frac{2\pi}{\omega_d}} e^{-\zeta\omega_n t_1}
\end{aligned} \tag{4.26}$$

Then, it follows that

$$\begin{aligned}
\frac{x_1}{x_2} &= e^{\zeta\omega_n \frac{2\pi}{\omega_d}} \\
&= e^{\frac{2\pi\zeta}{\sqrt{1-\zeta^2}}}
\end{aligned} \tag{4.27}$$

We define a quantity called the *log decrement* by

$$\delta = \log_e \frac{x_1}{x_2} = \frac{2\pi\zeta}{\sqrt{1-\zeta^2}} \tag{4.28}$$

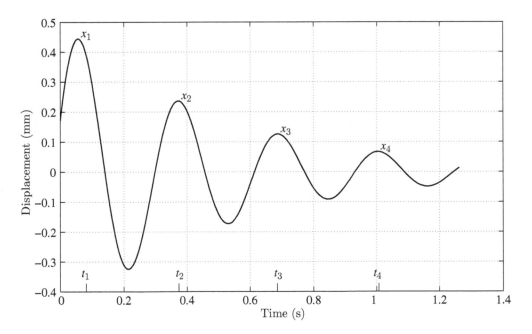

Figure 4.10: *Typical underdamped response and peak values*

that is a measure of the rate of decay of the oscillations due to damping. Then, the damping ratio, ζ, can be computed from

$$\zeta = \frac{\delta}{\sqrt{4\pi^2 + \delta^2}} \qquad (4.29)$$

If ζ is small, an approximate relation is

$$\delta \approx 2\pi\zeta \qquad (4.30)$$

In practice, in order to reduce the effect of measurement error on the damping estimate, we carry through the above process for n oscillations, where n is typically 10 or larger. Then,

$$\delta = \frac{1}{n} \log_e \frac{x_1}{x_{n+1}} \qquad (4.31)$$

An excellent way to verify the accuracy of equations is to use dimensional analysis. Obviously, every term in the equation has to have the same dimensions. For example, in

$$m\ddot{x} + c\dot{x} + kx = F \qquad (4.32)$$

every term has the dimension of force, or $[M\ L\ T^{-2}]$, where $[M]$ stands for mass, $[L]$ for length, and $[T]$ for time. Now, work out the dimension of c and k and make sure that they are correct. In the following equation,

$$I_o\ddot{\theta} + c\dot{\theta} + k\theta = T \qquad (4.33)$$

every term has the dimension of moment. Again, verify this [what are the dimensions of k and c in this equation, and why are they different from the k and c you had earlier?]

Example 4.1 Vertical oscillations in ships

Ships and boats suffer many kinds of oscillations in water. Figure 4.11 shows a highly simplified model for vertical oscillations. The ship shown has an area of cross section A, and under static conditions, floats at a height h. Using $x(t)$ to denote its motion from this static equilibrium, derive the differential equation of motion (EOM). Assume that there is linear viscous damping from the water. What are the natural frequency of oscillation and the damping ratio?

Kinetics

The forces acting on the ship are its weight W, the buoyancy force F_B, and the viscous damping force F_d, as shown in the FBD, Fig. 4.12.

Now, from Archimedes principle, the buoyancy force follows to be

Air

Water

Figure 4.11: *Floating body*

$$F_B = \rho A g(h - x) \qquad (4.34)$$

where ρ is the density of water, h is the submerged height when it is not moving, and x represents the displacement from that static position. The damping force is given to be

$$F_d = c\dot{x} \tag{4.35}$$

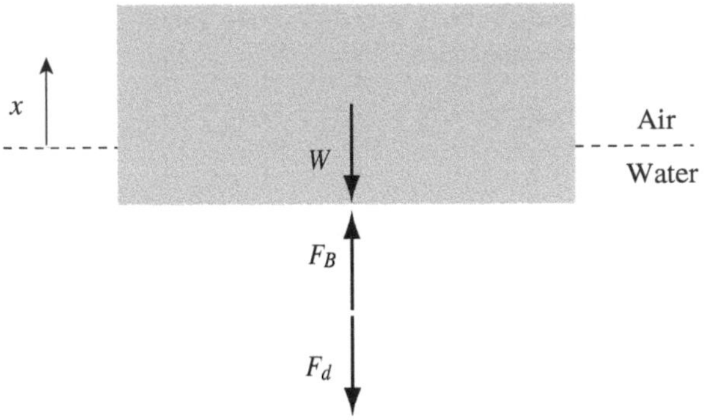

Kinematics

The acceleration is, simply

$$a = \ddot{x} \tag{4.36}$$

Newton's laws

Modeled as a particle, second law gives us

$$\vec{F} = m\vec{a}$$
$$m\ddot{x} = F_B - W - F_d$$
$$m\ddot{x} + c\dot{x} + \rho Agx = \rho Agh - mg \tag{4.37}$$

Although the EOM appears to be nonhomogeneous, it is actually homogeneous if we consider the fact that it should be in static equilibrium. From these considerations, it follows that the buoyancy force should equal weight. Or,

$$\rho Agh = mg \tag{4.38}$$

Hence, the EOM becomes

$$m\ddot{x} + c\dot{x} + \rho Agx = 0 \tag{4.39}$$

Then, the natural frequency of oscillation is

$$\omega_n = \sqrt{\frac{\rho Ag}{m}} \tag{4.40}$$

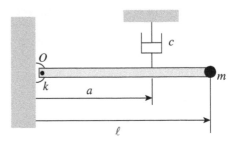

Figure 4.12: *FBD for a floating body*

— Example 4.2 Damped rod example

A rigid massless rod of length ℓ carries a mass m at one end, and is connected to the wall by a torsional spring of stiffness k, as shown in Fig. 4.13. A damper with viscous damping coefficient c is connected to the rod at a distance a from the wall.

1. Assume small displacements and derive the EOM. It may be assumed that the torsional spring is connected to the rod in such a way that it does not exert a moment when the rod is horizontal.
2. What is the maximum value of the damping coefficient, c, for underdamped motion?
3. What is the damped natural frequency?

Figure 4.13: *A damped rod*

Kinetics

Figure 4.14 shows the FBD of the system.

If we assume that the displacements are small, we can use linearized expressions. Then,

$$M_s = k\theta \tag{4.41}$$

$$F_d = ca\dot{\theta} \tag{4.42}$$

where k is the torsional spring stiffness and c is the viscous damping coefficient.

The sum of all the forces acting on the system is

$$\vec{F} = R_x\vec{i} + (R_y - W - F_d)\vec{j} \tag{4.43}$$

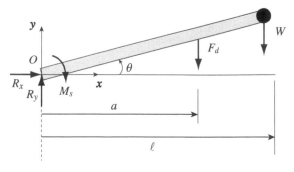

Figure 4.14: *FBD for the damped rod*

Kinematics

In this part of the analysis process, we use geometry and certain assumptions (that the rod is rigid) to derive expressions for velocities and accelerations. C in the equations represents the mass center of the body, which can be assumed to be at the location of the particle mass.

$$\vec{r}_C = \ell\left(\cos\theta\vec{i} + \sin\theta\vec{j}\right) \tag{4.44}$$

$$\vec{v}_C = \frac{d\vec{r}_C}{dt} = \ell\dot{\theta}\left(-\sin\theta\vec{i} + \cos\theta\vec{j}\right) \tag{4.45}$$

$$\vec{a}_C = \frac{d\vec{v}_C}{dt} = -\dot{\theta}^2\left(\cos\theta\vec{i} + \sin\theta\vec{j}\right) + \ell\ddot{\theta}\left(-\sin\theta\vec{i} + \cos\theta\vec{j}\right) \tag{4.46}$$

Also, since it is a rigid body, we should also determine its angular acceleration.

$$\alpha = \ddot{\theta} \tag{4.47}$$

Newton's laws

For a rigid body moving in two dimensions, Newton's second law is as follows.

$$\vec{F} = m\vec{a}_C \tag{4.48}$$

$$M_C = I_C\alpha \quad \text{or} \quad M_O = I_O\alpha \tag{4.49}$$

Again, only three of the above four equations will turn out to be independent. That is of course as it should be, since we have only three unknown quantities, namely, $\theta(t)$, $R_x(t)$, and $R_y(t)$.

Looking at the the above equations we now realize that, if we are interested only in $\theta(t)$, and do not need to compute the support reactions, we can simply use the moment equation about O. Then, assuming small displacements

$$-F_d a - W\ell - M_s = I_O\ddot{\theta} \tag{4.50}$$

where I_O is the effective mass moment of inertia of the system about O and is equal to $m\ell^2$. Reorganizing, we get the following EOM.

$$m\ell^2\ddot{\theta} + ca^2\dot{\theta} + k\theta = -mg\ell \tag{4.51}$$

Note that, since we did not use the force equations at all, we need not have computed the linear acceleration of the center of mass of the system and could have saved ourselves that much algebra. However, if we were interested in designing the hinge of the rod so that it could stand up to the dynamic forces, we *would* need to use the force equations to be able to compute R_x and R_y.

In standard form, the EOM becomes

$$\ddot{\theta} + 2\zeta\omega_n\dot{\theta} + \omega_n^2\theta = -\frac{g}{\ell} \tag{4.52}$$

where

$$\omega_n = \sqrt{\frac{k}{m\ell^2}} \tag{4.53}$$

$$\zeta = \frac{ca^2}{2\ell}\frac{1}{\sqrt{km}} \tag{4.54}$$

To determine the maximum value of the damping coefficient for which we would have underdamped motion, we set $\zeta = 1$ in the above equation. The damped natural frequency would be computed from $\omega_d = \omega_n\sqrt{1 - \zeta^2}$.

4.2 Energy

Let us revisit the subject of the energy of vibration that we started to discuss in Chapter 2. We already know the potential and kinetic energy expressions. The energy dissipated due to linear viscous damping is given by

$$\mathcal{D} = \frac{1}{2}c\,[\dot{x}(t)]^2 \tag{4.55}$$

which can be calculated by a straightforward substitution of velocity.

Let us do some numerical experimentation to get a sense of typical energy dissipation. Suppose that, we start the system off with an initial bout of energy (either through displacement or velocity or both—it does not really matter). Then, we look at how the system energy (defined as the sum of potential and kinetic energy) drops off with time. The actual numbers do not really matter here; hence, we are representing everything in normalized terms; for example, we can divide all energies by the initial energy, and do the same thing for the displacement. The number that does matter is the linear viscous damping ratio (ζ), which is a nondimensional quantity, and an extremely important one.

Figure 4.15 shows the decay of the total system energy with time; also shown on the graph is the displacement decay to help you correlate the energy variation with the actual motion. The damping ratio for this calculation is 0.1. Note how quickly the energy decays: it has dropped by about 90% after just two cycles. Also to be noted is the $2\omega_n$ dependence on time, as we discussed in Section 3.3.

Figure 4.16 shows the energy dissipation for the three values of damping (0.1, 0.4, and 0.8). Note the steep change from 0.1 to 0.4. The x-axis is really time, but represented in terms of the undamped period for ease of comparison in normalized terms. For example, after one undamped period of oscillation (a little different from its true damped period), the energy has dropped by 72% for $\zeta = 0.1$, by 99.5% for $\zeta = 0.4$, and by 99.98% for $\zeta = 0.8$!

Exercise 4.4

Verify Fig. 4.15 and Fig. 4.16. Use $m = 20\,\text{kg}$ and $k = 10\,\text{kN/m}$.

Use the closed form solution for $x(t)$ and derive an expression for the energy decay, which should help you put some mathematical oomph behind the statements made here.

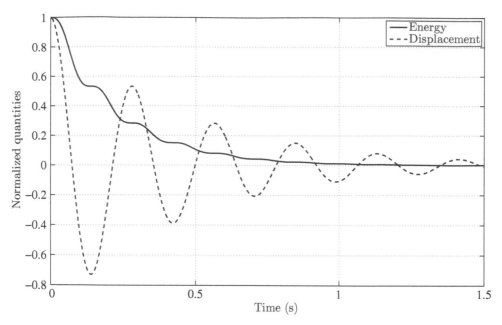

Figure 4.15: *Energy and displacement in a damped oscillator for $\zeta = 0.1$*

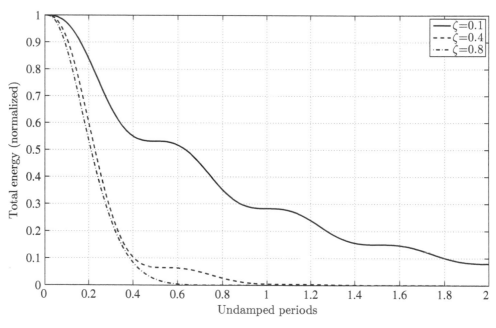

Figure 4.16: *Energy dissipation in a damped oscillator for varying values of damping*

4.3   Review Questions

1. When you slide a wooden block on a table, what kind of damping does it experience?

2. What kind of damping arises from cyclic variation of stress in materials?

3. What kind of damping does a submarine experience moving under water?

4. Is viscous friction always linear? Explain knowing what you know from your fluid mechanics class.

5. Define critical damping coefficient. How does it depend upon the mass of the system? On stiffness?

6. Given a system, how does its damped behavior change if its damping coefficient is doubled?

7. Given a system, how does its damped behavior change if its mass is doubled?

8. Given a system, how does its damped behavior change if its stiffness is doubled?

9. Given that an empty car has certain properties, how would its damped behavior change when it is fully loaded?

10. What is the *qualitative* change when the damping ratio is changed from a value less than 1 to a value greater than 1?

11. For what value of ζ does the exponential solution assumption for the ODE solution break down? What do you do in that case?

12. Why is critical damping desirable for many applications? Can you think up a few such applications?

13. Why is it difficult to achieve perfect critical damping in practice?

14. What happens to the speed of the response with excessive damping? Why?

15. If we connect the peak values in an underdamped response, what kind of a curve does it fall on? Why?

16. Indicate any one method of estimating damping experimentally.

17. Why is it a good idea to estimate damping from many cycles of vibration?

18. What is log decrement, and how is it related to the damping ratio?

19. At what rate does the energy drop in damped vibration? How does this compare with the rate of decrease of the displacement?

Problems

4.1 For each of the following systems, indicate if the system is stable or unstable, and if it is undamped, underdamped, critically damped, or overdamped.

    (a) $3\ddot{x} + 2\dot{x} + 10x = 0$

    (b) $2\ddot{x} + 5\dot{x} - 20x = 0$

    (c) $\ddot{x} + 95x = 0$

    (d) $\ddot{x} + 10\dot{x} + 25x = 0$

    (e) $\ddot{x} + 25\dot{x} + 5x - 3\sin 30t = 0$

4.2 Derive an expression for the time at which the free response of a single-degree-of-freedom (SDOF) damped system reaches a maximum value. Also find an expression for the maximum response.

4.3 (a) Derive the differential EOM for a simple pendulum of length ℓ with a bob of mass m in a viscous medium. The resulting damping coefficient is c. Assume small motions about the vertical equilibrium position.

(b) A simple pendulum is found to vibrate at a frequency of 0.5 Hz in vacuum and 0.45 Hz in a viscous fluid medium. Find the damping constant if the mass of the bob of the pendulum is 1 kg.

4.4 A thin plate of area A and weight W is attached to the end of a spring and is allowed to oscillate in a viscous fluid, as shown in Fig. 4.17. The idea is to experimentally estimate the coefficient of viscosity of the fluid. The period of oscillation of the system in air is τ_1, and when immersed in the fluid, it is τ_2. Derive an expression for μ, the coefficient of viscosity.

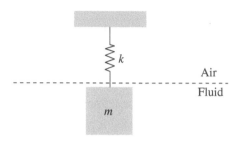

Figure 4.17: *Plate-spring system*

4.5 Consider a falling body in the atmosphere. As discussed earlier in Chapter 2 (page 14), if there is no friction, it would continue to accelerate until it reached the earth. Now, to be more realistic, let us consider the frictional resistance offered by the atmosphere. We will assume that the frictional force is proportional to the velocity. This is indeed a good assumption and an approximate formula for a falling sphere was discovered by the mathematician, Stokes, for the viscous damping coefficient, $c = 6\pi\mu R$, where μ is the coefficient of viscosity, and R is the radius of the sphere.

(a) Derive the EOM (ignoring buoyancy).

(b) Solve the equation and determine the steady-state value of the velocity. This is called the *terminal velocity.*

(c) ⚷ Now, include buoyancy (see Problem 2.24 on page 39) and solve the EOM. To do this part, assume that the falling object is a sphere of aluminum of radius 2.5 cm. You will have to look up other properties in a handbook.

4.6 Derive the differential EOM for the damped system shown in Fig. 4.18. Assume that the rod is itself massless (except for the point mass m shown). Determine the natural frequency of damped oscillation and the critical damping coefficient.

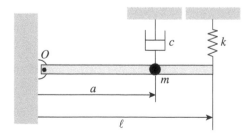

Figure 4.18: *A damped system, Problem 4.6*

4.7 As shown in Fig. 4.19, a plate of weight W rests on two counter-rotating wheels, a distance ℓ apart. The coefficient of friction between the wheels and the plate is μ.

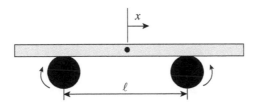

Figure 4.19: *Plate on rotating wheels*

(a) Derive the EOM for small displacements.

(b) If $\mu = 0.1$, determine the distance between the wheel centers so that the period of undamped motion is 1 s.

(c) If a linear viscous damper is added, what is the new EOM? If the damping coefficient is 30 Ns/m, and the mass is 2 kg, is the free motion underdamped or overdamped?

4.8 A shock absorber has a damping ratio of 0.8 when new, and 0.5 after years of driving in Philadelphia. How does the ratio of successive amplitudes of the car compare for the two situations? Use MATLAB to verify your answer.

4.9 A 2-kg piston supported on a helical spring vibrates freely with a natural frequency of 125 cpm (cycles per minute). When oscillating with an oil-filled cylinder, the frequency of free oscillation is reduced to 120 cpm. Determine the damping constant c.

4.10 If the ratio of successive amplitudes of damped-free vibration is 2:1, what is the logarithmic decrement and damping ratio?

4.11 The amplitude of vibration of an oscillating system was found to reduce to 1/30 its initial value after 10 cycles. Estimate the damping ratio of the system.

4.12 A system was observed to have a natural frequency of 5 rad/s and had very little damping in it. When some damping was added to it, it was found to vibrate at 4 rad/s.

(a) What is the damping ratio?

(b) If the mass in the original system is then reduced to one third its original value, what will be the damped natural frequency of the system?

4.13 Figure 4.20 shows an experimentally recorded response of a vibrating system. Assuming that it can be modeled as an SDOF system, estimate the damping ratio and undamped natural frequency of the system.

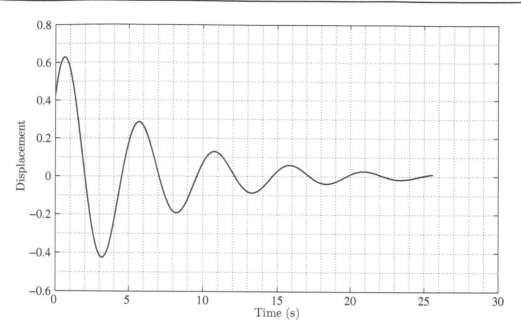

Figure 4.20: *An experimentally recorded vibration data*

4.14 A shock absorber design must be limited to 10% overshoot, when displaced from equilibrium and released. Determine the necessary damping ratio. (Overshoot is the maximum displacement the system experiences beyond its static equilibrium position.)

4.15 A rectangular door has height h, thickness d, width ℓ, and mass m. It is fitted with an automatic door closer, which consists of a torsional spring with a modulus k. Determine the necessary damping to critically dampen the return swing of the door. Use the numerical values $h = 2$ m, $d = 40$ mm, $\ell = 0.75$ m, $m = 36$ kg, and $k = 10$ Nm/rad. If the door is opened 90° and released, how long will it take until the door is within 1° of closing? Trick question: how long before it closes *completely*?

4.16 Find the equivalent damping coefficient of the configurations shown in Fig. 4.21 and Fig. 4.22.

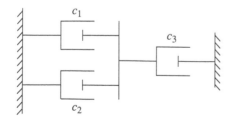

Figure 4.21: *Damper configuration*

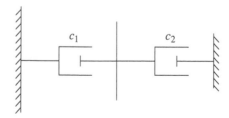

Figure 4.22: *Dampers in series*

4.17 Consider again the problem of the magnetic crane considered earlier (Problem 3.27).

Figure 4.23: *A magnetic crane carrying a car*

(a) Damping effects have been added, as shown in Fig. 4.23. Calculate the damped natural frequency of the system.

(b) Obtain the complete solution for the given initial conditions, $k = 35\,\text{kN/m}$, $c = 20\,\text{Ns/m}$, $m = 2000\,\text{kg}$, $y(0) = 0\,\text{m}$, and $\dot{y}(0) = 0.2\,\text{m/s}$.

4.18 Consider the damped mass-spring system shown in Fig. 4.24.

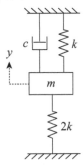

Figure 4.24: *A damped system, Problem 4.18*

(a) Obtain the EOM.

(b) Plot the response for the given values. $k = 110\,\text{N/m}$, $c = 1.5\,\text{Ns/m}$, $m = 2\,\text{kg}$, $y(0) = 0\,\text{m}$, and $\dot{y}(0) = -0.3\,\text{m/s}$.

4.19 Consider the horizontal pulley system shown in Fig. 4.25.

(a) Obtain the EOM.

(b) Find the natural frequency and obtain the response for $k = 60\,\text{N/m}$, $c = 3\,\text{Ns/m}$, $m = 0.65\,\text{kg}$, $y(0) = 0.1\,\text{m}$, and $\dot{y}(0) = -0.3\,\text{m/s}$. The mass of the pulley is negligible.

4.20 An inclined mass-pulley system is shown in Fig. 4.26.

(a) Obtain the EOM.

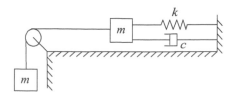

Figure 4.25: *A damped system with a pulley mechanism*

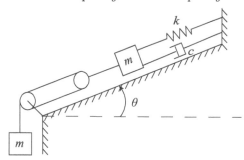

Figure 4.26: *An inclined mass-pulley system*

(b) Plot the response for the given values. $k = 95\,\text{N/m}$, $c = 4\,\text{Ns/m}$, $m = 1.35\,\text{kg}$, $y(0) = 0.2\,\text{m}$, and $\dot{y}(0) = 0\,\text{m/s}$. The masses of the pulleys are negligible.

4.21 Consider the selection of suspension springs and shock absorbers for an automobile. The mass of the vehicle is 1400 kg, and the desired damping ratio is 0.4. We would like the natural frequency to be in the range of 5–8 Hz. Determine the required stiffness of the springs and the viscous damping coefficient of the shock absorbers. The car is designed to carry a maximum of 300 kg of people and cargo. The static deflection (loaded or unloaded) cannot exceed 15 cm.

4.22 We need to select a suspension spring and a damper for an automobile. Its mass is 1500 kg; we would like the undamped natural frequency to be 10 Hz, and damping ratio to be 0.35.

(a) Determine the damped natural frequency, the stiffness coefficient, and the viscous damping coefficient of the shock absorbers.

(b) Four people with a total mass of 400 kg are sitting in the car. Determine the effective undamped and damped natural frequencies. Also, determine the new damping ratio.

4.23 The atomic force microscope (AFM) is a very useful tool for characterizing nano-sized structures and biological materials and processes. It has become indispensable for researchers and engineers working with nanomechanics. The essential element of the AFM is the probe, whose vibrational displacement is used as a sensor to measure extremely small structures and forces, and is shown schematically in Fig. 3.54. It is essentially a cantilever beam and can be modeled as an SDOF system. We looked at the undamped model in Chapter 3. Here, to be realistic, we add damping, as shown in Fig. 4.27. Derive the EOM, assuming that AB is a rigid body hinged at one end whose other end is coupled to a spring. For a more involved discussion on modeling the AFM probe you might want to read the article by Mahdavi et al. [17].

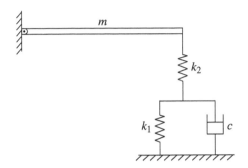

Figure 4.27: *Atomic force microscope model*

4.24 For the sample brake mechanism shown in Fig. 4.28, find the difference in the natural frequency for cases with and without the damper.

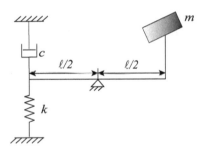

Figure 4.28: *Brake mechanism*

4.25 For the brake mechanism shown in Fig. 4.29, if the damper is attached as shown, find the natural frequency. Comment on the result when compared to the brake system from Problem 4.24.

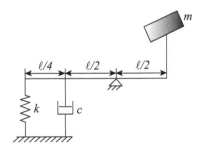

Figure 4.29: *Alternative brake mechanism*

4.26 A tire suspension system is shown in Fig. 4.30, with a torsional damper attached to it.

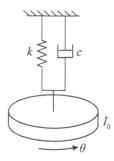

Figure 4.30: *Tire suspension*

Find the change in the natural frequency when the damper is added.

4.27 The system of mass m is supported by the massless rigid column of length ℓ and the spring-damper mechanism, as shown in Fig. 4.31. The horizontal displacement, x, of the mass is related to the rotating angle, θ, of the column and the length ℓ, of the column. Assume small displacements and derive the EOM of the system. Also, find the natural frequency of the system.

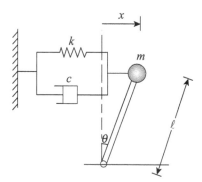

Figure 4.31: *Damped inverted pendulum system*

4.28 A free-vibrating mass-spring-damper system was built to experimentally characterize particle-produced damping [21]. The experimental setup of the system is shown in the figure and its equivalent model is illustrated in Fig. 4.32. This system has the form of a thrust damper in which a piston vibrates vertically in a damping medium. The free vibration damping is characterized by measuring its displacement decay. Derive the EOM of the equivalent system in which k is the stiffness, m the mass, c the damping coefficient, and x the displacement of the mass from its equilibrium position.

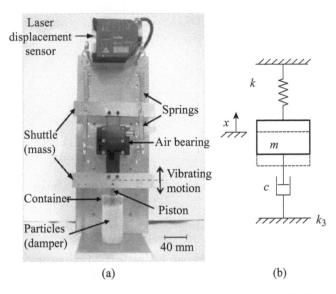

Figure 4.32: *Thrust damper (see Shah et al. [21])*

4.29 Consider an automobile modeled as a rigid rod of length ℓ and mass m, as shown in Fig. 4.33. Assume that the center of mass is at the center of the rod, and that the front and rear suspension properties are identical. Assume also that the car can only move up and down. (Of course, it can rotate too, but that would make it a *two*-DOF system! We will certainly revisit this problem in Chapter 7 with a more realistic model.) Derive the differential EOM.

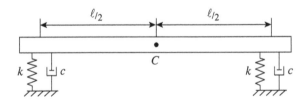

Figure 4.33: *Automobile modeled as a rigid rod on suspensions*

(a) Determine the undamped natural frequency, damping ratio, and the damped natural frequency for the following parameter values: $m = 1100\,\text{kg}$, $k = 40\,\text{kN/m}$, and $c = 6.5\,\text{kNs/m}$. What happens to these values when the car is loaded with an additional weight of 250 kg?

(b) Now consider a truck with a mass of 4000 kg. Suppose we want the truck to have the same damping ratio and natural frequency as the car in (a). Determine the suspension parameters k and c. How does its natural frequency and damping ratio change when it is loaded by an additional 1000 kg. Also, determine its static displacement when it is loaded.

4.30 Consider an automobile modeled as a rigid rod of length ℓ and mass m, as shown in Fig. 4.34. Assume that the center of mass is at a distance ℓ_1 from the left end of the rod. The suspension properties are selected so that the following is true: $k_1\ell_1 = k_2\ell_2, c_1\ell_1 = c_2\ell_2$. Assume also that the car can only move up and down. (Of course, it can rotate too, but that would make it a *two*-DOF system! We will certainly revisit this problem in Chapter 7 with a more realistic model.) Derive the differential EOM.

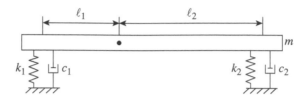

Figure 4.34: *Automobile modeled as a rigid rod on suspensions*

(a) Determine the undamped natural frequency, damping ratio, and the damped natural frequency for the following parameter values: $m = 1500\,\text{kg}$, $\ell_1 = 1.3\,\text{m}$, $\ell_2 = 1.7\,\text{m}$, $k_1 = 80\,\text{kN/m}$, and $c_1 = 8\,\text{kNs/m}$. What happens to these values when the car is loaded with an additional weight of 300 kg? Determine its static displacement.

(b) Now consider a nominally loaded truck with a total mass of 5000 kg and $\ell = 8\,\text{m}$ with a mass center that is located 55% from the left end when fully loaded. Suppose we want the truck to have the same damping ratio and natural frequency as the car in (a). Determine the suspension parameters k and c. Also, determine its static displacement.

4.31 A cable damper system is discussed in the article by Cheng et al. [22] and is shown schematically in Fig. 4.35. The cable is laid out in the horizontal direction with a chord length ℓ, a mass per unit span m, and flexural rigidity EI. An external viscous damper is transversely attached to the cable at a distance ℓ_d from one end of the cable. The damping coefficient of the viscous damper is denoted as c. Determine the natural frequency when $c = 0$, 5, and 10 Ns/m. Also, $\ell = 13.69$ m, EI $= 2.28\,\text{kNm}^2$, $m = 3.6\,\text{kg/m}$, $\ell_d = 3.5$ m. You might also want to investigate the effect of changing the point of attachment of the damper (ℓ_d) on the natural frequency and the damping ratio.

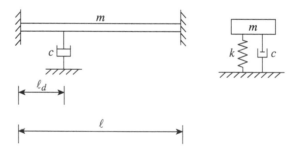

Figure 4.35: *Cable damper*

4.32 The vibrational response of seated human subjects is an important problem. One of the simplest models for it is to consider the human being as one lumped mass, and the human-seat interface as a spring and a damper [18]. The mass is 56.8 ± 9.4 kg, the spring stiffness is 75.5 ± 28.3 kN/m, and the damping coefficient is 3.840 ± 1.007 kN s/m. Determine the *range* of undamped and damped natural frequencies and damping ratios. We will look at more complicated models in the forthcoming chapters.

4.33 Derive the EOM, and find the natural frequency and damping ratio for the system shown in Fig. 4.36.

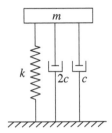

Figure 4.36: *Problem 4.34*

4.34 Derive the EOM, and find the natural frequency and damping ratio for the system shown in Fig. 4.37.

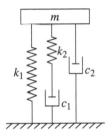

Figure 4.37: *Problem 4.34*

4.35 Derive the EOM, and find the natural frequency and damping ratio for the system shown in Fig. 4.38.

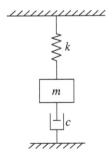

Figure 4.38: *Problem 4.35*

4.36 Derive the EOM, and find the natural frequency and damping ratio for the system shown in Fig. 4.39.

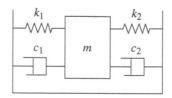

Figure 4.39: *Problem 4.37*

4.37 Derive the EOM, and find the natural frequency and damping ratio for the system shown in Fig. 4.40.

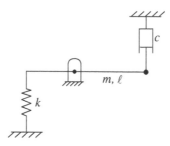

Figure 4.40: *Problem 4.37*

4.38 Derive the EOM, and find the natural frequency and damping ratio for the system shown in Fig. 4.41. Solve for the response.

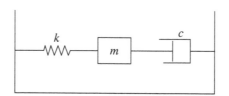

Figure 4.41: *Problem 4.38*

4.39 Derive the EOM, and find the natural frequency and damping ratio for the system shown in Fig. 4.42. Solve for the complete response.

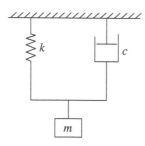

Figure 4.42: *Problem 4.40*

4.40 Derive the EOM, and find the natural frequency and damping ratio for the system shown in Fig. 4.43.

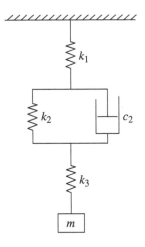

Figure 4.43: *Problem 4.40*

4.41 Derive the EOM, and find the natural frequency and damping ratio for the system shown in Fig. 4.44.

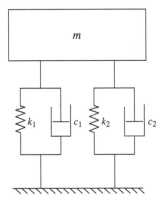

Figure 4.44: *Problem 4.41*

4.42 A uniform rod of mass m and length ℓ is hinged at one end; at the other end, a damper is attached, as shown in Fig. 4.45. Find the EOM of the system. Can you find a natural frequency for the system? Why or why not? Can the EOM be simplified in any way?

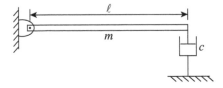

Figure 4.45: *Hinged rod with damper*

4.43 Prolonged, intensive exposure to hand-transmitted vibration could cause a series of disorders in the sensorineural, vascular, and muscular systems of the fingers, which are the major components of the hand-arm vibration syndrome. Wu et al. [23], hypothesize that the vibration power absorption density is a good measure for the vibration exposure intensity of the soft tissues of the fingers. The hand-finger is simulated by using a lumped parameter model combined with a two-dimensional finite element (FE) model, as shown in Fig. 4.46. The finger tip is modeled using a two-dimensional FE, whereas the effective mass of the hand finger is represented by the mass element m. The coupling between the finger tip and the hand finger is represented by the spring and damping element (k_1 and c_1). The contact between the finger tip and hand finger is simulated in the FE modeling, whereas the coupling between the hand and the vibrating plate is represented by a spring-damping unit (k_3 and c_3). The coupling between the hand, fore arm and ground is represented by using another spring-damper unit (k_2 and c_2). Find the EOM of the simplified model shown.

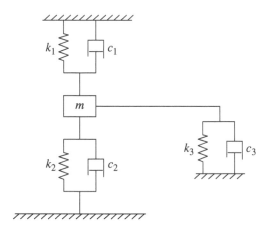

Figure 4.46: *Model for hand vibration*

5 SDOF Systems: Periodic Forced Response

Of course there's a lot of knowledge in universities: the freshmen bring a little in; the seniors don't take much away, so knowledge sort of accumulates.

—A. Lawrence Lowell

We consider the same single-degree-of-freedom (SDOF) system we had earlier, but subject it now to a force $F(t)$. Frequently in practice, the force is a complicated function of time, and we need to look at some typical scenarios and the consequent response of the system. In this chapter we will first focus on sinusoidal and periodic excitation; the next chapter will deal with other kinds of excitations.

5.1 Harmonic Excitation

The most common kind of excitation (in vibration, one often uses the word *excitation* to describe a force) for vibrating systems is harmonic. This is a consequence of periodic motions that are very common (by design) in most engineering systems.[1] In addition, it is important to study harmonic excitation, since the solution obtained here forms the mathematical basis for extending the solution to situations when the forces are *not* harmonic.

5.1.1 Equation of Motion

Consider the SDOF model again with a harmonic force, as shown in Fig. 5.1. Then, the equation of motion is

$$m\ddot{x} + c\dot{x} + kx = F_0 \sin \omega t \tag{5.1}$$

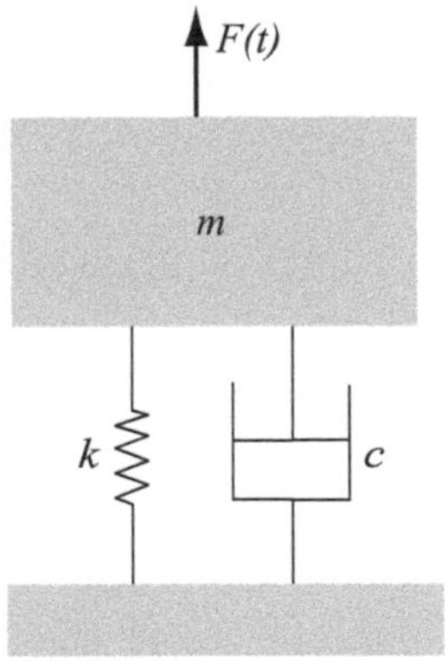

Figure 5.1: *A forced mass-spring-damper system*

---

[1] Rotating systems invariably lead to periodic excitation. How many rotating components or systems can you list from your knowledge of the engineering world?

where F_0 is the amplitude of the force, and ω is the frequency of excitation. It follows from the free body diagram (FBD), shown in Fig. 5.2, and from simple kinematics, $a = \ddot{x}$. Note that a more general harmonic force would be $F_0 \sin(\omega t + \phi)$. We will consider that case later. Next, we divide by m and write the equation in standard form:

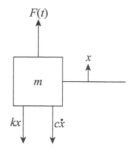

Figure 5.2: *FBD of the forced mass-spring-damper system*

$$\ddot{x} + 2\zeta\omega_n\dot{x} + \omega_n^2 x = \omega_n^2 \frac{F_0}{k} \sin\omega t \tag{5.2}$$

5.1.2 Solution

From the elementary theory of differential equations, the solution can be sought in the form

$$x(t) = x_c \cos\omega t + x_s \sin\omega t \tag{5.3}$$

Then,

$$\dot{x}(t) = -\omega x_c \sin\omega t + \omega x_s \cos\omega t \tag{5.4}$$
$$\ddot{x}(t) = -\omega^2 x_c \cos\omega t - \omega^2 x_s \sin\omega t \tag{5.5}$$

Substitution into the equation of motion and separation of the sine and cosine functions yield the following set of linear algebraic equations in the unknown coefficients.

$$\begin{bmatrix} -\omega^2 + \omega_n^2 & 2\zeta\omega\omega_n \\ -2\zeta\omega\omega_n & -\omega^2 + \omega_n^2 \end{bmatrix} \begin{bmatrix} x_c \\ x_s \end{bmatrix} = \begin{bmatrix} 0 \\ \omega_n^2 F_0/k \end{bmatrix} \tag{5.6}$$

Solving the set of algebraic equations, we get the following.

$$\begin{bmatrix} x_c \\ x_s \end{bmatrix} = \frac{1}{\Delta'} \begin{bmatrix} -\omega^2 + \omega_n^2 & -2\zeta\omega\omega_n \\ +2\zeta\omega\omega_n & -\omega^2 + \omega_n^2 \end{bmatrix} \begin{bmatrix} 0 \\ \omega_n^2 F_0/k \end{bmatrix} \tag{5.7}$$

where

$$\Delta' = \left(-\omega^2 + \omega_n^2\right)^2 + \left(2\zeta\omega\omega_n\right)^2 \tag{5.8}$$

Next, we define a convenient nondimensional parameter called the *frequency ratio*, r, that is a ratio of the forcing frequency to the undamped natural frequency of the system.

$$r = \frac{\omega}{\omega_n} \tag{5.9}$$

Using this, the unknown coefficients are given by

$$x_c = -\frac{2\zeta r}{\Delta} \frac{F_0}{k} \tag{5.10}$$

$$x_s = \frac{-r^2 + 1}{\Delta} \frac{F_0}{k} \tag{5.11}$$

where

$$\Delta = \left(-r^2 + 1\right)^2 + (2\zeta r)^2 \tag{5.12}$$

Hence, the final solution can be written in the form

$$x(t) = x_0 \sin{(\omega t - \phi)} \tag{5.13}$$

where

$$x_0 = \sqrt{x_c^2 + x_s^2}, \quad \text{and} \quad \phi = \tan^{-1}\left(\frac{-x_c}{x_s}\right) \tag{5.14}$$

substituting and simplifying, the nondimensional amplitude of motion is given by

$$\bar{x}_0 \stackrel{\text{def}}{=} \frac{x_0}{F_0/k} = \frac{1}{\sqrt{(-r^2 + 1)^2 + (2\zeta r)^2}} \tag{5.15}$$

and the phase angle is given by

$$\phi = \tan^{-1}\left(\frac{2\zeta r}{-r^2 + 1}\right) \tag{5.16}$$

$\bar{x}_0$, the ratio of displacement amplitudes given by Eq. (5.15), is the dynamic response amplitude divided by the static displacement that the system would experience if it were subjected to a static force of magnitude F_0 (instead of the harmonic force). Hence, this nondimensional ratio represents the effect of the dynamics on the system and is called the *dynamic magnification factor*.

Figure 5.3 shows the dynamic magnification factor, and Fig. 5.4 shows the phase lag ϕ between the force and the displacement; both are shown as functions of the frequency ratio and the damping ratio. Together, these figures constitute what is called the *frequency response* of the SDOF system.

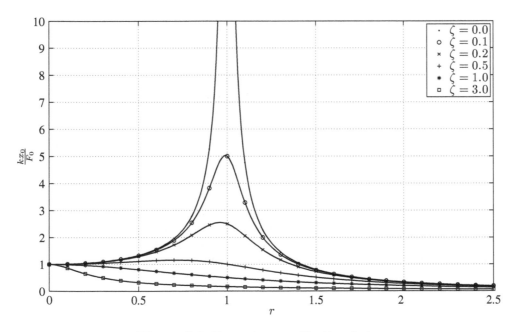

Figure 5.3: *Dynamic magnification factor*

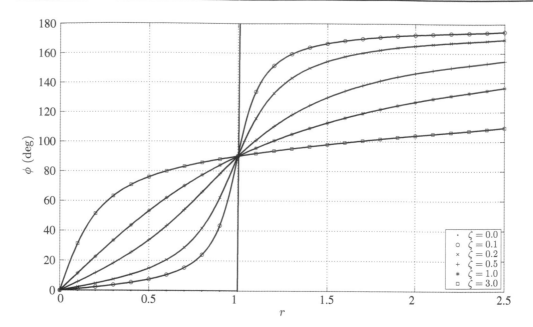

Figure 5.4: *Phase lag in harmonic excitation*

The frequency response figures should be carefully studied, since much of the design of vibrating systems is based on them. We, hence, enumerate below some important observations and implications to real systems that follow from this figure.

Exercise 5.1

Derive Eq. 5.15 following the procedure outlined in Section 5.2.1 providing all details.

5.1.3 Observations

- $\bar{x}_0 = 1$ for $r = 0$ for all values of ζ. In other words, for small excitation frequencies (in relation to the natural frequency of the system), the dynamic amplitude approaches the equivalent static amplitude. Or, the dynamic effects in the system could then be safely ignored. This is the case for many structural systems such as buildings.[2]

- $\bar{x}_0 \to 0$ for $r \gg 1$. For high frequencies of excitation, the response amplitude tends to zero for all values of damping. This is of course good news, since we now have a scheme by which we could reduce vibration in a machine or structure, namely, alter its mass and stiffness in such a way that the frequency ratio becomes a large number. (Not really! Check out the think box that follows).

- $\bar{x}_0$ becomes very large for $r = 1$. In fact, our analysis predicts that if $\zeta = 0$, the system response becomes unbounded. If there is any amount of damping in the system however, the response is bounded, but can get very large—large enough to destroy engineering systems; $\bar{x}_0 = 1/(2\zeta)$ for $r = 1$.[3] This phenomenon is called *resonance* and should be always considered in the design of any vibrating system.

---

[2]Earthquakes and wind-loaded tall buildings are important exceptions to this.

[3]You should also remember that we are using a linear model to make these predictions, and for large vibration amplitudes, the nonlinearities in the system would raise their ugly heads, and our model (and hence, its prediction) would no longer be accurate.

- The true maximum value of the amplitude can be found by differentiating Eq. (5.15) and setting it equal to zero. It occurs at the *resonant* frequency ratio,

$$r_r = \sqrt{1 - 2\zeta^2} \tag{5.17}$$

The corresponding maximum value is

$$\bar{x}_{0,\text{max}} = \frac{1}{2\zeta\sqrt{1 - \zeta^2}} \tag{5.18}$$

It also follows that this maximum exists only if $\zeta < 1/\sqrt{2}$.

- Although the true maximum is as described above, usually, especially if the damping ratio is very small, the maximum value is evaluated at $r = 1$, as described earlier.
- Each of the points of the curve in Fig. 5.3 corresponds to a sinusoidal response at the corresponding frequency. The only exception is for $\zeta = 0$, $r = 1$. Then, the system response is given by

$$x(t) = -\frac{F_0}{k}\frac{1}{2}\omega_n t \sin(\omega t) \tag{5.19}$$

which is a steadily increasing sinusoidal oscillation, as shown in Fig. 5.5.

- For small values of damping, in the vicinity of resonance, the response is very sensitive to changes in the frequency ratio. For example, if $\zeta = 0.01$, at $r = 1$, the response amplitude is $\bar{x}_0 = 50$, but at $r = 1.1$, $\bar{x}_0 = 4.7$. This fact then gives us a way to reduce the vibration by changing k or m, which would change the natural frequency, ω_n, that would in turn change r.
- At or near resonance, the response is also very sensitive to changes in ζ. For example, at $r = 1$ and for $\zeta = 0.01$, $\bar{x}_0 = 50$; but for $\zeta = 0.05$, $\bar{x}_0 = 10$. In contrast to this, for higher values of frequency, the response is relatively insensitive to changes in damping. For example, at $r = 3$

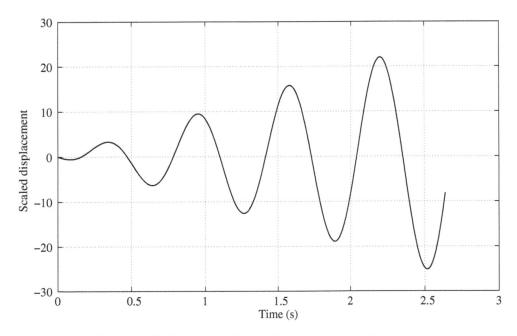

Figure 5.5: *Response of an undamped system at resonance*

and for $\zeta = 0.01$, $\bar{x}_0 = 0.125$. If we increase ζ to 0.05, the amplitude drops to 0.1249, which is a very small drop. If we increase the damping significantly to $\zeta = 1$, the response would be $\bar{x}_0 = 0.1$, which is a drop of 20% for increasing the damping by 100 times! This is confirmed by Fig. 5.6, which shows the nondimensional amplitude versus the damping ratio for the two values of r.

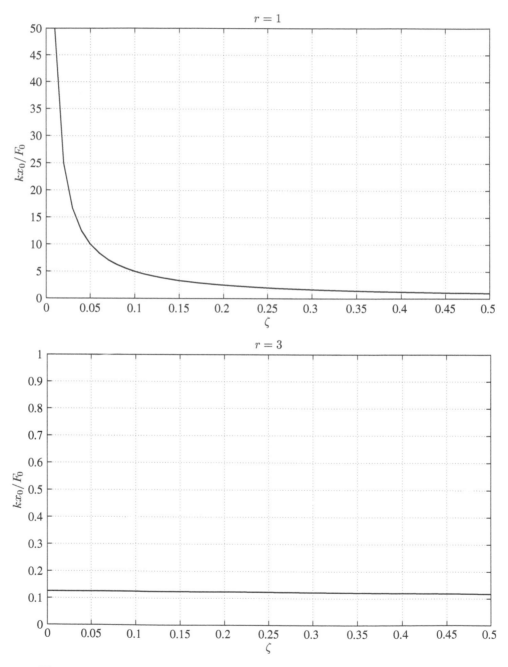

Figure 5.6: *Response amplitude versus damping ratio for $r = 1$, and $r = 3$*

- If $\zeta < 1/\sqrt{2}$, the dynamic magnification ratio is more than 1 until the resonant frequency. In fact, r has to get well past 1 before it will drop to a value below 1. Recall that a value greater than 1 means that the dynamic displacement amplitude is larger than the equivalent static displacement.

- From the phase figure it is clear that the response "keeps up" with the force at low frequencies; however, as the forcing frequency is increased, the phase lag becomes more pronounced, reaching 90° at resonance, and 180° for large frequencies. This means that at high frequencies, when the force is maximum in one direction, the displacement of the mass is maximum in the other direction![4] This is really a consequence of the fact that there is mass in the system and it cannot respond to the force *pronto* and takes its own sweet time doing it.

- Note that the phase angle is 90° at resonance irrespective of damping. This fact can be used to determine the natural frequencies of complicated systems experimentally: as the forcing frequency is gradually increased, the natural frequency is the place where the phase angle goes through 90°. This will work even for overdamped systems where we will not see any resonance in the amplitude of vibration.

Exercise 5.2

Determine the solution of an undamped system at resonance to get Eq. (5.19) and verify Fig. 5.5.

Exercise 5.3

Find the frequency at which the amplitude peaks, and show that it does not exist unless $\zeta < 1/\sqrt{2}$.

Based on the response amplitude of the system we stated that we can reduce the vibration of the structure to practically nil by increasing the frequency ratio r to a number way past a value of 1. This, alas, is not true. The reason goes back to something we have been stressing all along. *Never forget* that the results we predict depend strongly on the mathematical model we solve, which in turn depends on the assumptions we make when we derive the model. In this case, we have assumed that the real behavior of the system can be adequately and precisely explained by modeling it with *one* degree of freedom. If that were not true—in fact, it is hardly ever true—then, what do you think the frequency response curve would look like? Do you think real systems have only one natural frequency, and one resonant peak? If you dare to peek ahead, you will discover the answer to this in Chapter 8.

---

[4]Think about how counterintuitive this is. Doesn't dynamics just blow your mind?!

We made many comments that the response is more or less sensitive at some frequencies or damping ratios. The right way to do this is to derive a *sensitivity coefficient*. For example, the sensitivity of the displacement amplitude to damping would be defined by

$$\frac{\partial \bar{x}_0}{\partial \zeta}$$

Now, from the solution, determine this value, and prove our assertions stated in the observations with respect to sensitivity at various frequencies.

5.1.4 Phase Angles

It is important to note that the phase angle is determined from the ratio of two quantities

$$\phi = \tan^{-1}\left(\frac{2\zeta r}{-r^2 + 1}\right) \tag{5.20}$$

Even if we restrict the angle to a range of $-180°$ to $+180°$, since the inverse tangent is multi-valued, the answers turn out to be incorrect if care is not exercised. For example, if the numerator and denominator are both positive, or both negative, the ratio would turn out to be the same, and the phase angle obtained from a calculator would be indistinguishable. However, considering the sign of the numerator and the denominator separately would ensure correct calculations. All computer packages provide suitable strategies for doing this, often by using a function called `atan2`. Whether you are using a computer or a calculator, pay attention to the signs, and remember that the numerator is essentially a 'sine', and the denominator is a 'cosine' value, and then apply the knowledge you have about their signs in the four quadrants.

5.1.5 General Harmonic Forcing

Next, let us consider a more general case of the harmonic force

$$F(t) = F_c \cos \omega t + F_s \sin \omega t \tag{5.21}$$

Note that this expression can be written in a different form using simple trigonometry

$$F(t) = F_0 \sin(\omega t + \alpha) \tag{5.22}$$

where

$$F_0 = \sqrt{F_c^2 + F_s^2} \quad \text{and} \quad \phi = \tan^{-1}(F_c/F_s) \tag{5.23}$$

The EOM becomes

$$m\ddot{x} + c\dot{x} + kx = F_c \cos \omega t + F_s \sin \omega t$$
$$= F_0 \sin(\omega t + \alpha) \tag{5.24}$$

Proceeding along the same lines as before, it is straightforward to show that the response will be

$$x(t) = x_0 \sin(\omega t + \alpha - \phi) \tag{5.25}$$

where the amplitude, x_0 and the phase lag, ϕ are the same as before.

Exercise 5.4

Follow all the solution steps to show that the response to a general harmonic force, $F_0 \sin(\omega t + \alpha)$ leads to the same displacement amplitude and the same phase lag as derived earlier.

5.2   Complex Representation

Often, it is algebraically much easier to solve problems using a complex representation. Learning such an approach will also help you familiarize yourself with many concepts that are explored in system dynamics, rotor dynamics, and control theory. Let us consider again the forced harmonic equation.

$$\ddot{x} + 2\zeta\omega_n\dot{x} + \omega_n^2 x = \omega_n^2 \frac{F_0}{k} \sin\omega t \tag{5.26}$$

Next, we *artificially* rewrite the excitation as a complex function.

$$\ddot{x} + 2\zeta\omega_n\dot{x} + \omega_n^2 x = \omega_n^2 \frac{F_0}{k} \exp(i\omega t) \tag{5.27}$$

Note that the true excitation (say, $\sin\omega t$) would be obtained as the imaginary part of the complex function. Hence, our logic is that, if we can solve this complex equation, which will lead to a complex displacement, we could extract the imaginary part of it and treat that as the true solution.

Now, the solution becomes quite easy (much easier than the procedure we followed earlier by assuming sines and cosines); here, we seek a solution in the form

$$x(t) = x_0 \exp(i\omega t) \tag{5.28}$$

and substitute into the differential equation. This will lead (almost by inspection!) to

$$x_0 = \frac{F_0}{k} \frac{\omega_n^2}{[\omega_n^2 - \omega^2] + i\,[2\zeta\omega\omega_n]} \tag{5.29}$$

This can be further simplified using the frequency ratio r.

$$\frac{x_0}{F_0/k} = \frac{1}{[1 - r^2] + i\,[2\zeta r]} = H(i\omega) \tag{5.30}$$

The above function (often with the symbol, $H(i\omega)$ or $G(i\omega)$), is called the (nondimensional) frequency response function of the system being analyzed.

Since it is a complex function, it has a magnitude and a phase. The magnitude is given by

$$|H(i\omega)| = \frac{1}{\sqrt{(-r^2 + 1)^2 + (2\zeta r)^2}} \tag{5.31}$$

and the phase is given by

$$\angle H(i\omega) = -\tan^{-1}\left\{\frac{2\zeta r}{1 - r^2}\right\} \tag{5.32}$$

Now, the displacement amplitude can be written as

$$\frac{x_0}{F_0/k} = H(i\omega) = |H(i\omega)| \exp(-i\phi) \tag{5.33}$$

where

$$\phi \overset{\text{def}}{=} -\angle H(i\omega) \tag{5.34}$$

and the time-varying displacement is

$$x(t) = \frac{F_0}{k} |H(i\omega)| \exp\left(i\left(\omega t - \phi\right)\right) = x_0 \exp\left(i\left(\omega t - \phi\right)\right) \tag{5.35}$$

The above expression is clearly a complex function, which should give us pause. We are after all in the real world with real "feelable" forces and real sensible displacements! Of course, the truth is that it is all just a convenient stratagem, and we will have to extract its imaginary part to get the true displacement. That turns out to be

$$x(t) = \frac{F_0}{k} |H(i\omega)| \sin\left(\omega t - \phi\right) \tag{5.36}$$

which of course is the same expression we had earlier (Eq. 5.13).

Exercise 5.5

Consider the excitation, $F(t) = F_0 \cos\omega t$, and carry out the complex solution process as we did here for $\sin\omega t$; of course, you know what you should get for the answer.[5]

The complex representation also lends itself nicely to a pictorial analysis that provides more insight compared to purely algebraic derivations we have been dealing with. Representing the complex displacement on a complex plane (Fig. 5.7), it is clear that $x(t)$ would simply rotate with a frequency ω as time increases; this is what is often termed a *phasor*.[6]

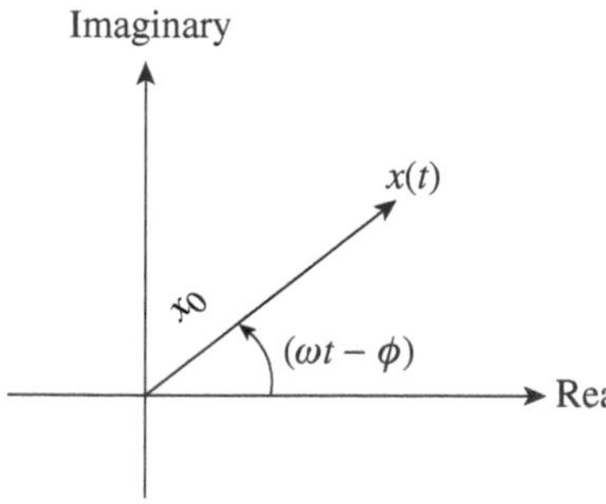

Given the complex form of $x(t) = x_0 \exp\left(i\left(\omega t - \phi\right)\right)$, the velocity, is given by

$$\dot{x} = i\omega x_0 \exp\left(i\left(\omega t - \phi\right)\right)$$

Figure 5.7: *Complex displacement as a phasor*

which would be a phasor at right angles to the displacement. Similarly, acceleration would be coincident with the displacement but opposite in direction

$$\ddot{x} = -\omega^2 x_0 \exp\left(i\left(\omega t - \phi\right)\right)$$

---

[5] $x(t) = \frac{F_0}{k} |H(\omega)| \cos\left(\omega t - \phi\right)$. Shame on you for checking!
[6] Not to be confused with the phaser from the Star Trek TV episodes!

This is shown in Fig. 5.8. You need to imagine these phasors spinning in counterclockwise direction as the system executes harmonic motion.

It is interesting to take this analogy further, and to represent the forces as well on the complex plane. To do this, let us rewrite the equation of motion.

$$m\ddot{x} + c\dot{x} + kx = F_0 \sin \omega t \quad (5.37)$$

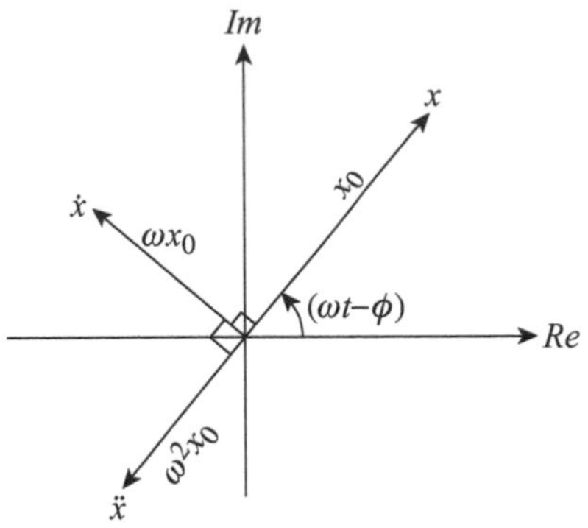

This now becomes

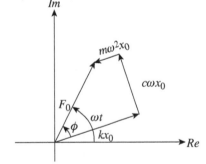

Figure 5.8: *Complex phasor diagram of displacement, velocity, and acceleration*

$$-m\omega^2 x_0 \exp\left(i\left(\omega t - \phi\right)\right) + ic\omega x_0 \exp\left(i\left(\omega t - \phi\right)\right) + kx_0 \exp\left(i\left(\omega t - \phi\right)\right) = F_0 \exp\left(i\omega t\right) \quad (5.38)$$

Simplifying, this becomes an algebraic equation if we divide out the common multiplier $\exp\left(i\omega t\right)$.

$$kx_0 e^{-i\phi} + ic\omega x_0 e^{-i\phi} - m\omega^2 x_0 e^{-i\phi} = F_0 \quad (5.39)$$

Now, we can look at this equation as an equilibrium equation, and put each of them on the phasor diagram, as shown in Fig. 5.9. This is often called a *force diagram*. Again, it is instructive to think about what is happening to each of these phasors as the system undergoes harmonic motion.

Let us examine these phasor diagrams further for different ranges of forcing frequencies. For small frequencies ($\omega \ll \omega_n$, or $r \ll 1$), the phase lag is small, and the harmonic force is nearly balanced by the static force due to the spring with only small contributions from the damper and the "inertia force" ($m\omega^2 x_0$). This can be seen clearly in Fig. 5.10. Next, as we increase the forcing frequency and approach resonance, the phase lag approaches 90° and we have the situation shown in Fig. 5.11; now, F_0 exactly balances the damping force $c\omega x_0$, and the inertial forces, $m\omega^2 x_0$ are balanced by the elastic force, kx_0. As the frequency is increased further, the phase angle becomes nearly 180° and the inertia force becomes dominant, as seen in Fig. 5.12.

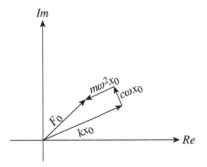

Figure 5.9: *Complex phasor diagram of the forces in dynamic equilibrium*

Figure 5.10: *Complex phasor diagram of the forces for $r \ll 1$*

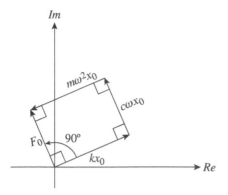

Figure 5.11: *Complex phasor diagram of the forces for* $r = 1$

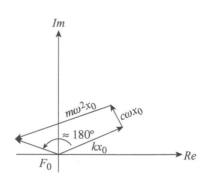

Figure 5.12: *Complex phasor diagram of the forces for* $r \gg 1$

5.3   Total Response

Now that we have studied both free and forced responses, we can put them together. In other words, the total response of a linear vibrating system will be a a simple addition of the two responses. For example, for the SDOF underdamped oscillator with harmonic excitation,

$$
\begin{aligned}
x(t) &= x_{\text{homogeneous}}(t) + x_{\text{particular}}(t) \\
&= x_{\text{free}} + x_{\text{forced}} \\
&= C e^{-\zeta \omega_n t} \sin(\omega_d t + \phi_h) + x_0 \sin(\omega t - \phi)
\end{aligned}
\tag{5.40}
$$

where C and ϕ_h would be evaluated from the initial conditions. x_0 and ϕ are of course given by specific expressions that depend on ω, F_0, and the other system parameters. Clearly, there is a one-to-one correspondence between what you learn in the differential equation theory about homogeneous and particular solutions, and what we call free and forced responses in the vibration theory.

Coming back to the practical system, what this means is that any oscillator has two components to its response: free and forced. The first is in response to what we do to it just before time zero; i.e., if we just disturbed the oscillator and walked away, that would be the free vibration. The forced response, on the other hand, is what the system exhibits in response to a persistent force. The real response of the system is then what it does in response to some disturbances that will keep happening (not just at time zero), and a continuous "thumping" it gets from a variety of sources. If you think of a car as an example, the free response would be the vibration you would feel because it went over a pothole (something that would die after a certain amount of time), and the forced response would be the continuous vibration you would feel as it travels over a rough road, responding to its roughness and undulations.

The free response to a given disturbance will die eventually; you know that already from Chapter 4. We hence often refer to it as *transient vibration*. The response that does not die (unless the force does) would be called *steady-state vibration*. Because the two are simply additive (a consequence of the linearity assumption), we do not bother to write this out every time. If we are interested in analyzing the free vibration, we do it; if we are interested in looking at the forced vibration, we will do it separately. We do know that the true response is a sum, but rarely would we put them together as we did in Eq. (5.40); we just did it here to make a point.

In numerical simulation, it is difficult to separate the two responses; i.e., you will always get the free response even if you were just looking for the steady-state response. A typical trick in numerical simulation is to wait until the transient response dies out, and to look at the response only after that time period. Of course, theoretically, it would take forever to die; in practice, beyond about four times the time constant (i.e., $4/\zeta\omega_n$ for an underdamped system, and $4/s_1$ for an overdamped system, where s_1 is the characteristic root with the smaller absolute value),[7] we can assume that the transient response has died, and that what is left is the steady-state vibration.

Figure 5.13 shows the total response of an SDOF system excited by a harmonic force; the results were obtained numerically using a program similar to Listings 2.5 and 2.6 shown on page 32; it is plotted on top of the analytical forced response solution that we obtained in this chapter. Note that the two start to coincide after about 2.5 s. For this system, the parameters were as follows: $\zeta = 0.25$, $\omega_n = 7\,\text{rad/s}$. Hence, the settling time would be four times the time constant, or $4/\zeta\omega_n$, which is about 2.3 s. That is *at least* how long we would need to wait before we can assume that the solution being obtained numerically is substantially the steady-state solution.

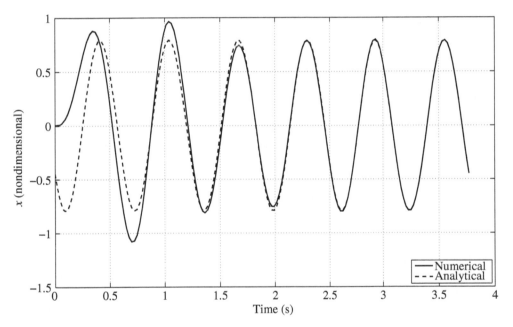

Figure 5.13: *Numerical response of an SDOF system with harmonic excitation*

We mentioned waiting four time constants for transients to die out. Assume an underdamped system and prove analytically that it is a reasonable assumption.

---

[7]We will discuss this in more detail in Chapter 6 when we want to look at transient response, but this is a good rule of thumb for now.

5.4   Where Does Damping Come From?

OK, it is time to go back and look closely at the origin of the dissipative mechanisms with a little more scrutiny. To reiterate what we discussed in Chapter 4, damping can be essentially defined as the irreversible dissipation of mechanical energy from a vibrating elastic body. So, clearly from a macroscopic point of view, we know that damping occurs. Unfortunately, at the microlevel, it is not all that obvious.

Damping is believed to result from a variety of energy loss mechanisms such as:

- conversion of mechanical energy to heat through internal friction within the elastic body. This is often due to a hysteretic phenomenon (and hence, nonlinear);
- the conversion of mechanical energy to heat through friction caused by the rubbing of one component of the elastic body against another;
- the transfer of mechanical energy from the vibrating elastic body to the adjacent structural components;
- the transfer of mechanical energy from the vibrating elastic body to the environment through acoustic radiation;
- the conversion of mechanical energy to heat through a viscous response either inherent in the system or subsequently added to the system (such as lubrication).

The energy dissipation mechanisms listed above are very complex and dependent upon a great number of factors including specifically, the composition of the elastic body, the crystallinity, the geometry, the temperature, the initial strain, the amount of any pre-load, the interrelationship between the vibrating body and other bodies, the amplitude and frequency of vibration, and the amount of viscosity. Due to the variety of dissipative mechanisms and the internal complexity of those mechanisms, it is extremely difficult to accurately predict (from analysis and computation) the damping effect of a given material and a given situation. What we have looked at so far in this book is really only the last item in the list enumerated above.

All this stuff is not to impose a generous dose of humility on you just when you thought you had a good handle on this "damping thing" and were feeling pretty confident (although it is difficult to argue against the virtue of humility!). The principal idea here is not to discourage you, but to alert you to the the state of the art thinking, and to warn you that damping is still a pretty big unknown in the field of vibrations, simply because of the incredible complexity mainly at the micromechanical level. That said, do not forget that we can still design vibrating systems, and add damping to reduce vibrations, and, in general, obtain well-functioning components and systems. Plus, we always have experimental ways of estimating and characterizing damping at the *macro* level, and that suffices most of the time.

Granted that phenomenologically, the mechanism may be poorly understood. However, even if we (i.e., the global "we, the engineers and scientists") cannot relate the micromechanics exactly to macroscopic properties, we can still characterize damping using the gross expressions that give us reasonable verisimilitude with reality.[8] The most important aspect of damping is clearly that it leads to loss of energy and is chiefly responsible for eventually decimating all vibration. Hence, we pursue a strategy that is common throughout the field of engineering: we try to match our model with the *observed* effect, especially, the one that is most important to us. And for realistic systems

---

[8]This approach is by no means anomalous in science. The underlying phenomena for the intensely practical experiences of gravity and magnetism are still not understood very well, but we can obviously design and work with real systems using empirically derived descriptions.

with damping, that critical observed effect is the energy dissipation. Here, we will look at a couple of specific kinds of damping mechanisms that we *can* analyze based on the energy concepts. It is hence important to be able to estimate the loss of energy due to the various damping mechanisms, and it is especially useful to evaluate this for cyclic or periodic oscillations. First, we will evaluate this for something we know very well: linear viscous damping, and then extend it to other kinds of damping mechanisms.

5.5   Energy Dissipated for Harmonic Motion

We briefly looked at the energy dissipated for transient free vibration in Chapter 4; let us say, it was mildly interesting, and marginally useful! But, when we consider harmonically excited systems, the energy dissipated *per cycle* becomes a very important quantity that can help deal with even nonlinear damping mechanisms.

Let us consider again a linear viscously damped oscillator undergoing cyclic motion. The displacement is then

$$x = x_0 \cos\left(\omega t - \phi\right) \tag{5.41}$$

The work done per cycle is, by definition, from Eq. (2.58),

$$W_d = \oint F_d \, dx = \oint F_d \, \dot{x} \, dt \tag{5.42}$$

The damping force is given by (under assumptions of linearity)

$$F_d = c\dot{x} \tag{5.43}$$

Then,

$$W_d = \oint c\dot{x}^2 \, dt = c\omega^2 x_0^2 \int_0^T \sin^2\left(\omega t - \phi\right) \, dt \tag{5.44}$$

where $T = 2\pi/\omega$ is the period of oscillation. The above integral simplifies to

$$W_d = \pi c \omega x_0^2 \tag{5.45}$$

Note that the work done (or energy dissipated) is a linear function of the frequency of oscillation; in other words, higher frequency vibration results in higher dissipation. In addition, it is proportional to square of the amplitude.

For an interesting geometric interpretation, consider that the damping force can be written as

$$F_d = c\dot{x} = c\omega\sqrt{x_0^2 - x(t)^2} \tag{5.46}$$

This leads to the following equation

$$\left(\frac{F_d}{c\omega x_0}\right)^2 + \left(\frac{x}{x_0}\right)^2 = 1 \tag{5.47}$$

which represents an ellipse in the (F_d, x_0) plane (Fig. 5.14). This is often called the hysteresis curve, whose area represents the energy dissipated per cycle.

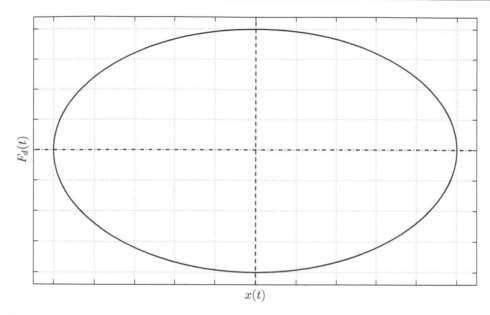

Figure 5.14: *Viscous damping ellipse showing the damping force versus displacement*

Exercise 5.6

Derive the expression for the work done, Eq. (5.45) and the ellipse equation, Eq. (5.47). How does the shape of the ellipse change as the frequency is increased, but if the amplitude does not change?

Two related definitions are common in industrial practice. *Specific damping capacity* is defined as the ratio of energy dissipated per cycle to the peak potential energy. A *loss coefficient*, η, is defined as this ratio per radian, and would hence be the specific damping capacity divided by 2π. For linear viscous damping, it is easy to show that the loss factor is given by

$$\eta = 2\zeta \frac{\omega}{\omega_n} \tag{5.48}$$

Exercise 5.7

Derive the expression for loss factor for linear viscous damping, Eq. (5.48).

5.5.1 Equivalent Viscous Damping

The cyclical energy dissipation leads us to a handy way of working with all kinds of damping mechanisms with nonlinear expressions, which would be very painful to analyze by our conventional methods. The idea is to find the energy dissipated per cycle and to equate it to that dissipated by an equivalent (linear) viscously damped system. Using the expressions derived above, we would compute equivalent damping coefficient from

$$c_{\text{eq}} = \frac{W_d}{\pi \omega x_0^2} \tag{5.49}$$

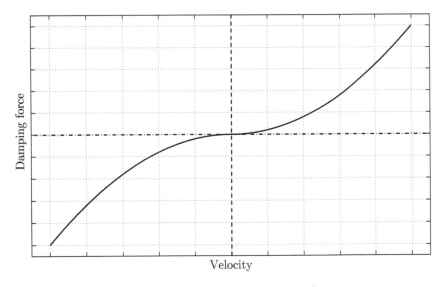

Figure 5.15: *Nonlinear damping force*

For example, consider a damping force shown in Fig. 5.15 given by

$$F_d = a\dot{x}^2 \, \text{sign}(\dot{x}) \tag{5.50}$$

which would be a force generated by a turbulent fluid around a vibrating body (a is some constant coefficient).

For harmonic motion, the work done is

$$W_d = \oint a\dot{x}^3 \, \text{sign}(\dot{x}) \, dt \tag{5.51}$$

With $x(t) = x_0 \sin \omega t$, we are led to

$$W_d = \frac{8}{3} a\omega^2 x_0^3 \tag{5.52}$$

From this, the equivalent *linear* viscous damping coefficient turns out to be

$$c_{\text{eq}} = \frac{8}{3\pi} a\omega x_0 \tag{5.53}$$

Figure 5.16 shows the actual force F_d for a purely harmonic motion (at some frequency); compared with it is the ellipse we would get if we used the equivalent linear viscous damping coefficient we just derived. Notice how the curves are quite different (it is no longer an ellipse), but that their areas are the same (which is after all how we derived the equivalent system).

Now, we could use this damping coefficient and solve all kinds of problems treating the nonlinear damping as if it were linear viscous damping. It is important to remember that this definition was derived assuming harmonic vibration, and is really only accurate for that situation. Of course, we will end up using it for nonharmonic vibration as well as a convenient engineering approximation. See Problem 5.33 for an investigation into the accuracy of doing this.

Note that using the energy (defined by the area) does not tell the whole story; the exact shape is indeed important, but this information is lost when we come up with an equivalent viscous damping

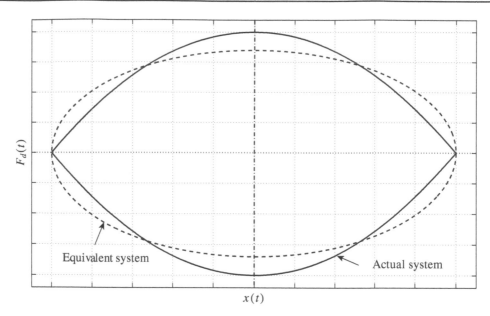

Figure 5.16: *Nonlinear damping force versus displacement for a cycle of motion; the dashed elliptical curve is based on the equivalent linear system*

and proceed with the analysis. Motivated students might want to read the book by Lazan [24] or Nashif et al. [25] to sate their appetite for an intensive study on this fascinating topic.

Exercise 5.8

Carry out the necessary numerical calculations to generate Fig. 5.16. Use $a = 200$, $\omega = 200$, $x_0 = 0.05$, and $\phi = 0.2$ in consistent SI units.

5.6   Coulomb Damping

As discussed very briefly in Chapter 4, Coulomb damping (or dry friction) occurs between sliding surfaces, and is approximately given by the following.

$$F_d = \begin{cases} +F_{d0} & \text{if } \dot{x} > 0 \\ -F_{d0} & \text{if } \dot{x} < 0 \end{cases} \tag{5.54}$$

where F_{d0} is assumed to be a constant force independent of velocity, usually equal to μN, where μ is the coefficient of sliding friction, and N is the normal force.

Now, consider a mass-spring oscillator subject to this force and an external harmonic excitation, as shown in Fig. 5.17. Even though the system is now nonlinear (thanks to the nonlinear damping), we assume that the system experiences a simple harmonic response with the same frequency as the excitation frequency. This is clearly an approximation, and something we cannot justify, or even come up with error bounds for our approximation

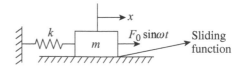

Figure 5.17: *Mass-spring system subject to Coulomb friction*

without a proper nonlinear analysis. (Now that your curiosity is aroused, check out a really old paper [26] for interesting and insightful answers to this very important problem).

Let us assume that it is a decent approximation (let me assure you that it is indeed so), and proceed for now. So,

$$x = x_0 \sin \omega t, \quad \dot{x} = \omega x_0 \cos \omega t \tag{5.55}$$

Next, we evaluate the work done over a cycle like we did for the linear viscous force.

$$W_d = \oint F_d \, dx = \int_0^T F_d \, \dot{x} \, dt \tag{5.56}$$

We substitute for the force and velocity and separate out the domains when the force is positive or negative, based on the sign of $\dot{x}(t)$ (see Fig. 5.18).

$$W_d = F_{d0} \, x_0 \left\{ \int_0^{\pi/2} \cos \omega t \, d(\omega t) - \int_{\pi/2}^{3\pi/2} \cos \omega t \, d(\omega t) + \int_{3\pi/2}^{2\pi} \cos \omega t \, d(\omega t) \right\}$$

$$= 4 F_{d0} \, x_0 \tag{5.57}$$

Using the concept of equivalent viscous damping, we get the following expression

$$c_{eq} = \frac{4 F_{d0}}{\pi \omega x_0} \tag{5.58}$$

Note that the equivalent damping factor is a function of the amplitude of vibration as well as the excitation frequency.

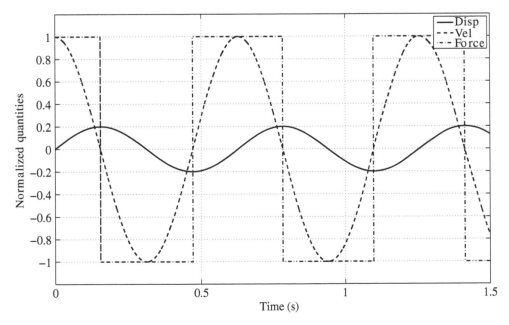

Figure 5.18: *Displacement, velocity, and Coulomb friction for harmonic motion*

The equivalent linear viscous damping ratio would then be

$$\zeta_{eq} = \frac{c_{eq}}{c_c} = \frac{2F_{d0}}{\pi r x_0} \tag{5.59}$$

Next, let us take this a step further, and use this viscous damping ratio as if it were a linear system, and substitute into the forced response expression, Eq. (5.15), to get the amplitude. Of course, this is a major approximation, since we now have our equivalent viscous damping ratio being dependent on the *amplitude* of vibration x_0. (Read that again to make sure you understand the implication.)

After some algebra, it leads to the following expression for the amplitude of vibration.

$$x_0 = \frac{F_0}{k} \frac{\sqrt{1 - \frac{16}{\pi^2}\left(\frac{F_{d0}}{F_0}\right)^2}}{(1 - r^2)} \tag{5.60}$$

Clearly, this is invalid unless

$$\frac{F_{d0}}{F_0} < \frac{\pi}{4} \tag{5.61}$$

Hence, along with all the other conditions, we also need this to be true (in other words, the frictional force needs to be small) for our approximate answer to be valid.

Exercise 5.9

Verify Eq. (5.57).

Exercise 5.10

Derive the amplitude equation, Eq. (5.60) along with the condition, Eq. (5.61).

It is interesting to test the validity of this approximation for a couple of cases. How do we do it without knowing any nonlinear analysis? It is quite simple really, and we already learned that we have the computer as a very powerful tool at our disposal: we use numerical integration. (Look at Listing 2.5 on page 31 and Listing 2.6 on page 32 to refresh your memory.) Note that you will have to define the friction force using a `sign` function, whether you use MATLAB or any other numerical tool. Also, note that the numerical analysis will include transients, and to compare the steady-state solution with the approximate analytical method, we need to wait until the transients die out, as discussed in Section 5.3.

Figure 5.19 shows the numerically simulated response of the oscillator with $F_{d0}/F_0 = 0.1$, and for $r = 0.1$ (Case 1), and Fig. 5.20 for $r = 0.9$ (Case 2). Figure 5.21 (Case 3) and Fig. 5.22 (Case 4) show similar results for a large value of the frictional force: $F_{d0}/F_0 = 0.5$. In Case 1, the response is indeed nearly sinusoidal (as it would be for the linear case), with an error of about 6%. Case 2 has an error of 22%, and the response shows beating, since the excitation and the natural frequency[9]

---

[9]We have been abusing the word, "natural frequency" quite a bit. It is only defined for linear systems, and really has no meaning for nonlinear systems. For now, just think of it as the natural frequency of the undamped system, which, in this case, is indeed linear.

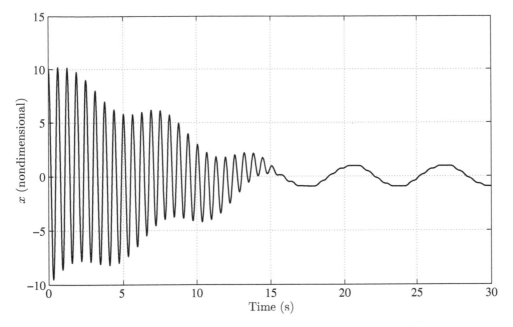

Figure 5.19: *Response of oscillator with Coulomb friction; Case 1*

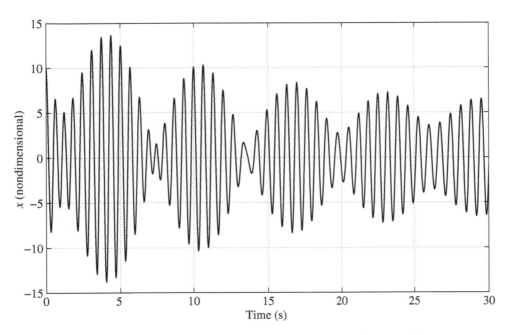

Figure 5.20: *Response of oscillator with Coulomb friction; Case 2*

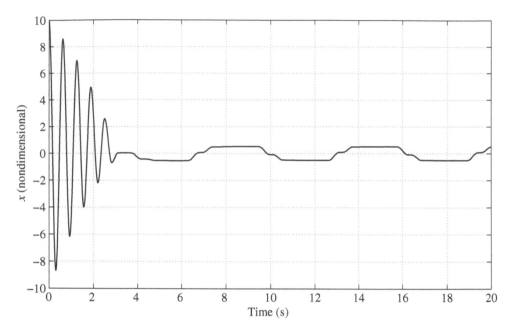

Figure 5.21: *Response of oscillator with Coulomb friction; Case 3*

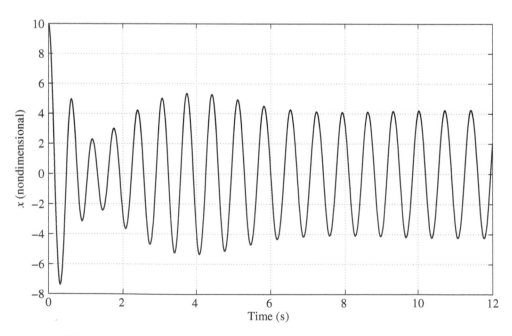

Figure 5.22: *Response of oscillator with Coulomb friction; Case 4*

are close to each other. Case 3, which is the case of a large frictional force, shows strong departures from the equivalent linear assumption with an error of 55%. Case 4 results in an error of only 6% as the response is mostly sinusoidal, and hence agrees more closely with the assumptions we made.

5.7 Material Damping

Material damping refers to energy dissipation that occurs within the confines of the micro- or macro structures of the material in vibration. It is also often called internal friction, internal damping, or hysteretic damping. The phenomena are quite varied and complicated, and are still the subject of research. We cannot, however, ignore this effect, as it contributes to a significant amount of damping in real engineering structures and machine components. Hence, in this book, we will not delve into all kinds of details; we will just try to get an idea of what it is all about and how we might characterize it approximately.

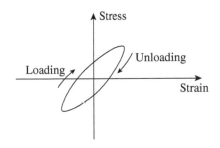

Even standard engineering materials are not perfectly elastic as we usually assume. When subjected to cyclic stress (in other words, when the structure is subject to periodic vibration), this inelastic behavior often takes the form of a hystere-

Figure 5.23: *Hysteresis loop due to material damping*

sis loop, as shown in Fig. 5.23. The stress-strain curves are different for loading and unloading, as shown in the figure. The net result is that there is a loss of energy, or *dissipation* that is equal to the area enclosed by the hysteresis loop.

Without getting into the details, we can reasonably estimate that the energy dissipated is proportional to the square of the amplitude of vibration. Then,

$$W_d = Ax_0^2 \tag{5.62}$$

where A is a proportionality constant. Again, equating this with the energy dissipated by linear viscous damping,

$$c_{eq} = \frac{A}{\pi\omega} = \frac{\eta k}{\omega} \tag{5.63}$$

where we have introduced a new nondimensional quantity, η by

$$\eta = \frac{A}{\pi k} \tag{5.64}$$

η (notionally same as the viscous loss factor introduced earlier) is called the structural damping factor in the context of material damping. Again, we can compute an equivalent damping ratio by using

$$\zeta_{eq} = \frac{c_{eq}}{c_c} = \frac{\omega}{2\omega}\eta \tag{5.65}$$

We can then substitute this value of the damping ratio into the harmonic response expression to get the response of the system with structural damping

$$x_0 = \frac{F_0}{k}\frac{1}{[(1-r^2)^2 + \eta^2]^{1/2}} \tag{5.66}$$

Also, we can substitute the equivalent damping coefficient back into the equation of motion for an interesting result.

$$m\ddot{x} + c_{eq}\dot{x} + kx = F_0 \sin \omega t \tag{5.67}$$

This leads to

$$m\ddot{x} + k\left[x + \frac{\eta}{\omega}\dot{x}\right] = F_0 \sin \omega t \tag{5.68}$$

Recognizing that in harmonic vibration, $\dot{x} = i\omega x$ written as a phasor, we can rewrite this equation as follows.

$$m\ddot{x} + k\left[1 + j\eta\right]x = F_0 e^{i\omega t} \tag{5.69}$$

In other words, the structural damping factor can be considered to make the effective stiffness a *complex* quantity. This is in fact the most common and easiest way to take structural damping into account.

The solution to the above equation can be found easily by assuming an exponential solution

$$x(t) = \tilde{x}_0 e^{i\omega t} \tag{5.70}$$

This leads to

$$\tilde{x}_0 = \frac{F_0}{k} \frac{1}{-r^2 + (1 + i\eta)} = \frac{F_0}{k} \frac{1}{(1 - r^2) + i\eta} \tag{5.71}$$

In real terms, the displacement becomes

$$x(t) = x_0 \sin \left(\omega t - \phi\right) \tag{5.72}$$

with the amplitude,

$$x_0 = \frac{F_0}{k} \frac{1}{\sqrt{(1 - r^2)^2 + \eta^2}} \tag{5.73}$$

and a phase lag of

$$\phi = \tan^{-1}\left(\frac{\eta}{1 - r^2}\right). \tag{5.74}$$

Example 5.1 Response of automobile engine on engine mounts

Consider an automobile engine supported on elastomeric engine mounts. They can be considered to be linearly elastic with hysteretic damping, and the engine mass may be modeled as an SDOF particle. The disturbing forces come from the rotation of the crankshaft and can be assumed to be harmonic with a frequency equal to the rotational speed of the crankshaft. Let $m = 300\,\text{kg}$, $k = 300\,\text{kN/m}$, and $\eta = 0.3$. The disturbing force may be assumed to be $1\,\text{kN}$, and the speed varies between 0 and $5000\,\text{rpm}$. Determine the vibrational response of the engine.

The natural frequency is

$$\omega_n = \sqrt{\frac{k}{m}} = 31.63\,\text{rad/s} \tag{5.75}$$

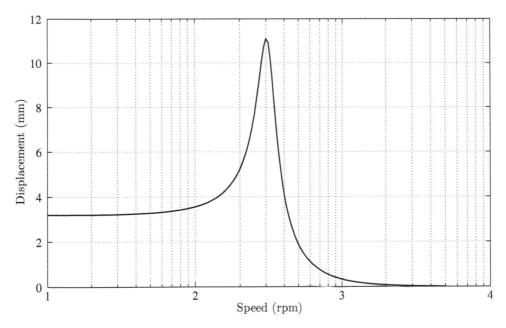

Figure 5.24: *Amplitude of engine vibration on elastomeric mounts*

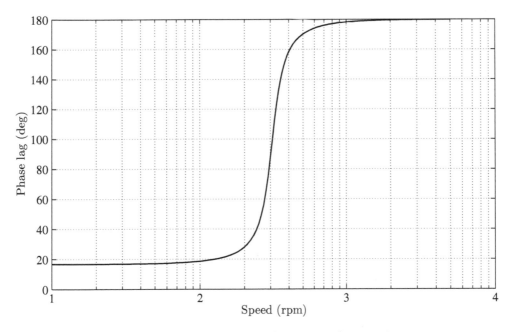

Figure 5.25: *Phase lag of engine vibration on elastomeric mounts*

and the range of forcing frequency is $0 < \omega < 524 \, \text{rad/s}$. Carrying out the calculations from Eq. (5.73), we find that the peak response is about 11 mm at 302 rpm. Figures 5.24 and 5.25 show the amplitude and phase, respectively.

5.8 Vibration Isolation

There are two fundamental problems in vibration isolation.

1. We would like to isolate delicate instruments and such from surrounding vibration. Examples are home CD players, computer components, packaging of anything, and people in automobiles. This problem is called *base excitation* or *support excitation*.

2. Most machinery generate vibrational forces that we would like to keep from transmitting to the surrounding environment. This is the so called *vibration transmission* problem. Examples are engine mounts in an automobile or aircraft engine, support systems for manufacturing equipment, etc.

The two items listed above are very different problems, but we will see here that they happen to lead to the same expression and similar conclusions.

5.8.1 Base Excitation

Consider an SDOF model supported on a moving base (with a displacement, $u(t)$), as shown in Fig. 5.26. Figure 5.27 shows the FBD of the mass isolated from its surroundings.

Note that the FBD is no different from the situation when the base is not moving. However, the forces due to the spring and the damper depend on relative motion between their ends and are given by

$$F_s = k(x - u), \quad F_d = c(\dot{x} - \dot{u}) \qquad (5.76)$$

Kinematics is straightforward and the acceleration is given by

$$a = \ddot{x} \qquad (5.77)$$

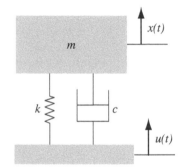

Figure 5.26: *Base excitation model*

Note that the forces depend on relative displacements, but the acceleration must still be absolute as specified by Newton's laws.

The equation of motion is then given by

$$m\ddot{x} = -F_s - F_d$$
$$= -k(x - u) - c(\dot{x} - \dot{u}) \qquad (5.78)$$

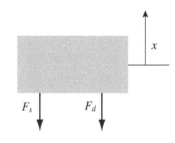

Figure 5.27: *Base excitation model*

Reorganizing,

$$m\ddot{x} + c\dot{x} + kx = c\dot{u} + ku \qquad (5.79)$$

Note that the base *displacement* has become a source of an *excitation* (force) on the mass.

This is as far as we can go without specifying the nature of the base displacement. For now, let us assume that it is harmonic. We will have a chance to revisit this problem when the base motion is not harmonic.

$$u(t) = u_0 \sin \omega t \qquad (5.80)$$

The EOM then becomes

$$m\ddot{x} + c\dot{x} + kx = c\omega u_0 \cos \omega t + k u_0 \sin \omega t \tag{5.81}$$

This is the form of the general harmonic force that we had earlier (Eq. 5.24), where

$$F_c = c\omega u_0 \quad \text{and} \quad F_s = k u_0 \tag{5.82}$$

Solution is then given by

$$x(t) = x_0 \sin(\omega t + \alpha - \phi) \tag{5.83}$$

where

$$\frac{x_0}{F_0/k} = \frac{1}{\sqrt{(-r^2+1)^2 + (2\zeta r)^2}} \tag{5.84}$$

Now,

$$\begin{aligned}
\frac{F_0}{k} &= \frac{1}{k}\sqrt{F_c^2 + F_s^2} \\
&= u_0 \sqrt{\left(\frac{c\omega}{k}\right)^2 + 1} \\
&= u_0 \sqrt{1 + (2\zeta r)^2}
\end{aligned} \tag{5.85}$$

Combining the above two equations

$$\boxed{\frac{x_0}{u_0} = \sqrt{\frac{1 + (2\zeta r)^2}{(-r^2+1)^2 + (2\zeta r)^2}}} \tag{5.86}$$

Also,

$$\alpha = \tan^{-1}\left(\frac{F_c}{F_s}\right) = \tan^{-1}(2\zeta r) \tag{5.87}$$

The phase difference between $x(t)$ and $u(t)$ is then,

$$\begin{aligned}
\beta &\stackrel{\text{def}}{=} \alpha - \phi \\
&= \tan^{-1}(2\zeta r) - \tan^{-1}\left(\frac{2\zeta r}{-r^2+1}\right)
\end{aligned} \tag{5.88}$$

Figure 5.28 shows the normalized response amplitude as a function of r and ζ, and Fig. 5.29 shows the phase lag β. Listing 5.1 shows the MATLAB implementation of a function to determine the amplitude ratio. There are several interesting observations that can be made from these figures.

Listing 5.1: *MATLAB code for solving the base excitation problem*

```
function [magnitude , phase]=tr (frequency_ratio , damping ratio)
%       tr .m
%       return non-dim transimissibility ratio given a frequency ratio (r)
%       and damping_ratio (zeta)
%       for a sdof system
%       same as base excitation problem
```

```
if (nargin<2)
        damping_ratio=0;
end;

A1=(1−frequency_ratio .^2);
A2=2*damping_ratio*frequency_ratio;
A3=1+(2*damping_ratio*frequency_ratio).^2;

magnitude=sqrt (A3./(A1.^2+A2.^2));
phase=atan2(A2,A1);
```

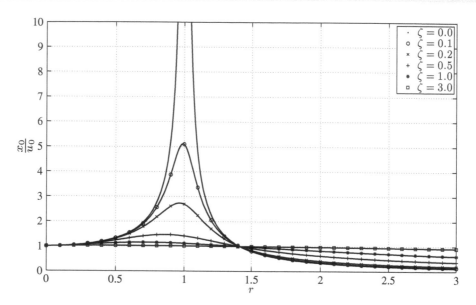

Figure 5.28: *Amplitude ratio as a function of frequency ratio and damping ratio*

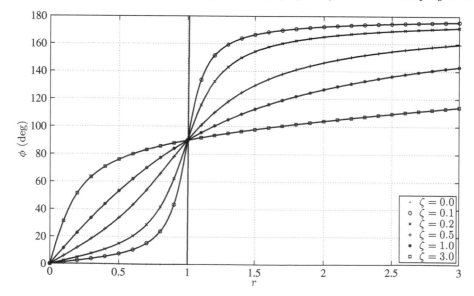

Figure 5.29: *Phase lag as a function of frequency ratio and damping ratio*

Observations

- $\frac{x_0}{u_0} \approx 1$ for $r \approx 0$; for small frequencies, the displacement amplitude of the mass is about the same as the base amplitude. Also, they are in phase. In other words, the system behaves like a static system.

- $\frac{x_0}{u_0} \uparrow$ near $r = 1$; again, we see the effect of resonance and this would be the worst possible situation. Note again how even small amounts of damping can reduce the amplitudes by a large amount in the resonant region.

- At $r = 1$ exactly, the response amplitude is given by

$$\left(\frac{x_0}{u_0}\right)_{r=1} = \left[\frac{1+(2\zeta)^2}{(2\zeta)^2}\right]^{1/2} \tag{5.89}$$

Clearly, the smaller the damping ratio, the larger the amplitude of vibration at resonance. This, however, is not necessarily its peak value.

- It stands to reason that we designed a spring-damper isolation system in order to reduce the vibration, and hence would demand that the ratio x_0/u_0 be less than 1. In a system without the isolators, a value of 1 is in fact what it would be. However, looking at the figure, we realize that we *have* to have $r > \sqrt{2}$ to achieve the needed $x_0/u_0 < 1$.

- A strange (but true!) result is that, for regions of frequencies with $r > \sqrt{2}$, the amplitude of vibration is *larger* for larger damping. Or, we are actually hurting our design if we add large amounts of damping. In fact, at these frequencies, having no damping is the best situation! Or is it? Think a little before you read further.

- Before you conclude that having no damping would be the ideal final design, let us look at a practical system. Suppose that we had to design such an isolation system (in other words, select a spring and a damper) for a delicate instrument, we would not only need to make sure that the response amplitude is small at the operating frequency, but also ensure that if it does experience resonance, it comes through unscathed. In fact, the system would need to *get to the operating speed*, and this invariably means that we have to go through the resonant speed. A lightly damped system would get destroyed easily under resonant conditions. Hence, the only way of preventing that is to have sufficient damping.

The design criteria for the typical isolation system can then be summarized as follows.

- Choose large enough damping to ensure passage through resonance without damage; but do not use any more than that.

- Select an operating frequency ratio (r) large enough to meet the vibration specifications under normal operating conditions.

The first of these criteria can be used to select ζ, and the second to select r, and therefore, k.

Exercise 5.11

Determine the maximum value of the amplitude ratio, x_0/u_0, and the frequency ratio at which it occurs.

Why would adding damping make the situation worse at higher frequencies? Does it have anything to do with energies? It appears that the damper is working against us—so does it have something to do with phase angles? Why the magic number $\sqrt{2}$ that makes all the difference?

Example 5.2 Automobile on suspensions

An automobile traveling at a constant speed v on a road can be modeled as an undamped spring-mass system on a road with sinusoidal roughness (Fig. 5.30). Derive an expression for the amplitude of the mass as a function of vehicle speed, and determine the most unfavorable speed.

The springs of an automobile are compressed 75 mm under its own weight. The road profile can be approximated by a sine wave of amplitude 55 mm and wavelength of 12 m. Assume a ζ of 0.8 for a new car, and 0.4 for an old one.

This is a problem in base excitation as illustrated in Fig. 5.30.

$$m\ddot{x} + c\dot{x} + kx = c\dot{u} + ku \tag{5.90}$$

where

$$u(t) = u_0 \sin \omega t \tag{5.91}$$

with

$$\omega = \frac{2\pi v}{\lambda} \tag{5.92}$$

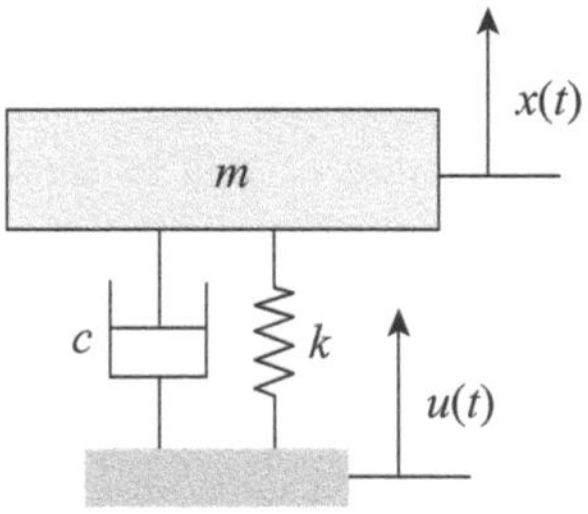

Figure 5.30: *Automobile on a suspension*

where v is the forward speed of the car (assumed constant), and λ is the wavelength of the road roughness.

The solution is

$$x(t) = x_0 \sin (\omega t + \alpha - \phi) \tag{5.93}$$

where

$$\frac{x_0}{u_0} = \sqrt{\frac{1 + (2\zeta r)^2}{(-r^2 + 1)^2 + (2\zeta r)^2}} \tag{5.94}$$

and

$$\alpha = \tan^{-1}(2\zeta r), \quad \phi = \tan^{-1}\left[\frac{2\zeta r}{-r^2 + 1}\right] \tag{5.95}$$

Under static conditions,

$$mg = k\Delta \tag{5.96}$$

This leads to the natural frequency,

$$\omega_n = \sqrt{\frac{k}{m}} = \sqrt{\frac{g}{\Delta}} \tag{5.97}$$

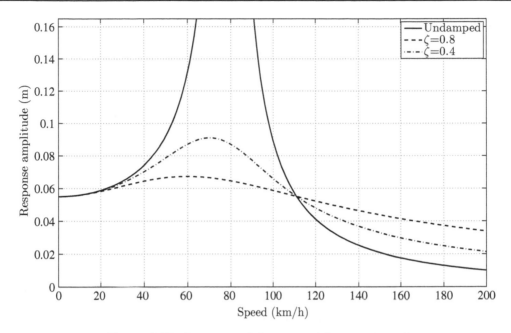

Figure 5.31: *Response of the automobile versus speed*

Substituting numerical values, $\omega_n = 11.4368$ rad/s. Note that a speed range of zero to 200 km/h (say) will yield a range of excitation frequency, $\omega = 0$ to 29 rad/s. This corresponds to $0 < r < 2.5434$.

Next, we substitute the numerical values into Eq. (5.94) to get the response versus speed for different values of damping ($\zeta = 0$, 0.8, 0.4). The response is shown in Fig. 5.31; note how the undamped response is better than the damped response beyond a speed of 115 km/h. It also follows that, to get a low vibrational amplitude you need to go as fast as possible![10]

5.8.2 Force Isolation

This is the second case when we would like to design mounts for machinery that would otherwise transmit excessive forces to the surrounding environment. Examples of such machinery abound: automotive engines, aircraft engines, refrigeration compressors, air-conditioning units, generators, turbine shafts, lathes and such manufacturing equipment, and so on.

The essential problem is shown in Fig. 5.32; the machine is generating a harmonic force and is mounted to the ground on springs and dampers. Figure 5.33 shows the FBD of the system, which is no different from the case considered earlier of a mass subject to a harmonic force.

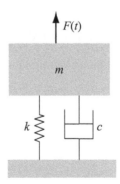

Figure 5.32: *Machine on mounts*

---

[10]We suggest that you keep this figure with you the next time you travel in your car or motorcycle to present as scientific evidence in support of speeding, should you be stopped by a traffic police officer!

Recall that the governing equation is

$$m\ddot{x} + c\dot{x} + kx = F(t) = F_0 \sin \omega t \quad (5.98)$$

and that the solution is

$$x(t) = x_0 \sin (\omega t - \phi) \quad (5.99)$$

where

$$\frac{x_0}{F_0/k} = \frac{1}{\sqrt{(-r^2 + 1)^2 + (2\zeta r)^2}} \quad (5.100)$$

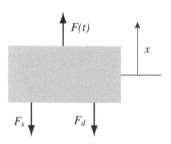

Figure 5.33: *FBD of machine on mounts*

Now, what we are interested in is not the displacement of the machine itself, but the force that is transmitted to the ground. Figure 5.34 shows the forces transmitted: clearly, since the machine is connected to the ground through the spring and damper, it is these elements that apply the forces to the ground.

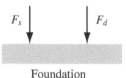

Figure 5.34: *Forces transmitted to the foundation*

Note however that this figure is *not* an FBD, since an FBD of the ground would need to include all the forces acting on it. We do not need an FBD, since we are not going to apply Newton's second law to it. We just need the transmitted force, which is clearly given by

$$\begin{aligned} F_T(t) &= kx + c\dot{x} \\ &= kx_0 \sin (\omega t - \phi) + c\omega x_0 \cos (\omega t - \phi) \\ &= \left[(kx_0)^2 + (c\omega x_0)^2 \right]^{1/2} \sin (\omega t - \phi - \alpha) \end{aligned} \quad (5.101)$$

where α is some phase angle that we are not particularly interested in calculating. However, we *are* interested in the amplitude of the transmitted force,

$$\begin{aligned} F_{T0} &= \left[(kx_0)^2 + (c\omega x_0)^2 \right]^{1/2} \\ &= x_0 \left[k^2 + (c\omega)^2 \right]^{1/2} \\ &= kx_0 \left[1 + (2\zeta r)^2 \right]^{1/2} \end{aligned} \quad (5.102)$$

We next define a nondimensional quantity called the *transmissibility ratio* (TR) by

$$\begin{aligned} \text{TR} &= \frac{\text{Amplitude of force transmitted}}{\text{Amplitude of disturbing force}} \\ &= \frac{F_{T0}}{F_0} \\ &= \frac{kx_0}{F_0} \left[1 + (2\zeta r)^2 \right]^{1/2} \end{aligned} \quad (5.103)$$

Substituting for x_0, we get

$$\boxed{\text{TR} = \sqrt{\frac{1 + (2\zeta r)^2}{(-r^2 + 1)^2 + (2\zeta r)^2}}} \quad (5.104)$$

which is the *same* expression as we obtained for the nondimensional ratio of displacement amplitudes in the case of base excitation (Eq. (5.86) on page 145). Hence, although it is a very different problem here (with ratios of forces instead of displacements), all the detailed conclusions we derived on page 147 apply here. In particular, note that adding damping at higher frequencies can actually transmit larger forces; we still need adequate damping however to survive the ravages of resonance, as explained there.

Example 5.3 Transmissibility

A machine has a mass of 500 kg and has an operating speed of 800 rpm. A vibration isolation system has to be designed so that the following specifications are met:

- $TR = 0.15$ (or less) at the operating speed.
- $TR < 1.25$ when the machine is accelerated up to the operating speed from zero.

Determine the effective spring stiffness and the viscous damping coefficient of the vibration isolation system.
   This is a problem in vibration transmission.

$$m\ddot{x} + c\dot{x} + kx = F_0 \sin \omega t \tag{5.105}$$

The solution is

$$x(t) = x_0 \sin(\omega t - \phi) \tag{5.106}$$

where

$$\frac{x_0}{F_0/k} = \sqrt{\frac{1}{(-r^2 + 1)^2 + (2\zeta r)^2}} \tag{5.107}$$

and

$$\phi = \tan^{-1}\left[\frac{2\zeta r}{-r^2 + 1}\right] \tag{5.108}$$

The transmitted force is $F_T = c\dot{x} + kx = F_{T0} \sin(\omega t + \alpha - \phi)$. This leads to TR, the transmissibility ratio.

$$TR = \frac{F_{T0}}{F_0} = \sqrt{\frac{1 + (2\zeta r)^2}{(-r^2 + 1)^2 + (2\zeta r)^2}} \tag{5.109}$$

   Given data: $m = 500$ kg, $\omega = 800$ rpm $= 83.78$ rad/s. It is also given that the maximum TR can be 1.25, which happens when $r \approx 1$. Hence,

$$1.25 = \sqrt{\frac{1 + 4\zeta^2}{4\zeta^2}} \tag{5.110}$$

Solving, $\zeta = 0.6667$.
   Next, we use the fact that TR is 0.15 at the operating speed to determine r, the frequency ratio.

$$0.15 = \sqrt{\frac{1 + (2\zeta r)^2}{(-r^2 + 1)^2 + (2\zeta r)^2}} \tag{5.111}$$

where we use the value of ζ just computed and determine r from a quadratic equation in r^2. This leads to $r = 8.9319$. Next, $\omega_n = \omega/r = 9.3794$ rad/s. The required stiffness is computed from $k = \omega_n^2 m = 43,986$ N/m. Similarly, the damping coefficient is computed from $c = 2\zeta\sqrt{km} = 6253$ Ns/m.

5.9   Rotating Unbalance

Engineering systems frequently contain rotating parts. Examples are driveshafts and crankshafts in automobiles, turbine shafts in generators, compressors, aircraft engines, propeller shafts, centrifuges, helicopter rotors, and so on. A consequence of rotation is a rotating unbalance effect that results from a difference between the center of mass and the geometric center of the rotating part. This will normally lead to an excitation that is harmonic and, of course, that means vibration that we have to contend with.

Figure 5.35: *Jeffcott rotor model*

A simple model for a rotating shaft is the so-called Jeffcott rotor model shown in Fig. 5.35. It consists of a rigid disk in the center of a massless flexible shaft. The shaft is supported at its endpoints on bearings that we will model as rigid supports. A cross-sectional view of the disk is shown in Fig. 5.36. C is the mass center and P is the geometric center of the disk. As the disk rotates, the line PC will spin about

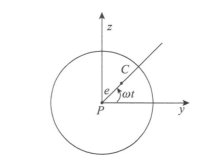

Figure 5.36: *Cross-sectional view of the disk*

P with an angular velocity ω. The distance PC is called the unbalance eccentricity and is usually denoted by e. In practice, this eccentricity is very small (order of mils, or thousands of an inch). This should prompt you to ask "What is the big deal then?" or "Why worry about small inaccuracies?" We can only answer that question after developing a mathematical model for the system; so just hang on to your hat.

Kinematics

The position of the center of mass is given by

$$y_C = y + e \cos \omega t \qquad (5.112)$$
$$z_C = z + e \sin \omega t \qquad (5.113)$$

where (y, z) are the displacement coordinates of P. The acceleration of the center of mass is then

$$\vec{a}_C = \ddot{y}_C \vec{j} + \ddot{z}_C \vec{k}$$
$$= \left[\ddot{y} - \omega^2 e \cos \omega t\right] \vec{j} + \left[\ddot{z} - \omega^2 e \sin \omega t\right] \vec{k} \qquad (5.114)$$

Kinetics

An FBD of the disk is shown in Fig. 5.37. The shaft acts like a spring in two dimensions with spring stiffness k; we have also assumed some damping in the system with a viscous damping coefficient c.

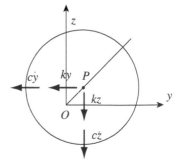

Figure 5.37: *FBD of the disk*

Equation of motion

Applying Newton's second law,

$$\vec{F} = m\vec{a}_C \qquad (5.115)$$

we get

$$m\vec{a}_C = -k(y\vec{j} + z\vec{k}) - c(\dot{y}\vec{j} + \dot{z}\vec{k}) \tag{5.116}$$

Substituting for the acceleration and simplifying, the equations of motion turn out to be

$$m\ddot{y} + c\dot{y} + ky = me\omega^2 \cos\omega t \tag{5.117}$$

$$m\ddot{z} + c\dot{z} + kz = me\omega^2 \sin\omega t \tag{5.118}$$

Clearly, this system has two degrees of freedom (DOF) and does not really belong in this chapter. On the other hand, note that the two DOF are completely decoupled and we can treat the problem as two independent single-DOF systems. This model was used only to illustrate the way a rotating unbalance leads to harmonic excitation; we will not attempt to solve both these equations.

This is a good time to answer the question raised in the beginning of this section: "If e is very small in practice, why bother with all this?" If you look at the above equations, you will see that the *amplitude* of the force is itself dependent on the speed of rotation and is in fact proportional to its square. In other words, for higher speeds of rotation the unbalance force can get to be very large, in turn leading to large amplitudes of vibration. With most of modern technology pushing for higher speeds in almost every kind of machinery, it stands to reason that we have to deal with this problem.

We will now solve only one of these equations, say the one in y. Also, the model we used above does not reflect the fact that the mass of the system is not necessarily the one involved with the rotating unbalance. In other words, m in the L.H.S. of the equations is not the same as the m in me on the R.H.S. To illustrate this, consider the problem of balancing the wheels of your car that you have to carry out when you get a new tire. In modeling that problem m would be the mass of the wheel (perhaps, 25 kg), and the mass of the rotating unbalance, m_0, would be the mass of the lead weight that is placed on the rim in an effort to reduce the unbalance (perhaps, 100 g).

Hence, the equation we are going to solve is

$$m\ddot{y} + c\dot{y} + ky = m_0 e\omega^2 \cos\omega t \tag{5.119}$$

The solution can be written by inspection by substituting $m_0 e\omega^2$ for F_0 in Eq. (5.15) (on page 121). Simplification leads to a nondimensional amplitude $\bar{x}_0$.

$$\boxed{\bar{x}_0 \overset{\text{def}}{=} \frac{mx_0}{m_0 e} = \frac{r^2}{\sqrt{(-r^2 + 1)^2 + (2\zeta r)^2}}} \tag{5.120}$$

Figure 5.38 plots the above expression as a function of the frequency ratio, r, and the damping ratio ζ.

Observations

- For many (practical and technical) reasons, light flexible shafts is what we would like in most rotating systems. This corresponds to a small ω_n. For a given ω, this corresponds to a large r. Hence, we would like machinery to operate at the far right end of the graph as much as possible.
- $\bar{x}_0 \approx 0$ for small r. For low frequencies, the *amplitude* of the force is very small, and hence the response amplitude is very small as well. (Contrast this with the response to $F_0 \sin\omega t$ on page 121.) This means that for slow-running shafts, we may not have to worry about the effect of rotating unbalance.

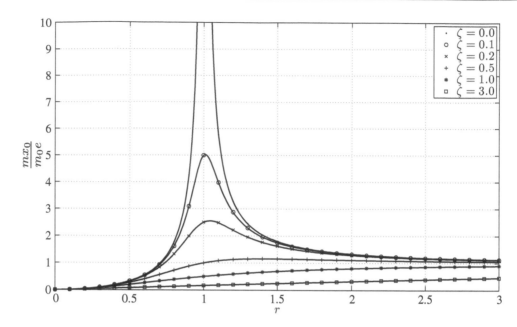

Figure 5.38: *Unbalance response of an SDOF system*

- $\bar{x}_0 \uparrow$ for $r \approx 1$. This is again resonance and should be avoided as much as possible. Also, if we wish to operate at a frequency (which, in this case, is the same as the speed of rotation) higher than the resonant frequency, we will have to have sufficient damping to be able to get past it without damage.

- $\bar{x}_0 \approx 1$ for large r. Hence, if we expect to operate at a large r, for a given unbalance $(m_0 e)$, we would want a large m.

Example 5.4 Compressor rotor

The compressor rotor shown in Fig. 5.39 consists of a rigid disk of mass m mounted at the end of a mass-less cantilever beam of radius r and length ℓ. It was found that the vibration at resonance was 2.5 mm. What should be the running speed of the rotor for the vibration at that speed (i.e., the running speed) to be less than 1.6 mm? The known parameters of the system are as follows: $m = 2.5$ kg, $E = 2.1 \times 10^{11}$ N/m^2, $\ell = 1$ m, $r = 5$ cm, $c = 1600$ Ns/m.

In nondimensional terms, the amplitude of vibration is given by

$$\frac{mx_0}{m_0 e} = \frac{r^2}{\sqrt{(-r^2 + 1)^2 + (2\zeta r)^2}} \tag{5.121}$$

Bearing

Compressor

Figure 5.39: *Compressor rotor*

At resonance $(r = 1)$, this becomes

$$\frac{mx_0}{m_0 e} = \frac{1}{2\zeta} \tag{5.122}$$

Hence, taking the ratio and substituting the two amplitudes (at resonance and at the running speed),

$$\frac{1.6}{2.5} = \frac{\sqrt{(-r^2 + 1)^2 + (2\zeta r)^2}}{2\zeta r^2} \tag{5.123}$$

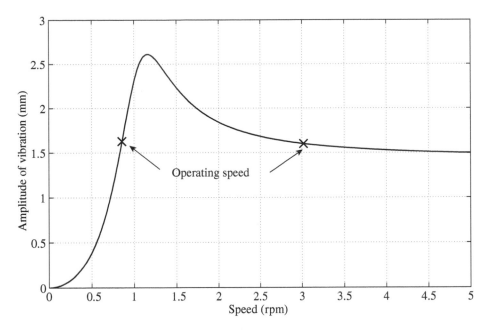

Figure 5.40: *Compressor rotor response*

The equivalent stiffness of a cantilever beam at the end is given by

$$k = \frac{3EI}{\ell^3} \tag{5.124}$$

where $I = \pi r^4/4$. Using the numerical values, we get $\omega_n = \sqrt{k/m} = 1112.2$ rad/s. Also, $\zeta = c/(2\sqrt{km}) = 0.2877$. We substitute this value of ζ into Eq. (5.123), yielding a quadratic equation

$$0.1916r^4 + 1.6689r^2 - 1.0000 = 0 \tag{5.125}$$

Solving, $r = 0.8046$ or 2.8396.

This gives us two possible running speeds of $\omega = r\omega_n = 895$ rad/s or 3158 rad/s. In terms of rpm, that is 8,545 or 30,159. Figure 5.40 shows the response amplitude versus speed. Note how the peak amplitude is in reality a little more than 2.5 mm, since we approximated the resonance to be at $r = 1$. Also, note the two possible operating speeds that will give the same amplitude of vibration.

5.10   Periodic Excitation

We will again consider an SDOF oscillator subject to a force; we know its equation of motion to be

$$m\ddot{x} + c\dot{x} + kx = F(t) \tag{5.126}$$

Now, we consider forces that are not harmonic, but are still periodic (for example, see Fig. 5.41). Such forces are quite common in manufacturing, automotive, and aerospace applications. From Fourier theory we know that such functions can always be written in the form of an infinite series

$$F(t) = \frac{A_0}{2} + \sum_j A_j \cos j\omega t + \sum_j B_j \sin j\omega t \tag{5.127}$$

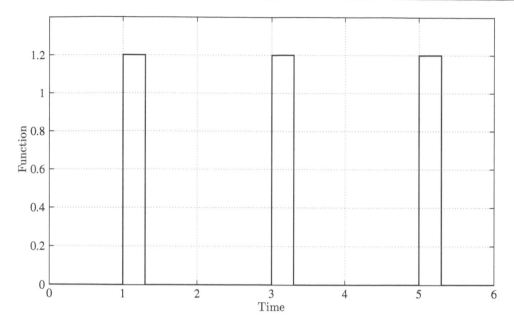

Figure 5.41: *A periodic function that is not harmonic*

where ω is the fundamental frequency of the periodic function; the period is related to it by

$$\tau = \frac{2\pi}{\omega} \tag{5.128}$$

The *Fourier coefficients* are evaluated from the following expressions (see Appendix A for an overview of Fourier series).

$$A_0 = \frac{2}{\tau} \int_{-\tau/2}^{\tau/2} F(t)\, dt \tag{5.129}$$

$$A_j = \frac{2}{\tau} \int_{-\tau/2}^{\tau/2} F(t) \cos j\omega t\, dt \tag{5.130}$$

$$B_j = \frac{2}{\tau} \int_{-\tau/2}^{\tau/2} F(t) \sin j\omega t\, dt \tag{5.131}$$

Hence, the equation of motion becomes

$$m\ddot{x} + c\dot{x} + kx = \frac{A_0}{2} + \sum_j A_j \cos j\omega t + \sum_j B_j \sin j\omega t \tag{5.132}$$

Note that we can consider the system to be acted on by an infinite number of forces, one constant, one with a frequency of ω, another with a frequency 2ω, and so on. Since the system is linear, we can use the principle of superposition to get the response to each of these forces that we can add eventually to get the total response. Let us denote the response to the constant force $x_0(t)$, the response to the jth cosine force $x_{jc}(t)$, and the response to the jth sine force $x_{js}(t)$. Then, we can write three classes of equations as follows.

$$m\ddot{x}_0 + c\dot{x}_0 + kx_0 = \frac{A_0}{2} \tag{5.133}$$

$$m\ddot{x}_{jc} + c\dot{x}_{jc} + kx_{jc} = A_j \cos j\omega t, \quad j = 1, 2, \dots \tag{5.134}$$

$$m\ddot{x}_{js} + c\dot{x}_{js} + kx_{js} = B_j \sin j\omega t, \quad j = 1, 2, \dots \tag{5.135}$$

Once we solve each of these equations, we can get the total response as follows by superposition.

$$x(t) = x_0 + \sum_j x_{jc} + \sum_j x_{js} \tag{5.136}$$

We can now solve each of these equations easily, since we have solved identical equations in the earlier sections. The first equation yields a constant solution, and the last two are harmonic excitations with a frequency of $j\omega$, where j could be any integer. Their solutions are found by substituting $j\omega$ for ω in Eqs. (5.15) and (5.16).

$$x_0 = \text{constant} \tag{5.137}$$

$$x_{jc} = x_{jc0} \cos(j\omega t - \phi_j) \tag{5.138}$$

$$x_{js} = x_{js0} \sin(j\omega t - \phi_j) \tag{5.139}$$

where the amplitudes are as follows.

$$x_0 = \frac{A_0}{2k}$$

$$x_{jc0} = \frac{A_j}{k} \frac{1}{\left\{ \left[1 - \left(\frac{j\omega}{\omega_n} \right)^2 \right]^2 + \left[2\zeta \frac{j\omega}{\omega_n} \right]^2 \right\}^{\frac{1}{2}}}$$

$$x_{js0} = \frac{B_j}{k} \frac{1}{\left\{ \left[1 - \left(\frac{j\omega}{\omega_n} \right)^2 \right]^2 + \left[2\zeta \frac{j\omega}{\omega_n} \right]^2 \right\}^{\frac{1}{2}}} \tag{5.140}$$

The phase angle is given by

$$\phi_j = \tan^{-1} \left[\frac{2\zeta \frac{j\omega}{\omega_n}}{1 - \left(\frac{j\omega}{\omega_n} \right)^2} \right] \tag{5.141}$$

Putting them all together, the solution is

$$x(t) = \frac{A_0}{2k} + \sum_j x_{jc0} \cos(j\omega t - \phi_j) + \sum_j x_{js0} \sin(j\omega t - \phi_j)$$

$$= \frac{A_0}{2k} + \sum_j \{ x_{jc0} [\cos j\omega t \cos \phi_j + \sin j\omega t \sin \phi_j] + x_{js0} [\sin j\omega t \cos \phi_j - \cos j\omega t \sin \phi_j] \}$$

$$= \frac{A_0}{2k} + \sum_j [x_{jc0} \cos \phi_j - x_{js0} \sin \phi_j] \cos j\omega t + \sum_j [x_{jc0} \sin \phi_j + x_{js0} \cos \phi_j] \sin j\omega t \tag{5.142}$$

Hence, the solution is in the form

$$x(t) = \frac{a_0}{2} + \sum_j a_j \cos j\omega t + \sum_j b_j \sin j\omega t \tag{5.143}$$

In other words, like the force, the displacement is also periodic with a fundamental period, $\tau = \frac{2\pi}{\omega}$. (What else did we expect from a *linear* system?) Note that there are now several (in fact, infinite) frequencies where resonance can occur; i.e., whenever, $j\omega = \omega_n$, we can have resonance.

Root mean square value

In order to characterize the magnitude of the response, we often use the root mean square (RMS) value of the response. Recall (!) that the general definition of the RMS value of a function is

$$x_{\text{RMS}} = \sqrt{\overline{x^2}} = \sqrt{\lim_{T \to \infty} \frac{1}{T} \int_0^T x^2(t)\, dt} \tag{5.144}$$

For a sine wave, $x(t) = x_0 \sin \omega t$, the wave is periodic, and the integration limit just becomes the period.

$$\overline{x^2} = \frac{x_0^2}{T} \int_0^T \frac{1}{2} \left(1 - \cos 2\omega t\right)\, dt = \frac{1}{2} x_0^2 \tag{5.145}$$

Hence, the RMS value would be $x_0/\sqrt{2}$.

For the Fourier series approximation, using the above result in Eq. (5.143), the mean square value can be calculated as follows.

$$
\begin{aligned}
\overline{x}^2 &= \frac{1}{T} \int_0^T \left[\frac{a_0}{2} + \sum_j a_j \cos j\omega t + \sum_j b_j \sin j\omega t \right]^2 dt \\
&= \frac{1}{T} \int_0^T \frac{a_0^2}{4}\, dt + \frac{1}{T} \int_0^T \left[\sum_i \sum_j a_i a_j \cos i\omega t \cos j\omega t \right] dt \\
&\quad + \frac{1}{T} \int_0^T \left[\sum_i \sum_j b_i b_j \sin i\omega t \sin j\omega t \right] \\
&\quad + \frac{2}{T} \int_0^T \left(\frac{a_0}{2} \sum_j a_j \cos j\omega t \right) dt + \frac{2}{T} \int_0^T \left(\frac{a_0}{2} \sum_j b_j \sin j\omega t \right) dt \\
&\quad + \frac{2}{T} \int_0^T \left[\sum_i \sum_j a_i b_j \cos i\omega t \sin j\omega t \right] \\
&= \left(\frac{a_0}{2} \right)^2 + \frac{1}{T} \int_0^T \sum_j a_j^2 \cos^2 j\omega t\, dt + \frac{1}{T} \int_0^T \sum_j a_j^2 \sin^2 j\omega t\, dt \\
&= \left(\frac{a_0}{2} \right)^2 + \frac{1}{2} \sum_j a_j^2 + \frac{1}{2} \sum_j b_j^2
\end{aligned}
\tag{5.146}
$$

The RMS value would of course be the square root of the above value

$$x_{\text{RMS}} = \sqrt{ \left(\frac{a_0}{2} \right)^2 + \frac{1}{2} \sum_j a_j^2 + \frac{1}{2} \sum_j b_j^2 } \tag{5.147}$$

--- **Example 5.5 Response to a rectangular wave** ---

Determine the response of a mass-spring-damper system to a rectangular wave shown in Fig. 5.42.

First, we determine the Fourier coefficients of a rectangular wave. It is easier to replot the force in terms of ωt, as shown in Fig. 5.43.

Then, the Fourier coefficients are given by

$$
A_0 = \frac{1}{\pi} \int_0^{2\pi} F(\omega t)\, d(\omega t)
$$

$$
= \frac{1}{\pi} \int_0^{\pi} (+F_0)\, d(\omega t) + \frac{1}{\pi} \int_\pi^{2\pi} (-F_0)\, d(\omega t)
$$

$$
= F_0 (\pi - \pi) = 0 \tag{5.148}
$$

$$
A_j = \frac{1}{\pi} \int_0^{2\pi} F(\omega t) \cos j\omega t\, d(\omega t)
$$

$$
= \frac{1}{\pi} \int_0^{\pi} (+F_0) \cos j\omega t\, d(\omega t) + \frac{1}{\pi} \int_\pi^{2\pi} (-F_0) \cos j\omega t\, d(\omega t)
$$

$$
= \frac{F_0}{\pi} \left[+\frac{\sin j\omega t}{j} \bigg|_0^{\pi} - \frac{\sin j\omega t}{j} \bigg|_\pi^{2\pi} \right] = 0 \tag{5.149}
$$

$$
B_j = \frac{1}{\pi} \int_0^{2\pi} F(\omega t) \sin j\omega t\, d(\omega t)
$$

$$
= \frac{1}{\pi} \int_0^{\pi} (+F_0) \sin j\omega t\, d(\omega t) + \frac{1}{\pi} \int_\pi^{2\pi} (-F_0) \sin j\omega t\, d(\omega t)
$$

$$
= \frac{F_0}{\pi} \left[+\frac{-\cos j\omega t}{j} \bigg|_0^{\pi} + \frac{\cos j\omega t}{j} \bigg|_\pi^{2\pi} \right]
$$

$$
= \frac{2F_0}{j\pi} [1 - \cos j\pi] \tag{5.150}
$$

Note how the cosine term (A_j) went to zero because it is an *odd function*. What! Odd and even functions seem to be familiar terms you heard in a mathematics class, but you did not think it was necessary to retain them after the exams, and everything has been lost in the mists of memory? Well, it is time to review (see Appendix A). Note also that the even angles of the sine series also disappear (for example, $\sin 2\omega t$).

Back to the problem. Now, the force can be written as an infinite Fourier series

$$
F(\omega t) = \frac{4F_0}{\pi} \left[\sin \omega t + \frac{1}{3} \sin 3\omega t + \frac{1}{5} \sin 5\omega t + \cdots \right] = \frac{4F_0}{\pi} \sum_{j=1}^{\infty} \frac{1}{2j+1} \sin [(2j+1)\omega t] \tag{5.151}
$$

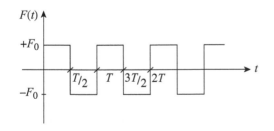

Figure 5.42: *A force in the form of a rectangular wave*

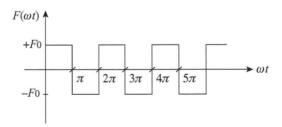

Figure 5.43: *Rectangular wave force replotted in terms of ωt*

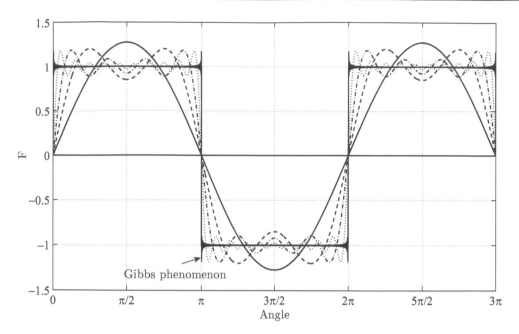

Figure 5.44: *Fourier series approximation of a rectangular pulse for 1, 3, 7, 20, and 500 terms*

One quick question to ask (us being practical minded engineers) is: "What is a suitable approximation for infinity?" Clearly, to be practical, we will need to truncate this series. An easy (although not mathematically rigorous) way to check this is to plot the approximations by successively including additional terms. For example, if we included only one term, it would be a sine wave; addition of one more term would include $\sin 3\omega t$, and so on. Figure 5.44 shows a few of the approximations going all the way to 500 terms (I hope you are surprised by the result—take a look at the figure, and think a little before reading further).

Clearly, the approximation gets better with additional terms, but only up to a point. As we keep increasing the number of terms, the agreement does get better in the continuous part of the function; however, it gets worse where we have a discontinuity in the function. If we had thought clearly, this should have been obvious from the beginning. In a Fourier series approximation, the sines and cosines are all continuous, and we had no right to expect them to somehow approximate a discontinuous function exactly at all points. This problem of exaggerated fits at points of discontinuity is called Gibbs phenomenon, and should be guarded against. For this case, 20 terms seems to be a reasonable compromise.

Now that we have established a Fourier series approximation of the force, we are ready to determine the response of our damped oscillator. Substituting in Eq. (5.140), we can determine the Fourier coefficients of the response as follows.

$$x_0 = 0$$
$$x_{jc0} = 0$$
$$x_{js0} = \frac{F_0}{k}\frac{2}{j\pi}[1 - \cos j\pi]\ \frac{1}{\left\{\left[1 - \left(\frac{j\omega}{\omega_n}\right)^2\right]^2 + \left[2\zeta\frac{j\omega}{\omega_n}\right]^2\right\}^{\frac{1}{2}}} \tag{5.152}$$

Figures 5.45 and 5.46 show the amplitude x_{js0} of the response for the first four terms for an underdamped and an over damped case, respectively. First, in the underdamped case, note the shift of peaks corresponding to the resonant frequency for that respective harmonic. Note also the falling amplitude values with increasing order that make the higher order terms less and less relevant. In the overdamped case, naturally there is no resonant peak for any of them; however, the amplitude values still fall for increasing orders.

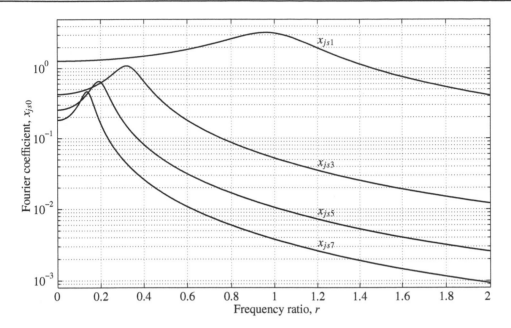

Figure 5.45: *Amplitudes x_{js0} plotted versus r for $\zeta < 1$; note the semilog scale*

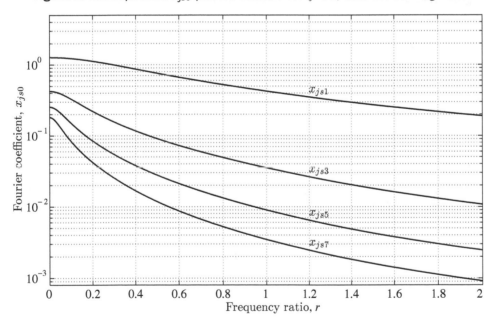

Figure 5.46: *Amplitudes x_{js0} plotted versus r for $\zeta > 1$; note the semilog scale*

Again, note that

$$x(t) = \frac{x_0}{2} + \sum_j x_{jc0} \cos j\omega t + \sum_j x_{js0} \sin j\omega t$$

$$= \sum_j x_{js0} \sin j\omega t \tag{5.153}$$

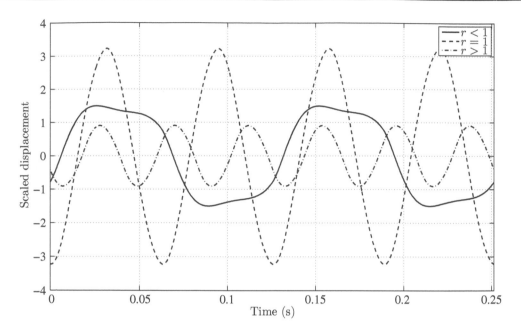

Figure 5.47: *Response of the damped oscillator to the rectangular force for ζ < 1*

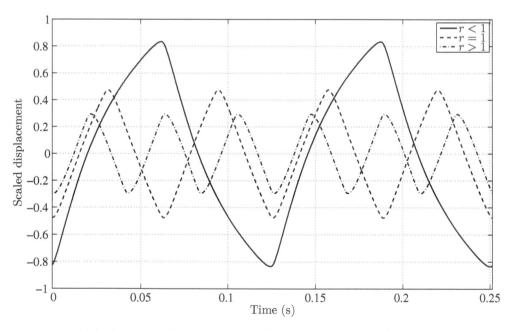

Figure 5.48: *Response of the damped oscillator to the rectangular force for ζ > 1*

Figure 5.47 shows the response of the system with a damping ratio of 0.5, and for frequency ratios of 0.5, 1, and 1.5. Figure 5.48 shows the response for $\zeta = 1.5$. Note how the amplitude, frequency, and *shape* of the response changes.

The RMS value can also be computed and is shown in Fig. 5.49 for four values of damping. Note again the resonant bumps at fractions of the natural frequency. If we reduce damping to a very small value ($\zeta = 0.01$), we

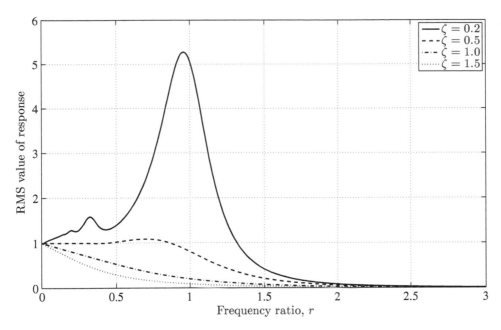

Figure 5.49: *RMS value of the response for various values of damping*

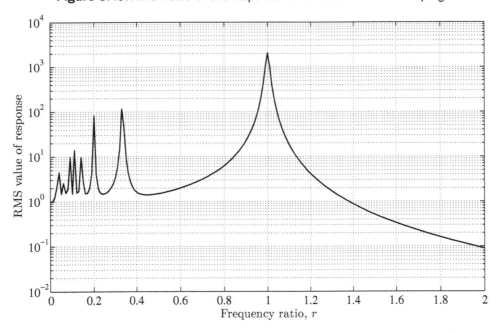

Figure 5.50: *RMS value of the response for very low damping ($\zeta = 0.01$)*

can clearly see the *subharmonic*[11] resonances; see Fig. 5.50, which is plotted on a logarithmic scale to emphasize the effect.

---

[11]subharmonic, as the name implies, is a fraction of the harmonic, or the fundamental forcing frequency; *superharmonic* response would be at a multiple of the harmonic.

Response spectrum

Another interesting way to look at the results is by using the concept of a response spectrum. This is where we plot the amplitudes in the individual harmonics versus the harmonic in a bar chart. The easiest way to understand this is by means of the example we just discussed. Let us first plot the force spectrum; this would mean plotting the magnitude of the terms from Eq. (5.151), as done in Fig. 5.51. Note how the force amplitudes drop off with increased harmonic. Hence, we expect the response also to drop off with increased harmonic (Why is that? Because, in linear systems, remember that the amplitude of the displacement is directly proportional to the amplitude of the force). Let us see if this surmise is correct.

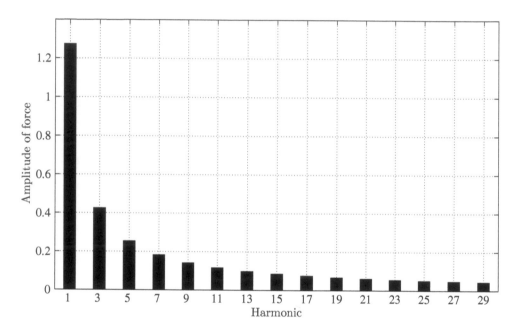

Figure 5.51: *Force spectrum for the rectangular force*

Let us plot the response spectrum by using the coefficients from Eq. (5.152). We will also do this for a range of excitation frequencies, which will help us examine the response critically. First, we use $r = 1$, or what we may call the resonant frequency, for the case when it is underdamped ($\zeta = 0.2$); the result is shown in Fig. 5.52. Clearly, the first harmonic dominates the response overwhelmingly; in fact, we could truncate the Fourier series after the first term, and be satisfied with the accuracy we derive from doing so. Before we get over confident however, let us look at $r = 0.33$; this is shown in Fig. 5.53. Note how the third harmonic response is no longer negligible; this is of course because of the term $j\omega/\omega_n$ in the denominator with $j = 3$. So, clearly, we would be at fault if we ignored the terms 3, 5, and 7 in this case. How about $r = 0.2 = 1/5$; this is shown in Fig. 5.54. Now, the fifth term is no longer small as we might expect, but many of the other terms are comparable in magnitude as well.

So, our surmise a paragraph ago was not quite right: even though the force amplitude drops off with increasing harmonics (look at Fig. 5.51 again), depending on the excitation frequency, the response amplitude may or may not do so. The moral of this story is hence that, in the case of periodic excitation, we need to be very careful with how we approximate things, as resonant excitation now can happen at infinite (practically, let us say, "many") frequencies.

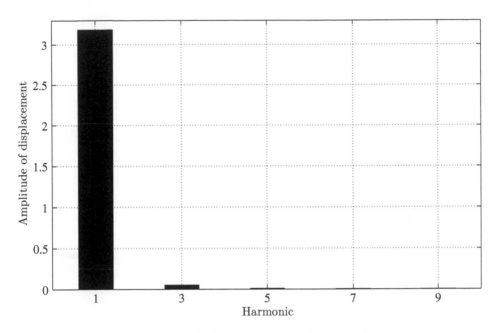

Figure 5.52: *Response spectrum for r = 1*

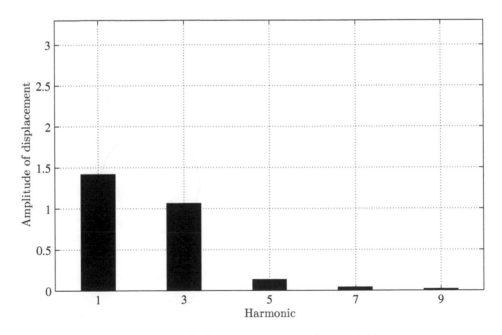

Figure 5.53: *Response spectrum for r = 0.33*

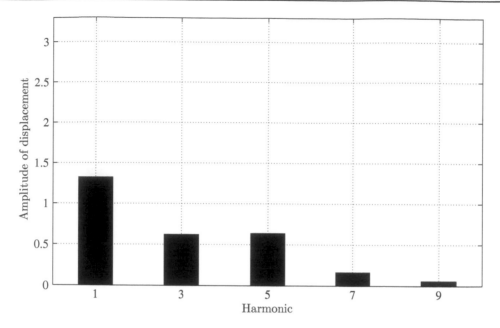

Figure 5.54: *Response spectrum for r = 0.2*

5.11 Forces in Reciprocating Engines

Reciprocating engines comprise a big application topic, which will not be discussed in detail here. Still, let us look at a simple model that will at least help us understand the origin of vibratory forces. Let us consider a vertical single cylinder [27, 16] engine, as shown schematically in Fig. 5.55. The entire engine is normally mounted on engine mounts; we will simply model the engine as an SDOF mass-spring system with a disturbing force coming from the engine dynamics. The first step is to come up with an adequately accurate description of the forces.

The underlying mechanism is what is called the slider-crank mechanism (something you probably studied in a mechanism or machinery dynamics class), and is shown in Fig. 5.56. The slider

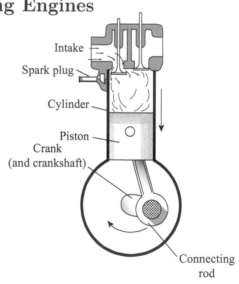

Figure 5.55: *Single cylinder engine schematic (see Obert [28])*

Figure 5.56: *Slider-crank mechanism*

is subject to forces (due to combustion in the case of the internal combustion engine), and the connecting rod connecting the slider to the crankshaft results in a rotation of the crank.

If the rotation of the crankshaft is denoted by θ, the linear displacement of the piston from when $\theta = 0$ is denoted by x, and the rotation angle of the connecting rod is denoted by α; they are related by the geometric constraints as follows:

$$x = \ell\,(1 - \cos\alpha) + r\,(1 - \cos\theta) \tag{5.154}$$
$$r\sin\theta = \ell\sin\alpha \tag{5.155}$$

where r is the crank radius and ℓ is the connecting rod length. From this it follows that

$$\cos\alpha = \sqrt{1 - \sin^2\alpha} = \sqrt{1 - \frac{r^2}{\ell^2}\sin^2\theta} \tag{5.156}$$

Note that these relations are purely kinematic (i.e., they have nothing do with forces).

Usually, the connecting rod length is much larger than the crank radius. Hence,

$$\ell \gg r \tag{5.157}$$

Therefore, we can use the binomial series expansion (What! You do not remember binomial expansion? Look at Appendix A for a refresher).

$$\cos\alpha = \sqrt{1 - \frac{r^2}{\ell^2}\sin^2\omega t} = 1 - \frac{r^2}{2\ell^2}\sin^2\omega t + \frac{r^2}{4\ell^2}\sin^4\omega t + \cdots \tag{5.158}$$

Next, we assume that the motion of the crankshaft is uniform at a constant rate ω. This is not true in general especially because of the variations of the gas torque, but is a reasonable approximation for the analysis that we intend to do here. Then,

$$\theta = \omega t, \quad \dot\theta = \omega \tag{5.159}$$

Substituting this into the expression for the motion of the piston, we can find its displacement, velocity, and acceleration (kinematically).

$$x(t) = r\,(1 - \cos\omega t) + r\left\{\frac{1}{2}\frac{r}{\ell}\sin^2\omega t - \frac{1}{4}\left(\frac{r}{\ell}\right)^2\sin^4\omega t + \cdots\right\} \tag{5.160}$$

$$\dot{x}(t) = r\omega\sin\omega t + r\omega\left\{\frac{1}{2}\frac{r}{\ell}(2\sin\omega t\cos\omega t) - \frac{1}{4}\left(\frac{r}{\ell}\right)^2(4\sin^3\omega t\cos\omega t) + \cdots\right\}$$

$$= r\omega\sin\omega t + r\omega\left\{\frac{1}{2}\frac{r}{\ell}\sin 2\omega t - \left(\frac{r}{\ell}\right)^2\left(\frac{1}{16}\sin 2\omega t - \frac{3}{32}\sin 4\omega t\right) + \cdots\right\} \tag{5.161}$$

where we have used trigonometric formulae to simplify things. Note that, since r/ℓ is small, $(r/\ell)^2$ is smaller still, and the subsequent terms not listed are even smaller than that. This justifies the approximation we are making by dropping the higher order terms. Differentiating again,

$$\ddot{x} = r\omega^2\cos\omega t + r\omega^2\left\{\frac{r}{\ell}\cos 2\omega t - \left(\frac{r}{\ell}\right)^2\left(\frac{1}{8}\cos 2\omega t - \frac{3}{16}\cos 4\omega t\right) + \cdots\right\} \tag{5.162}$$

If we decide to drop after the first term (making a further approximation) we would have

$$\ddot{x} = r\omega^2\cos\omega t + r\omega^2\left\{\frac{r}{\ell}\cos 2\omega t\right\} \tag{5.163}$$

The crankshaft, the connecting rod, and the piston all have masses that are non-negligible. Hence, it is clearly a multi-DOF (and what is called a multibody) dynamics problem. One way to approximate this is to lump the mass of the connecting rod at two points, the crankpin, and the piston. Let the sum total of the masses at the crankpin be m_1, and that at the piston be m_2. Then, we can estimate a disturbing force coming from the "inertia force" associated with m_2 with

$$F_{m2} = -m_2 \ddot{x} \tag{5.164}$$

Similarly, the inertia force associated with m_1 is given by

$$F_{m1} = -m_1 \frac{d}{dt} (-r \cos \omega t) = -m_1 r \omega^2 \cos \omega t \tag{5.165}$$

Hence, the total unbalance force acting on the engine mass is given by the sum of these two

$$F(t) = r\omega^2 \left\{ [-(m_1 + m_2)] \cos \omega t - m_2 \left[\left(\frac{r}{\ell} \right) \cos 2\omega t + \left(\frac{r}{\ell} \right)^2 \left(\frac{1}{8} \cos 2\omega t - \frac{3}{16} \cos 4\omega t \right) + \cdots \right] \right\} \tag{5.166}$$

Clearly, the expression for the force turned out to be in the same form as the Fourier series we had to derive for other periodic functions. Hence, it follows that the disturbing force on the engine is periodic.

Without solving for the vibrational amplitude (which is difficult to do without all parameters including damping), we can still carry out a quantitative analysis about the nature of the excitation. Note that, if the forcing frequency, which is the rotational speed of the engine, coincides with the natural frequency, we will get resonance because of the first term in the force. In addition, since we have a 2ω term in the force, we also expect resonance when the rotational speed is close to $1/2$ the natural frequency. Then, so also $1/4$ and so on. Clearly, as the amplitude of the force drops off with increasing frequency, the successive harmonic amplitudes of the vibration are also going to be somewhat small. But we make this statement with some caution, remembering our periodic excitation example earlier in this chapter; resonance can change everything!

— **Example 5.6 Reciprocating engine vibration** —

Solve the single engine reciprocating engine vibration problem for the following numerical values.[12] The mass of the piston (m_p) is 3 kg, that of the connecting rod (m_c) is 1.2 kg, and the mass of the engine (m_e) is 120 kg. The effective stiffness of the support (k) is 1.1 MN/m; assume a linear viscous damping ratio of 0.2. The crank radius (r) is 22 cm, and the connecting rod length (ℓ) is 80 cm. The engine rpm of interest is 500.

As an approximation, we will lump 2/3 the mass of the connecting rod at the crank pin, and 1/3 with the piston. Then,

$$m_1 = \frac{2}{3} m_c = 0.8 \text{ kg}, \quad m_2 = m_p + \frac{1}{3} m_c = 3.4 \text{ kg} \tag{5.167}$$

The effective natural frequency is

$$\omega_n = \sqrt{\frac{k}{m_e}} = 95.74 \text{ rad/s} \tag{5.168}$$

The engine speed is 52.36 rad/s, which gives us frequency ratios of 0.55 and 1.09 for the excitation frequencies of ω and 2ω, respectively. Neglecting terms of order higher than $\left(\frac{r}{\ell} \right)^2$, the excitation force is of the form

$$F(t) = B_1 \cos \omega t + B_2 \cos 2\omega t \tag{5.169}$$

---

[12]The numbers for this problem are based on the treatment in the book by Timoshenko et al. [16].

where

$$B_1 = -(m_1 + m_2)r\omega^2 = -2{,}533\,\text{N} \tag{5.170}$$

$$B_2 = -m_2 r\omega^2 \left(\frac{r}{\ell}\right) = -564\,\text{N} \tag{5.171}$$

Hence, the response is given by

$$x(t) = x_{1c0}\cos(\omega t - \phi_1) + x_{2c0}\cos(2\omega t - \phi_2) \tag{5.172}$$

where

$$x_{1c0} = \frac{|B_1|}{k}\,\frac{1}{\sqrt{(1-r_1^2)^2 + (2\zeta r_1)^2}} = 3.1\,\text{mm} \tag{5.173}$$

$$x_{1c0} = \frac{|B_2|}{k}\,\frac{1}{\sqrt{(1-r_2^2)^2 + (2\zeta r_2)^2}} = 1.1\,\text{mm} \tag{5.174}$$

where $r_1 = \omega/\omega_n$ and $r_2 = 2\omega/\omega_n$. The phase angles are

$$\phi_1 = -163°, \quad \phi_2 = -66° \tag{5.175}$$

The response is shown in Fig. 5.57.

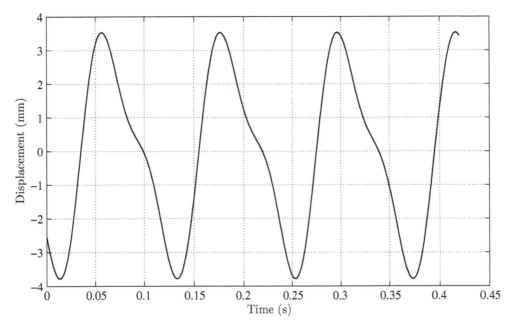

Figure 5.57: *Vibrational response of the single cylinder engine*

5.12 Review Questions

1. In harmonic excitation, how is the response frequency related to the excitation frequency? Is this always true?

2. What is the frequency ratio often used in harmonic excitation? How does it change with the excitation frequency?

3. Define the dynamic magnification factor. Why is it called so?

4. What is resonance, and under what conditions does it occur?

5. How does the amplitude at resonance depend upon the damping ratio?

6. What happens to the response amplitude for large excitation frequencies?

7. What is the amplitude at low excitation frequencies? Why?

8. What is the phase lag at low frequencies? Why?

9. What is the phase lag at resonance? What does it mean?

10. In what way does the phase lag at resonance depend on the damping ratio? Why?

11. What is the minimum value of damping above which a resonant peak is not observed? (*Hint*: it is not 1.)

12. Why is the response to harmonic excitation *not* sinusoidal for $r = 1$ and $\zeta = 0$? Can this occur in practice?

13. Under what conditions is the response amplitude most sensitive to changes in excitation frequency?

14. Under what conditions is the response amplitude most sensitive to changes in damping?

15. While operating well above the resonant frequency, how effective is addition of damping to reduce the vibration amplitudes? Explain.

16. How would you use the phase plot for estimating the natural frequency?

17. What is the complex frequency response function, and why is it complex?

18. What is a phasor, and how is it used in harmonic response solution?

19. From the phasor diagram, what do you learn about the size of the damping term at different frequencies?

20. In the phasor diagram, which force dominates for high frequencies?

21. Define transient and steady-state vibration.

22. In practice, when would you be most interested in transient vibration?

23. In practice, when would you be most interested in steady-state vibration?

24. In numerical simulation, how long would you wait for the transients to die out? Why?

25. What are the various sources of damping? How many of those have you experienced?

26. How does the energy dissipated per cycle for harmonic excitation with linear viscous damping depend on the frequency? On vibration amplitude?

27. What does the area enclosed by the hysteresis curve represent?

28. Define specific damping capacity and loss factor.

29. For linear viscous damping, how does the loss factor depend on the frequency ratio?

30. Given some arbitrary damping mechanism, how would you define an equivalent viscous damping coefficient?

31. What is the principal assumption in deriving the equivalent viscous damping coefficient?

32. For Coulomb friction, what are the various parameters and variables the equivalent viscous damping coefficient depend upon? Are you surprised by anything in this?

33. In structural damping, what happens to the effective stiffness? Why did the *damping* end up changing the *stiffness*?

34. What is the base excitation problem? Give some practical examples.

35. What is the vibration transmission problem? Give some practical examples.

36. In base excitation, what is the minimum frequency ratio for suitable design?

37. Is it a good idea to have a lot of damping in a vibration isolator? Explain.

38. Why would it not be a good idea to design a vibration isolator with zero damping? Explain.

39. Define transmissibility ratio.

40. Given a practical situation with excessive vibration, what are all the things you could do to reduce the response amplitude?

41. Why is consideration of unbalance in rotating machinery important even if the unbalance magnitude is very small?

42. In rotating unbalance, what is the response amplitude for low frequencies? For high frequencies?

43. In periodic excitation, when can resonance occur? Why?

44. Define the root mean square value of a periodic function.

45. What is Gibbs phenomenon and what limitation does it impose on the Fourier solution process for discontinuous functions?

46. What is the response spectrum for periodic response? Why is it useful?

47. In a reciprocating engine, what is the nature of the time dependence of the force on the engine mass?

48. We only considered a single cylinder engine in the text; what would happen with multicylinder engines?

Problems

5.1 Consider an SDOF damped system, Fig. 5.58, subject to a general harmonic force, $F(t) = F_0 \cos(\omega t - \alpha)$. Determine the response of the system, $x(t)$.

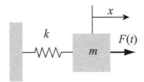

Figure 5.58: *Mass-spring-damper system*

5.2 For a damped spring-mass system subject to a harmonic force, determine the forcing frequency at which the amplitude of vibration is maximum. Also, find the value of the maximum amplitude.

5.3 A vibratory system with a natural frequency of 15 Hz is suddenly excited by a harmonic excitation at 8 Hz. What should the damping factor of the system be so that the system settles down within 5% of the steady-state amplitude in 200 ms?

5.4 A spring-mass system vibrates horizontally on a surface with coefficient of friction $\mu = 0.5$. When excited harmonically at 5 Hz, the amplitude of displacement of the mass is 10 cm. Determine the equivalent viscous damping. Assume $m = 30$ kg and $k = 5000$ N/m.

5.5 A spring-mass system is subjected to a harmonic force whose frequency is close to the natural frequency of the system. If the forcing frequency is 45 Hz and the natural frequency is 52 Hz, determine the period.

5.6 A spring-mass system with $m = 15$ kg and $k = 3000$ N/m is subjected to a harmonic force of amplitude 300 N and frequency ω. If the maximum amplitude of the mass is observed to be 150 mm, find the value of ω.

5.7 A spring-mass-damper system is subjected to a harmonic force. The amplitude is found to be 15 mm at resonance and 8 mm at a frequency 0.5 times the resonant frequency. Find the damping ratio of the system.

5.8 When a spring-mass-damper system is subjected to a harmonic force $F(t) = 8\cos 3t\,(\text{N})$, the resulting displacement is given by $x(t) = 2.1\cos 3t\,(\text{mm})$. Find the work done during the first 10 s.

5.9 A mass m is suspended by a spring of stiffness 5000 N/m and is subjected to a harmonic force with an amplitude of 150 N and a frequency of 8 Hz. The amplitude of the forced motion of the mass is observed to be 25 mm. Find the value of m.

5.10 A spring-mass system consists of a 15 kg mass and a spring with a stiffness of 3000 N/m. The mass is subjected to a harmonic force $F(t) = 15\cos\omega t$ N. Find the amplitude of the forced motion at the end of (i)$\frac{1}{2}$ a cycle (ii) $2\frac{1}{2}$ cycles.

5.11 A spring-mass system with $m = 10$ kg and $k = 4000$ N/m is subjected to a harmonic force having an amplitude of 100 N and an excitation frequency of ω. If the maximum amplitude of the mass is observed to be 100 mm, find the value of ω.

5.12 For a vibrating system $m = 15$ kg, $k = 3000$ N/m, and $c = 40$ Ns/m. A harmonic force of amplitude 230 N and frequency 15 Hz acts on the mass. If the initial displacement and velocity of the mass are 10 mm and 6 mm/s, respectively, find the complete solution.

5.13 A spring-mass system with $m = 100$ kg and $k = 5000$ N/m is subjected to a harmonic force of amplitude 250 N and frequency ω. If the maximum amplitude of the mass is observed to be 100 mm, find the value of ω.

5.14 A spring-mass-damper system is subjected to a harmonic force. The amplitude is found to be 20 mm at resonance and 10 mm at a frequency 0.9 times the resonant frequency. Find the damping ratio of the system.

5.15 A damped system is subjected to a harmonic force for which the frequency can be adjusted. It is determined experimentally that, at twice the resonant frequency, the steady amplitude is one-tenth of that which occurs at resonance. Determine the damping ratio for the system.

5.16 A spring-mass system is excited by a harmonic force. At resonance the amplitude is measured to be 0.58 cm. At 0.80 of resonant frequency, the amplitude is measured to be 0.46 cm. Determine the damping ratio ζ for the system.

5.17 In Problem 2.25, we considered a pendulum with a support that was moving (shown here again in Fig. 5.59). Now, assume that the angle θ remains small, and that the base motion is give by

$$x = A \sin \omega t$$

Determine $\theta(t)$.

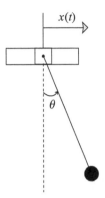

Figure 5.59: *Simple pendulum with base motion*

5.18 A particle of mass m is connected to a spring of stiffness k, as shown in Fig. 5.60. One end of the piston is given a motion $x_2 = A \sin \omega t$; the other end moves inside the cylinder with a viscous friction coefficient c. Determine the amplitude of the particle motion and its phase with respect to the piston.

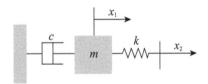

Figure 5.60: *Mass-spring-damper system*

5.19 A mass m is attached to one end of a rigid massless link ABC, as shown in Fig. 5.61. The link ABC can pivot at B. (Note that the angle ABC is always 90°.) The other end of the spring attached to point A is given a motion $y = A \cos \omega t$. Assume small motions and determine $\theta(t)$.

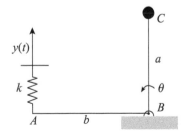

Figure 5.61: *Oscillating system*

5.20 Figure 5.62 shows a motor mounted on a bracket that is hinged about the point O and can be assumed to be perfectly rigid. The motor results in a harmonic force on the bracket. Determine the response amplitude of the rigid rod. The rod has mass m and length ℓ. The spring is mounted at the end, and the damper at the midpoint.

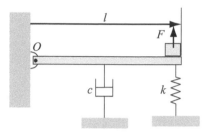

Figure 5.62: *Motor on a rigid bracket*

5.21 Consider a motor driving a load through a pair of gear trains, as shown in Fig. 5.63. The motor results in a torque excitation at twice the rotational speed Ω. Determine the forced response. Ignore backlash, slip, and friction. The motor inertia is 0.025 kgm², and the load inertia is 0.32 kgm². The motor shaft is 20 cm long, and the load shaft is 40 cm long. They each have an outer radius of 4.5 cm, and an inner radius of 2.5 cm, and the gear ratio is 10. They are both made of steel. The torque magnitude is 55 Nm, and $\Omega = 250$ rpm. Assume a combined damping ratio of 0.2.

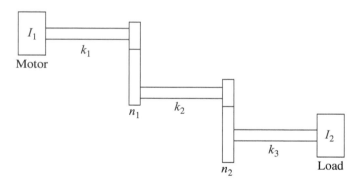

Figure 5.63: *Motor drive train*

5.22 An electric motor of mass 75 kg is to be mounted at the free end of a cantilever beam of length 5 m. The running speed of the motor is 1200 rpm, and the motor develops a harmonic force of amplitude 5000 N. The amplitude of vibration should be limited to 0.5 cm. Determine the necessary cross-sectional dimensions of the beam. Assume that it is a square cross section. The beam is made of steel with a Young's modulus of 2.1×10^{11} N/m².

5.23 The free vibration of an SDOF system is observed to decay from an amplitude of 20 mm to half that value in 10 cycles. Calculate the amplitude of vibration at resonance if the base is subjected to a harmonic motion 1 mm in amplitude.

5.24 A machine that is supported on four steel springs (with negligible damping) has resonance at 500 rpm. At 1300 rpm, the amplitude of motion is 0.5 mm. What would be the amplitude if the springs are replaced by four rubber isolators with an effective damping ratio of $\zeta = 0.3$?

Assume that the undamped natural frequency has not changed. How different are the two amplitudes? Why?

5.25 A spring-mass system is excited by a harmonic force. At resonance the amplitude is measured to be 9.2 mm. At 80% of the resonant frequency, the amplitude is measured to be 6 mm. Determine the damping ratio ζ for the system.

5.26 Find the equation of motion for the system shown in Fig. 5.64. Solve for the vibrational response.

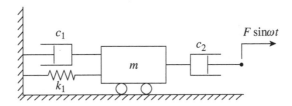

Figure 5.64: *Harmonically excited system*

5.27 A weight of 50 N is suspended from a spring of stiffness 4000 N/m and is subjected to a harmonic force of amplitude 60 N and frequency 6 Hz. Find the following.

(a) The extension of the spring due to the suspended weight.

(b) The static displacement of the spring due to the maximum applied force.

(c) The amplitude of forced motion.

5.28 A motor shaft, shown in Fig. 5.65, is subjected to a harmonic torque at varying speeds. Determine the response of the shaft at the location of the motor mass. $m = 30$ kg; the shaft is hollow and is made of steel, and has the following dimensions: length $\ell = 3$ m, outer radius $r_1 = 5$ cm, and inner radius $r_2 = 4.5$ cm. The torque amplitude is 10 kNm. The speed varies from 0 to 2000 rpm. Assume that the damping is negligible. If you wanted to limit the maximum vibration to 5°, what would be the damping that you would need to add?

Figure 5.65: *Motor shaft in torsion*

5.29 A pump mounted at the center of a simply supported beam causes a harmonic excitation Fig. 5.66. Determine the vibrational response at the location of the pump. $m = 22$ kg; the shaft is made of aluminum, and has the following dimensions: length, $\ell = 1.3$ m, rectangular width $b = 6$ cm, and height, $h = 2.5$ cm. The force has an amplitude of 5.6 N. The speed of the pump varies between 0 and 1200 rpm. Assume that the damping is negligible, and plot the vibrational response versus the speed. If the maximum vibration that can be tolerated is 1.0 cm, what are the various ways you could do it? Note that you cannot change the length of the beam, or the speed or excitation of the force. You need to provide a numerical answer here.

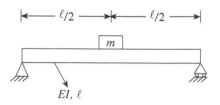

Figure 5.66: *Pump on a beam*

5.30 Consider a mass-spring slider on a horizontal surface that has Coulomb friction. The friction is given by μN, where μ is the coefficient of sliding friction, and N is the normal force. $m = 10$ kg, $k = 8000$ N/m, and $\mu = 0.15$.

(a) Determine the equivalent viscous friction coefficient.

(b) The system is subject to harmonic vibration with amplitude $F_0 = 25$ N and frequency that varies between 0 and 20 Hz. Determine the frequency response of the system assuming the equivalent linear viscous model.

(c) Solve the actual problem (with the nonlinear damping) numerically for three frequency values (say, 2, 4.5, and 10 Hz) and compare with the solution in (b). What are your conclusions?

5.31 A spring-mass system is subjected to a harmonic force whose frequency is close to the natural frequency of the system. The forcing frequency is 39.8 Hz and the natural frequency is 40 Hz. Solve for the response and plot it as a function of time. You should see a waxing and waning of the displacement called the *beating* phenomenon.

5.32 An aircraft engine is representable by a mass-spring model and is excited by a harmonic force. There is structural damping in the system estimated to be about $\eta = 0.1$. Determine the harmonic frequency response of the system and plot it. $m = 300$ kg, $k = 500$ kN/m, $F_0 = 1$ kN, and $0 < \omega < 1,000$ rad/s.

5.33 Consider a mass-spring system subjected to the nonlinear damping force we discussed on page 135; i.e.,

$$F_d = a\dot{x}^2 \, \text{sign}(\dot{x})$$

where a is a constant parameter. The equation of motion is then

$$m\ddot{x} + a\dot{x}^2 \, \text{sign}(\dot{x}) + kx = F(t) \tag{5.176}$$

Assume that $m = 20$ kg, $k = 15$ kN/m, $F_0 = 10$ kN, and $a = 200$ Ns^4/m^2 (note the units of a).

(a) Find the equivalent linear viscous damping coefficient using the energy dissipated per cycle; do it for a frequency of 20 rad/s. [Hint: Substituting equivalent damping into the linearly damped system frequency response equation will yield a quadratic equation for x_0 for this problem.]

(b) Use the equivalent damping ratio and determine the *free* response to a set of initial conditions: $x(0) = 0.05$ m, $\dot{x}(0) = 0.01$ m/s.

(c) Compare the above response to a numerically simulated response with the actual nonlinear force.

(d) Determine the response of the system to a harmonic force using the equivalent linear viscous damping coefficient. Do this for a range of frequencies from 0 to 60 rad/s, and $F_0 = 25$ N.

(e) Compare the harmonic response to the actual response obtained by numerical simulation at the following three frequencies: 10, 20, and 50 rad/s.

(f) What are your conclusions about the limits of validity of the approximation?

(g) ⚠⚠ Obtain the nonlinear frequency response of the system from numerical simulation, and estimate the amplitude at the forcing frequency. Compare with the linear frequency response you obtained in (d).

5.34 (a) During an earthquake, a single-story building frame in sunny California is subjected to a harmonic ground acceleration $\ddot{x}_g$, as shown in Fig. 5.67. The building may be modeled as a mass m supported on two beams with stiffness k. Determine the steady-state motion of the floor.

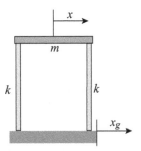

Figure 5.67: *Building in an earthquake*

(b) If the ground acceleration $\ddot{x}_g = 75 \sin \omega t$ mm/s², $m = 50{,}000$ kg, $k = 2.5 \times 10^8$ N/m, and $\omega = 16$ rad/s, determine the horizontal displacement of the floor.

5.35 Consider model of a grinding wheel as a solid steel cantilever shaft with a rigid mass at the end, as shown in Fig. 5.68. The shaft has length ℓ, Young's modulus E, and area moment of inertia I. Assume that its mass is negligible compared to that of the grinding wheel m. As the wheel rotates there is a rotating unbalance that leads to harmonic excitation on the system. Determine the forced response of the system; note that the vibrational displacement at the end would result in manufacturing errors, and would need to be minimized. Use the following numerical values: $\ell = 0.3$ m, radius of the shaft $= 2.5$ cm, $m = 15$ kg, unbalance magnitude $= 0.05$ kg m, and $\omega = 100$ rpm.

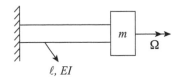

Figure 5.68: *Grinding wheel model*

5.36 An aircraft radio of mass 7.2 kg is to be isolated from engine vibrations. The engine speed varies between 1000 rpm and 2000 rpm. What static deflection must the isolators have for 90% isolation?

5.37 A CD player is to be supported on a vibration isolation system so that the ground vibration does not cause it to skip tracks. The ground vibration has an amplitude of 1 mm, and can occur at a frequency of 25 Hz. When it is playing, the amplitude of vibration should be less than 0.1 mm, and the maximum possible vibration (that can occur during resonance) should be limited to 3 mm. If the mass of the system is 5 kg, design an isolation system.

5.38 A microprocessor control unit has to be mounted on a set of isolation mounts under the hood of an automobile. No more than 125% of the vibration of the bracket it is mounted on should be experienced by the control unit itself. It has a mass of 300 g and the engine speed can be taken to range from 2000 to 5000 rpm. Design the mount.

5.39 A refrigerator unit of mass 45 kg is to be supported by four springs of stiffness k each. If the unit operates at 450 rpm, what should be the value of the spring constant if only 10% of the shaking force of the unit is to be transmitted to the supporting structure?

5.40 An isolation system has to be designed for a machine such that no more than 20% of the disturbing force gets transmitted to the surrounding area at its operating speed. It is also desired that, as the machine is accelerated to the running speed, no more than two times the disturbing force should get transmitted. The running speed of the machine is 15,000 rpm, and the effective mass of the machine is 1250 kg. Determine the parameters of the isolation system.

5.41 We would like to design the suspension parameters for an automobile. The road can be assumed to have a sinusoidal roughness with an amplitude of 1 cm and a wavelength of 5 m. The effective mass is 1200 kg. It is desired that the automobile vibration never exceeds 1.35 cm. In addition, over the entire range of cruising speeds of 75 to 100 kph we would like the vibration to be lower than 0.45 cm. Select the suspension parameters k and c. Recall that the ground roughness results in a sinusoidal input with $\omega = \frac{2\pi v}{\lambda}$.

5.42 A lathe creates a significant amount of dynamic forces that we would like to isolate from the surroundings. It has an effective mass of 550 kg and an operating speed of 1200 rpm. The specifications are as follows.

- The vibrational displacement of the tool piece on the lathe should never exceed 3.5 times the static displacement.

- The transmissibility ratio at the operating speed should not exceed 0.2.

Design a vibration mount for the lathe.

5.43 A cantilever beam with one end fixed at the wall was needed to be tested by exciting it at the free end by a sinusoidal force. It was decided to model the cantilever as a rigid mass m_1 with spring stiffness k_1 connected to the wall. In order to excite this by a shaker, a fixture was designed. The fixture can be modeled as a mass m_2 with stiffness k_2 and a damping constant c_2. The mass m_1 is bolted to this fixture m_2. The schematic of the system is shown in Fig. 5.69. Then the shaker head is subjected to a harmonic excitation $u = u_0 \sin \omega t$.

(a) Derive the equations of motion.

(b) Given $k_1 = 3000\,\text{N/m}$, $k_2 = 3250\,\text{N/m}$, $m_1 = 0.2\,\text{kg}$, $m_2 = 9.8\,\text{kg}$, $c_2 = 100\,\text{Ns/m}$, $u_0 = 0.02\,\text{m}$, and $\omega = 8\,\text{rad/s}$, solve for the magnitude and the acceleration of the system.

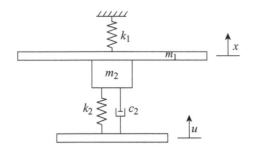

Figure 5.69: *Excitation test of a cantilever beam*

(c) Compute the magnitude of the force transmitted to the wall through the spring k_1.

(d) Compute the magnitude of the force transmitted to the shaker.

5.44 A precision grinding machine is supported on an isolator that has a stiffness of 1.2 MN/m and a viscous damping constant of 1.1 kNs/m. The floor on which the machine is mounted is itself subject to harmonic disturbance due to another machine in the vicinity. Find the maximum acceptable displacement of the floor if the amplitude of the grinding wheel vibration is to be less than 1 µm. Assume that the grinding machine is a rigid body with a mass of 600 kg.

5.45 A machine has a mass of 600 kg and has an operating speed of 800 rpm. During its operation it is subject to a harmonic force with amplitude 175 N. A vibration isolation system has to be designed so that the following specifications are met:

- The transmitted force amplitude should not exceed 50 N at the operating speed.

- The vibrational amplitude of the machine itself should never exceed 5 mm when it is accelerated up to the operating speed from zero.

Determine the effective spring stiffness and the viscous damping coefficient of the vibration isolation system.

5.46 An electric motor of mass 72 kg is mounted on an isolator block of mass 1000 kg, and the natural frequency of the total assembly is 3 Hz with a damping ratio of $\zeta = 0.10$. If there is an unbalance in the motor that results in a harmonic force of $F = F_0 \sin \omega t$, determine the amplitude of vibration of the block and the force transmitted to the floor with $F_0 = 125$ N, $\omega = 35$ rad/s.

5.47 Consider a reciprocating saw held by a human operator. As injuries due to vibration are a significant occupational hazard that should be avoided, it is very important to design the saw and its handles carefully. Assume that it can be modeled as a single-DOF system subject to a harmonic force from the motor. Design the isolation system so that the transmissibility ratio is never larger than 4.5, and lower than 0.5 at the operating speed. The speed range of operation is 50–300 cps, and the effective mass is 20 kg.

5.48 Consider the wing of an aircraft with an engine mounted on it, as shown in Fig. 5.70. The wing can be modeled as a uniform (OK, we are stretching the truth a little here!) cantilever beam of length ℓ, Young's modulus E, and area moment of inertia I. Assume that half its mass m_w can be lumped, along with the mass of the engine m_e. The engine generates a harmonic force at the running speed. Determine the forced response of the system.

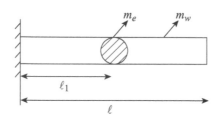

Figure 5.70: *An SDOF model of an aircraft wing*

5.49 The vibrational response of seated human subjects is an important problem. One of the simplest models for it is to model the human being as one lumped mass, and the human-seat interface as a spring and a damper [18]. The mass is 56.8 ± 9.4 kg, the spring stiffness is 75.5 ± 28.3 kN/m, and the damping coefficient is 3.840 ± 1.007 kN s/m. it is subjected to a harmonic base motion of amplitude 5 cm, and a frequency range of $0 < \omega < 20$ Hz. Determine the response for the *range* of given parameters. We will look at more complicated models in the forthcoming chapters.

5.50 A diesel-electric generator is mounted on a set of rubber isolators. The system exhibits an amplitude resonance peak of 5:1 at a running speed of 600 rpm. Above what speed would the transmission of the force be reduced to one half?

5.51 A turbine shaft can be modeled as a rigid disk at the center of a simply supported massless beam, as shown in Fig. 5.71. The running speed of the machine is ω, the mass of the disk is m, the shaft has length ℓ, and area moment of inertia I; it is made of a material with Young's modulus E. The turbine wheel has a rotating unbalance, $m_0 e$.

Figure 5.71: *Turbine shaft*

(a) Determine the most unfavorable speed (critical speed) of the system.

(b) Determine the amplitude of vibration at the running speed.

(c) If the shaft is made *thinner* by half, what would be the vibrational amplitude? Does that agree with your intuition?

(d) Suppose we add damping in the form of a fluid-film bearing that results in a damping coefficient of c, then what are the vibrational amplitudes?

(e) For one of the two sizes of the shaft, plot the vibrational response of the system as a function of the rotational speed (use MATLAB).

You may use the following numerical values: $m = 6$ kg, $E = 2.1 \times 10^{11}$ N/m^2, $\ell = 0.8$ m, $r = 30$ mm, $\omega = 12{,}000$ rpm, $m_0 e = 0.1$ kg mm, $c = 3500$ N s/m.

5.52 A variable speed electric motor, having an unbalance, is mounted on an isolator. As the speed of the motor is increased from zero, the amplitudes of vibration of the motor are observed to be 0.45 in. at resonance, and 0.09 in. well beyond resonance. Find the damping ratio of the system.

5.53 A turbine rotor runs at 50,000 rpm and has a mass of 5000 kg. The following two pieces of data were gathered in an experiment.

- As the rotor was accelerated to its operating speed, it was found to vibrate excessively at 30,000 rpm. The vibration amplitude was measured to be 6 mm.

- As the rotor's speed was further increased to well past its operating speed, the vibration became steady and was found to be 1 mm.

Estimate the vibrational displacement at the running speed.

5.54 A centrifugal pump with a mass of 60 kg and operating at 1000 rpm is mounted on six springs each of stiffness 6000 N/m. Find the maximum permissible unbalance in order to limit the steady-state deflection amplitude to 5 mm.

5.55 Derive an expression for the frequency response of a system with material damping excited by rotating unbalance.

5.56 Derive an expression for the base excitation of an SDOF system with material damping.

5.57 Consider the automobile supported on a spring and damper subject to base motion from a harmonic road roughness. Assume that there is structural damping in addition to viscous damping. Determine the vibrational response over a range of speeds. The parameters are as follows: $m = 1200$ kg, $k = 10$ kN/m, $\zeta = 0.35$ for viscous damping, $\eta = 0.1$ for structural damping, road surface roughness $= u_0 \sin\left(\frac{2\pi d}{\lambda}\right)$, with $u_0 = 0.1$ m, $\lambda = 1.2$ m, and speed range $= 0 < v < 100$ kph. Note that d is the forward displacement given by $d = vt$, for constant speed.

5.58 Consider a rotating shaft subjected to Coulomb friction due to dry rubbing at the bearing. The shaft is subjected to rotating unbalance excitation. Assume the following values: $m = 300$ kg, $k = 10^9$ N/m, unbalance $m_0 e = 15$ kg m, speed $0 < \omega < 30,000$ rpm, Coulomb friction coefficient $\mu = 0.15$.

(a) Determine the vibrational response of the rotor using an equivalent viscous damping coefficient. Plot it versus the speed of the rotor.

(b) Numerically simulate the response for three values of the speed and compare with the solution obtained in (a).

5.59 A compressor has a mass of 2000 kg and a running speed of 1800 rpm. A vibration isolation has to be designed for the system. The design specifications are as follows.

(a) The transmissibility ratio at the running speed should not exceed 0.25.

(b) The transmissibility ratio should not exceed 3.75 as the machine is sped up to the operating speed.

Design the isolation system.

5.60 This problem concerns a delicate instrument of mass 20 kg that has to be isolated from surrounding vibration in an aircraft. During normal flight, the vibration in the aircraft is harmonic with an amplitude of 5 mm, and a frequency of 50 Hz. For fault-free operation, it is required to hold the vibration of the equipment down to 1 mm. Also, any vibration larger than 12 mm is likely to destroy the equipment, and hence the design of the suspension should ensure that the vibration amplitude in the instrument is less than this value even under resonant conditions. Design a suitable vibration isolation system.

5.61 Consider a simple model for the vibration transmitted to the driver in an automobile. The seat can be modeled as a spring and a damper, and the driver as a particle mass. We wish to isolate the driver from the engine vibration transmitted through the chassis. Assume the following parameter values and select appropriate spring and damper values for the seat; note that the design has to be satisfactory for a *range* of values. $m = 50\text{--}100\,\text{kg}$, $\omega = 200\text{--}600\,\text{rad/s}$. Transmissibility should be less than 10%.

5.62 A machine with rotating unbalance has a mass of 300 kg and is supported on springs and dampers. The normal running speed of the machine is 4000 rpm. It is observed that as the machine is sped up to the operating speed from rest, it has a maximum vibrational amplitude of 1 mm. Also, for very large speeds, the vibrational amplitude is found to be 0.5 mm. Determine the vibrational amplitude at the operating speed.

5.63 A machine has an effective mass of 450 kg and a running speed of 2300 rpm. It is supported on four springs, each with a spring constant of 2.25 MN/m. The machine has a harmonic force acting on it that leads to a (modest) vibrational amplitude of 0.1 mm at the operating speed. The vibration of the machine, as it passes through resonance, was however too high and it was decided to add damping. Determine the damping coefficient of the damper you would need to add to keep the vibrational amplitudes to less than 0.5 mm at resonance. Also, find the new vibrational amplitude at the operating speed.

5.64 An industrial machine of mass 450 kg is supported on springs with a static deflection of 0.50 cm. If the machine has a rotating unbalance of 0.23 kg m, determine (a) the dynamical amplitude at this speed and (b) the force transmitted to the floor at 1200 rpm. Assume damping to be negligible.

5.65 Design a solid steel shaft supported in bearings that carries the rotor of a turbine at the middle. The rotor weighs 1100 N and delivers a power of 200 hp at 300 rpm. In order to keep the stress due to the unbalance in the rotor small, the critical speed of the shaft is to be made one-fifth of the operating speed of the rotor. The length of the shaft is to be made equal to at least 30 times its diameter.

5.66 Consider a one-DOF system excited by two supposedly identical motors. They are almost identical, but their frequencies are off by a small amount. This situation is quite common as the motor speeds can be different simply because of manufacturing or operational differences. In other words,

$$F_1 = F_{10} \sin \omega_1 t, \quad F_2 = F_{20} \sin \omega_2 t \tag{5.177}$$

Assume $F_{10} = F_{20}$, $\omega_1 = \omega$, $\omega_2 = 0.95\omega$, and let $\zeta = 0.2$, $k = 20$ kN/m, $m = 1.5$ kg, and $\omega = 200\,\text{rad/s}$.

(a) Determine the forced response of the system, and plot $x(t)$. Make sure you choose a long enough time interval to catch all the action. Note: this is very important; otherwise, you will miss the point.

(b) What you observe here is called a *beating phenomenon*, where you predominantly see two frequencies corresponding to the average of the two excitations as well as a difference between them. The slower frequency wave *modulates* the high-frequency response. Now, obtain an analytical approximation to prove this point (*Hint*: use elementary trigonometry). You can also demonstrate this with two tuning forks set at slightly different natural frequencies. In fact, tuning of a piano or a guitar involves essentially changing the frequency of the instrument until the beats are extinguished.

5.67 Determine the response of the SDOF oscillator to the periodic excitation shown in Fig. 5.72. Plot the response and determine the RMS value.

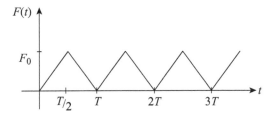

Figure 5.72: *Periodic force for Problem 5.67*

5.68 Determine the response of the SDOF oscillator to the periodic excitation shown in Fig. 5.73. Plot the response and determine the RMS value.

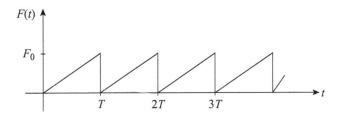

Figure 5.73: *Periodic force for Problem 5.68*

5.69 Determine the response of the SDOF oscillator to the periodic excitation shown in Fig. 5.74. Plot the response and determine the RMS value.

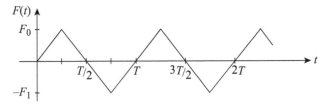

Figure 5.74: *Periodic force for Problem 5.69*

5.70 Determine the response of the SDOF oscillator to the periodic excitation shown in Fig. 5.75. Plot the response and determine the RMS value.

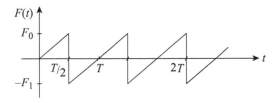

Figure 5.75: *Periodic force for Problem 5.70*

5.71 Determine the response of the SDOF oscillator to the half-sine periodic excitation shown in Fig. 5.76. Plot the response and determine the RMS value.

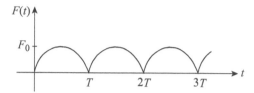

Figure 5.76: *Periodic force for Problem 5.72*

5.72 Determine the response of the SDOF oscillator to the half-cosine periodic excitation shown in Fig. 5.77. Plot the response and determine the RMS value.

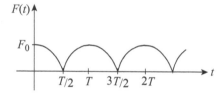

Figure 5.77: *Periodic force for Problem 5.72*

5.73 Determine the response of the SDOF oscillator to the periodic excitation shown in Fig. 5.78. Plot the response and determine the RMS value.

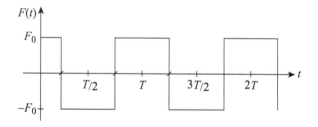

Figure 5.78: *Periodic force for Problem 5.73*

5.74 Determine the response of the SDOF oscillator to the periodic excitation shown in Fig. 5.79. Plot the response and determine the RMS value.

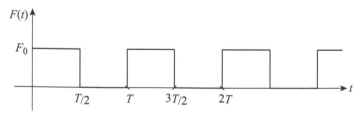

Figure 5.79: *Periodic force for Problem 5.74*

5.75 Determine the response of the SDOF oscillator to the periodic excitation shown in Fig. 5.80. Plot the response and determine the RMS value.

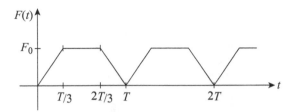

Figure 5.80: *Periodic force for Problem 5.76*

5.76 Consider the slider-crank mechanism in the reciprocating engine that we analyzed earlier in Section 5.11. There, we ignored all terms of order higher than r/ℓ. Now, derive the expression for the forces correct to the order of $\left(\frac{r}{\ell}\right)^2$.

5.77 Consider a single cylinder engine with an effective mass of 180 kg. The mass of the piston is 3.8 kg, and that of the connecting rod is 1.1 kg. The effective stiffness of the support is 2 MN/m. Assume a linear viscous damping ratio of 0.15. The crank radius is 16 cm, and the connecting rod length is 75 cm.

(a) Find the vibrational response of the engine at the nominal running speed of 600 rpm.

(b) Suppose we are interested in the entire range of engine speeds: $0 < \omega < 1000$ rpm. Find the amplitudes at ω and 2ω, and plot them versus the engine speed. Also, plot the RMS value.

5.78 Consider the same engine as in Problem 5.77. But this time, assume the more realistic material damping with $\eta = 0.25$.

(a) Find the vibrational response of the engine at 600 rpm.

(b) Suppose we are interested in the entire range of engine speeds: $0 < \omega < 1000$ rpm. Find the amplitudes at ω and 2ω, and plot them versus the engine speed. Also plot the RMS value.

5.79 Consider the problem of an automobile on a rough road discussed in this chapter on page 148. We solved the problem for a single harmonic excitation from the road. The truth however is that the road roughness often consists of not a single frequency, but a range of frequencies. Assume that the road roughness has the spectrum shown in Fig. 5.81. Assume the following parametric values: $m = 1500$ kg, $k = 100$ kN/m, and $\zeta = 0.35$.

(a) Determine the cumulative dynamic response for a speed of 100 kph. Plot the response versus time. What is the RMS value?

(b) Carry out the analysis for a range of speeds, and plot the RMS value versus speed.

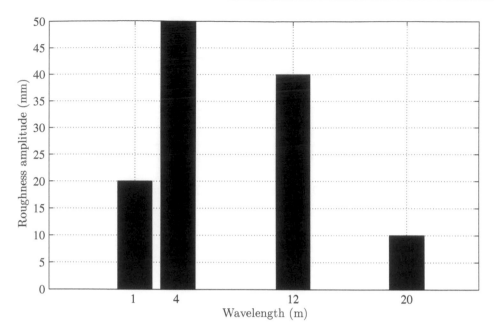

Figure 5.81: *Spectrum of road roughness*

Design Problems

5.1 Design of an Automobile Suspension System

In this problem you will design a suspension system for a typical automobile from a consideration of its ride quality. This is a continuing design problem that we will revisit when we learn about multiple-degree-of-freedom systems in a latter chapter.

The principal objective is to keep the passenger's sensation of discomfort below a certain level. The only source of vibration for the vehicle will be assumed to be road surface irregularities (thereby ignoring aerodynamic forces, engine vibrations, etc). There is no single right answer to this project, and there may be multiple design configurations that will satisfy the design objectives.

The following data is available at the outset.

Car body

Sprung mass of the car (m_s)	1200 kg
Unsprung mass of the car (m_{us})	100 kg
Distance between the axles (ℓ)	1.8 m
Distance of center of mass from the front axle (a)	0.7 m
Radius of gyration for pitching motion (r_g)	0.75 m

Tires

They are 165×13 radial-ply tires, assumed to be inflated to 30 psi. If inflated to 30 psi, they will each have a stiffness (k_t) of 100 kN/m, and a damping coefficient (c_t) of 3.43 kN s/m. At an inflation pressure of 15 psi, $k_t = 45$ kN/m, and $c_t = 4.45$ kN s/m. At intermediate pressures of inflation, the stiffness and damping values can be obtained by linear interpolation.

Roads

We will design and predict the performance of the car on two types of roads: (a) smooth highway and (b) a rough gravel surface. They will both have random roughness in general, but will be assumed here to be sine waves. The smooth highway has a wavelength of 1.9 m and a roughness amplitude of 1.65 mm. The gravel surface has a wavelength of 3.33 m and a roughness amplitude of 6 mm. The legal speed limits for the two roads are 100 kph and 40 kph, respectively. It will be assumed that we are designing cars to behave well for speeds below the legal speed limit only.

Suspension

The essential suspension design is left to you. However, there are some general guidelines as follows.

- Damping ratio for new shock absorbers will be 0.5 or so; for old absorbers, it will be 0.3 or so.
- The stiffness for each suspension spring for the given car is likely to be in the range of 5–100 kN/m (from other considerations).

Design specifications

The International Standard 2631 recommends the following comfort limits on vertical accelerations with the idea that a higher vibration is usually better tolerated if the trip is very short.

- Over a 1-hour period, a maximum of $0.1g$.
- Over a 8-hour period, a maximum of $0.02g$.

Note that g refers to acceleration due to gravity (9.81 in SI units).

In addition, it is of interest to minimize the pitching motion that might occur. Also, the maximum vibrational displacement amplitude (that may occur at any speed up to the speed limit) would have to be minimized. From other considerations, the static displacement of the suspension spring (under normal load) should not exceed 15 cm.

Model

Even simple analyses should ideally include seven degrees of freedom, as shown in Fig. 5.82. How-ever, we can gain a good understanding of the problem and get an acceptable design with simpler models. You will be analyzing the *same* problem based on progressively complex models.

In this chapter, we will use an SDOF model; this is the simplest model consisting of 1/4 the mass of the car (sprung and unsprung masses lumped together), on a suspension that consists of the tire and the spring in series. This is often called the quarter–model of the car (say, "Model 1"), and is a good starting point in this problem Fig. 5.83.

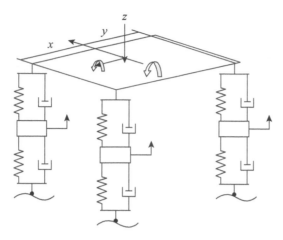

Figure 5.82: *Seven-DOF model for a car*

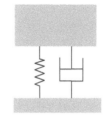

Figure 5.83: *Model 1*

Analysis and design procedure

Typically, you would start with an assumed value for the spring stiffness and damper, and analyze the problem. Then, check to see if the performance satisfies the design specifications listed earlier. Then, reanalyze the problem with a different value of stiffness and damping, and check to see if the vibration levels are lower or higher. An important parameter in your analysis is the speed of the car; in other words, all amplitudes and such need to be examined as a function of the speed of the car.

The kinds of analyses you will need to carry out are:

- determine the natural frequency;
- determine the forced response due to the road roughness.

The following points would be of interest (in addition to many more you can probably think of).

(a) Is there a particular value of stiffness for each suspension that would be very satisfactory from all considerations?

(b) How would the performance degrade with age? (What happens when the damping coefficient of the shock absorber goes down?)

(c) Are there certain specific speeds where the vibration is very high? Is one likely to travel on the roads at those specific speeds?

(d) Which inflation pressure is better for the tires from the (somewhat narrow) point of view of this project? Should the front and rear tires be inflated to different pressures?

(e) Does the model predict vibrational levels that exceed the recommended standards?

5.2 Design of a Test Rotor

Objective

In this problem you will be concerned with the design of a turbine test rotor. The problem is based on a recent design requirement of a turbomachinery company. The intended application of the rotor is in a test setup for studying blade vibrations in turbine wheels.

Design configuration

The rotor will consist of a rigid disk on a flexible shaft. Some of the parameters of the system are known; the essential design parameters that you will need to select are the shaft geometric properties such as diameter and length, and the bearing properties. Note that the shaft need not be uniform; i.e., its diameter could change along its axis. The rotor will be supported on two identical bearings that can be assumed to be simple supports. Figure 5.84 shows the overall configuration.

Figure 5.84: *Test rotor configuration*

The following numerical values are known.

m_d	mass of the disk	600 kg
ω	running speed	4000–5200 rpm
E	Youngs modulus	2.06×10^{11} Pa
G	Shear modulus	8.5×10^{10} Pa
ρ	material density	7850 kg/m^3
$m_0 e$	disk unbalance	15 kg mm

Design requirements

The following requirements are to be met.

- The running speeds should be greater than the first frequency.
- The unbalance response at the location of the disk at the running speed should not exceed 0.0254 mm (1 mil).
- The forces transmitted at the running speed should not exceed 150 N.
- The torsional natural frequency should be in the range of running speeds.

The design parameters should be within certain limits.

- Shaft diameter: $0.13 \text{ m} < d_s < 0.19 \text{ m}$.
- Shaft length: $1.8 \text{ m} < \ell < 2.2 \text{ m}$.
- Bearing stiffness: $0.2 \text{ MN/m} < k_b < 0.3 \text{ MN/m}$.
- Bearing damping coefficient: $1.5 \text{ kNs/m} < c_b < 2.5 \text{ kNs/m}$.

Analyses

The following analyses are required for each of the models.

- Natural frequency.
- Unbalance response of the system over a speed range from zero to twice the operating speed.
- Forces transmitted to the foundation through the bearings over a speed range from zero to twice the operating speed.
- Torsional natural frequency.

Procedure

Described below are two models for the *same* system. For each of them, select the design parameters to meet the design requirements and carry out the required analyses. We will look at more complicated models in Chapter 8.

Model 1 Ignore the mass of the shaft and model it as a flexible beam. Ignore the bearing stiffness (assume that they are rigid).

Model 2 Lump half the mass of the shaft with that of the disk. Note that this is still a SDOF model.

6 SDOF Systems: Nonharmonic Excitation

People who think they know everything are a great annoyance to those of us who do.

—Isaac Asimov

W hile it is true that a large percentage of vibration that occurs is due to persistent periodic—or harmonic—excitation, there are indeed a significant number of practical applications where the excitation is irregular. Examples abound in automotive, aerospace, and manufacturing applications. Also, testing of structures and systems can involve intentional excitation that is transient. In this chapter, these are the cases that we will study. We will learn two different techniques. One is the so-called convolution theorem, which is an important concept not only in vibrations, but also in the general dynamical systems theory as well as in completely different areas such as heat transfer.[1] The second approach is the Laplace transform, which is a very useful general technique of analyzing linear systems with nonharmonic (and also harmonic) excitations, especially if you follow up this chapter with a book in system dynamics or control theory.

6.1 Impulse Excitation

First, let us consider impulsive excitation of dynamic systems. Recall from elementary dynamics the definition of an *impulse*

$$\hat{F} = \int F \, dt \tag{6.1}$$

Note that the quantity called impulse is defined for *any* force, and would have units of Newton seconds (Ns) in SI units. Related to this is the concept of an impulsive force, which is a force that is typically large, and acts for a short time. Such forces often arise in collisions and impacts. In such cases, from an engineering point of view, the question arises: Considering that a collision between two bodies, for example, might last milliseconds, do we really care to know about the exact nature of the force in terms of its time dependence? Perhaps more important is the impulse, or in a sense,

---

[1]Of course, thermal people, belonging to quite a different species from us dynamics folks, call it often by a different name: Duhamel's integral.

Figure 6.1: *Pulses leading to an impulse function*

an integrated quantity of an unknown force, which can still help us determine the response of the system subsequent to the collision even if we cannot determine what exactly happens over the short time during the collision. This kind of logic motivates the definition of the mathematical impulse as follows.

An impulse function is best described by first considering a force in the form of a pulse, Fig. 6.1, of width b and height $1/b$ so that the impulse due to the force, or the area under the curve, is equal to 1. Now, suppose we shrink the width of the pulse by $1/2$ keeping the area constant (and still equal to 1). Then, we get the slimmer version shown in the second figure whose height is $2/b$. Then, we repeat to get the third figure, where the height is $4/b$ and the width is $b/4$. Suppose we continue this process until the width is very small. In the limit, this process leads to a force that is (theoretically) infinitely large, and lasts an infinitesimally small amount of time. This is what we call an *impulse*.

Of course, this concept of an impulse does not have to just apply to a force, but could be any quantity. The impulse is actually one of the functions belonging to a class of what are called *generalized functions*.

Let us define the impulse more formally. Also, to be general, let us define a *unit* impulse existing not at time zero, but at some specific time, τ, with the following set of statements.

$$\delta(t - \tau) = \begin{cases} 0 & \forall\, t \neq \tau \\ \text{undefined} & \text{for } t = \tau \end{cases} \qquad (6.2)$$

$$\int_0^\infty \delta(t - \tau)\, dt = 1 \quad 0 < \tau < \infty \qquad (6.3)$$

An impulse F of magnitude $\hat{F}_0$ would have an area of $\hat{F}_0$, and would hence be given by

$$F = \hat{F}_0 \delta(t - \tau) \qquad (6.4)$$

If we have to show it in a graph, we would do it as shown in Fig. 6.2. Note that if F has units of force (N), $\hat{F}_0$ would have units of Ns, and $\delta(t - \tau)$ has units of s^{-1}.

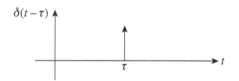

Figure 6.2: *An impulse occurring at $t = \tau$*

An important property of the impulse function can be deduced from a simple observation (a mathematically rigorous proof can be obtained as well, but we will stick to an intuitive understanding here). Suppose we have an arbitrary force $F(t)$, as shown in Fig. 6.8. Consider $F(t)\delta(t - \tau)$. Some thinking will tell you that it has the following properties.

$$F(t)\delta(t - \tau) = \begin{cases} 0 & \forall\, t \neq \tau \\ \text{undefined} & \text{for } t = \tau \end{cases} \qquad (6.5)$$

$$\int_0^\infty F(t)\delta(t - \tau)\, dt = F(\tau) \quad 0 < \tau < \infty \qquad (6.6)$$

Note that we now have a new way of defining the values of an *arbitrary* function of time in terms of the impulse function! That is, we can pick off the value of a function at a specific time (say τ) by using the impulse function on the left-hand side of Eq. (6.6). This property will come in handy shortly, so do not forget it.

A doublet function is often defined to specify a point moment. It is essentially an impulse followed by a negative impulse. Carry out some research to get its mathematical definition. What would be the integral of a doublet function?

6.1.1   Response to an Impulse

Consider a damped oscillator subject to an impulsive force, $\hat{F}_0\delta(t)$, whose impulse magnitude is $\hat{F}_0$ (Fig. 6.3). Recall that the impulse acts over a short amount of time (assumed here to be infinitesimal). There are then three regions of time to consider: the time before the impulse, during the impulse, and after the impulse. If the mass was not moving before the impulse (a common convenient assumption without loss of generality), we can ignore that time period. We can also ignore whatever motion happens during the action of the impulse because it is very small compared to the normal time span of the motion of the system. Hence, we can think of time zero as the time when the impulse has finished acting.

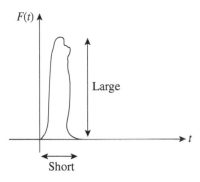

Figure 6.3: *An impulsive force*

With these assumptions it is quite easy to carry out the analysis. Clearly, from Newton's second law, we have impulse as being equal to change in momentum (if this surprises you, it is time to dig out your elementary dynamics book and read up the billiard ball problems that you used to love so much).

$$m\dot{x}(0) = \hat{F}_0 \quad \Rightarrow \dot{x}(0) = \frac{\hat{F}_0}{m} \tag{6.7}$$

Hence, we can think of the impulsive force as something that imparts an initial condition to the oscillator. After doing so, it vanishes; that is why, there will be no force on the right-hand side of the equation of motion. The other initial condition is

$$x(0) = 0 \tag{6.8}$$

OK; you understand why the initial velocity is what was given above. But why is the displacement initial condition zero? Doesn't the impulse have an effect on it? If that is correct (we can assure you that it *is* correct!), what does it tell us about the speed of response of dynamic systems? How would it change if the mass were very small?

Note how we have converted a forced response problem into a *free response* problem. Free unforced response is of course old hat to you by now, and the response can be determined to be

$$x(t) = \frac{\hat{F}_0}{m\omega_d}e^{-\zeta\omega_n t}\sin(\omega_d t + \phi) \tag{6.9}$$

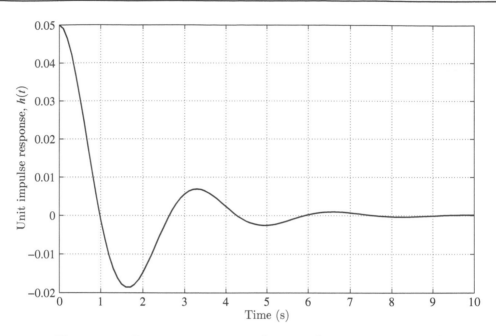

Figure 6.4: *Unit impulse response function (underdamped system)*

The response to a *unit* impulse is called the *impulse response function*, and is often denoted by $h(t)$ and is given by

$$h(t) = \frac{1}{m\omega_d} e^{-\zeta\omega_n t} \sin(\omega_d t + \phi) \tag{6.10}$$

where

$$\phi = \tan^{-1}\left[\frac{\sqrt{1-\zeta^2}}{\zeta}\right] \tag{6.11}$$

The response for an underdamped system ($\zeta = 0.3$) is shown in Fig. 6.4.

Exercise 6.1

Use appropriate initial conditions and derive the impulse response of an underdamped SDOF oscillator, and show that we get Eq. (6.9).

For an undamped system, the unit impulse response function simplifies to

$$h(t) = \frac{\hat{F}_0}{m\omega_n} \sin\omega_n t \tag{6.12}$$

which of course does not decay. The unit impulse response function is an important function that plays a critical role in experimental vibration.

Exercise 6.2

Use appropriate initial conditions and derive the impulse response of an overdamped SDOF oscillator.

There is another way to derive the unit impulse response. Consider the equation again.

$$m\ddot{x} + c\dot{x} + kx = \delta(t) \tag{6.13}$$

We integrate both sides of the equation over a time interval ϵ, and we let the time interval shrink to zero. Then,

$$\lim_{\epsilon \to 0} \int_0^\epsilon m\ddot{x}\, dt = \lim_{\epsilon \to 0} m\left[\dot{x}(\epsilon) - \dot{x}(0)\right] = m\dot{x}(0+) \tag{6.14}$$

$$\lim_{\epsilon \to 0} \int_0^\epsilon c\dot{x}\, dt = \lim_{\epsilon \to 0} c\left[x(\epsilon) - x(0)\right] = cx(0+) \tag{6.15}$$

$$\lim_{\epsilon \to 0} \int_0^\epsilon kx\, dt = \lim_{\epsilon \to 0} k\epsilon x(0) = 0 \tag{6.16}$$

Note that the last equation follows by mean value theorem. Following through all the steps, we will get the same initial condition, Eq. (6.7).

Example 6.1 The response of a diving platform

An Olympic swimmer, with a mass of 60 kg, jumps on a diving platform with a downward velocity of 6 m/s before he jumps off into the pool (Fig. 6.5). The platform is 3 m long, and has a rectangular cross section with a width (b) of 45 cm and a thickness (h) of 6 cm; it is made of an aluminum alloy with $E = 6.9 \times 10^{10}$ N/m^2, and density $\rho = 2800$ kg/m^3. Determine the subsequent response of the platform. (We know it might look cold hearted to focus on the platform rather than thinking of the swimmer who probably dived 20 m or more into a pool that must look very small from such a height! But, we, being coolly logical engineers, are interested in the robustness of the platform to keep it safe for this person and all such foolhardy people.)

We can model the platform as a cantilever beam with its mass being lumped at the end, as shown in Fig. 6.6. The stiffness is computed from the force-deflection characteristic of the cantilever beam with a force at the end

$$k = \frac{3EI}{\ell^3} = 6.210 \times 10^4 \text{ N/m} \tag{6.17}$$

where the area moment of inertia was computed from $bh^3/12$. The mass can be computed from the geometric properties.

$$m = \rho bh\ell = 227 \text{ kg} \tag{6.18}$$

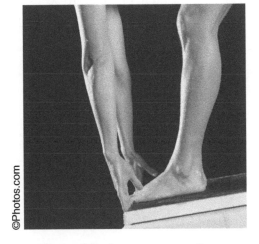

©Photos.com

Figure 6.5: *A venturesome diver*

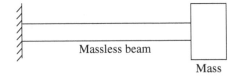

Massless beam

Mass

Figure 6.6: *Model of the diving platform*

Let us also assume that there is some typical structural damping with an equivalent damping ratio of 0.05 (we can always estimate this quite well from a simple log decrement testing on the platform). Hence, the undamped natural frequency of the beam is $\omega_n = 27.58$ rad/s, and the damped natural frequency is $\omega_d = 27.54$ rad/s.

The impulse delivered by the diver to the beam is simply the product of his mass and velocity, and is $\hat{F} = 360$ Ns. Hence, the response of the beam (more precisely, that of the end point of the beam) is given by

$$x(t) = \frac{\hat{F}}{m\omega_d} e^{-\zeta \omega_n t} \sin\left(\omega_d t + \phi\right) \tag{6.19}$$

where

$$\phi = \tan^{-1}\left[\frac{\sqrt{1 - \zeta^2}}{\zeta}\right] \tag{6.20}$$

Substituting numerical values, we get the response shown in Fig. 6.7, with a peak value of about 10 cm. Since it is a linear model, proportionally larger masses of the divers and higher velocities would increase the deflection proportionately.

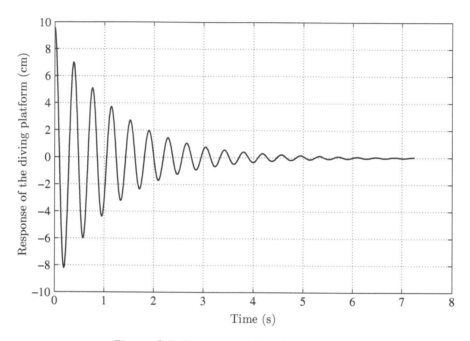

Figure 6.7: *Response of the diving platform*

6.2 Arbitrary Excitation

Suppose we have an arbitrary force (Fig. 6.8) acting on the system. What we mean by this is that the force's dependence on time cannot be specified in terms of well-known functions of time; note that this does not preclude the possibility that the force could have a simple functional description. The equation to be solved is then

$$m\ddot{x} + c\dot{x} + kx = F(t) \tag{6.21}$$

We could of course solve this numerically to determine $x(t)$, using the technique discussed in Chapter 2. However, it turns out that we can derive a more satisfying solution by using the concept developed in this chapter.

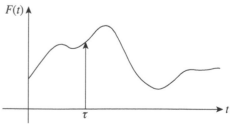

Figure 6.8: *An arbitrary force*

6.2.1   Convolution Theorem

For the force shown in Fig. 6.9, considering a length of time $\Delta\tau$ at time τ, we have an infinitesimal impulse of magnitude

$$\Delta\hat{F} = F(\tau)\Delta\tau \qquad (6.22)$$

This impulse, acting on the oscillator at time τ generates a response that is given by $\Delta\hat{F}h(t - \tau)$, where $h(t - \tau)$ is the unit impulse response derived earlier, and is also shown in Fig. 6.9. So, now we can think about the effect of the force on the system as due to a simple sum of all of these individual impulses[2]:

$$x(t) \approx \sum \Delta\hat{F}h(t - \tau) \qquad (6.23)$$

In the limit as as $\Delta\tau \to 0$, we shrink the time interval, and we get

$$x(t) = \int_0^t F(\tau)h(t - \tau)\,d\tau \qquad (6.24)$$

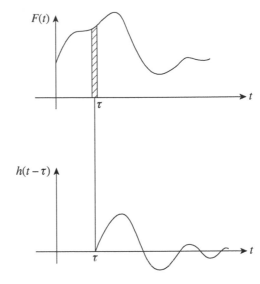

Figure 6.9: *Impulse response with an arbitrary force*

The above integral is called a convolution integral, and the result is called the convolution theorem. It is also called *Duhamel's integral.*

Come to think of it, it is quite remarkable that we have been able to write an explicit solution to the differential equation with an *arbitrary* right-hand side. On the other hand, we still need to evaluate an integral that will likely not yield to a nice closed-form expression. Nevertheless, as will be shown later, this is still a very useful result both from a conceptual and a practical viewpoint.

The convolution theorem is a very important result used in many fields. For example, you will see this in your controls and heat transfer classes as well. If it is the same theorem, it stands to reason that there must be a common characteristic between the very different systems. How and why is the temperature distribution similar to the vibrational response? And, different? You may need to do some digging to come up with an answer!

Exercise 6.3

Show that the convolution integral can be transformed to the following alternate form.

$$x(t) = \int_0^t F(t - \tau)h(\tau)\,d\tau \qquad (6.25)$$

---

[2]This is of course possible only because of the linearity of the system, and the principle of superposition.

In other words, the convolution is symmetric in $F(t)$ and $h(t)$. Oftentimes, this form of the convolution integral might be algebraically easier to apply.

6.2.2 Some More Test Functions

Often, we need to determine the response of vibrating systems to other kinds of standard generalized functions either because they approximate the real behavior of forces, or because we need a standard for specifying and comparing systems. We discuss some of them here.

Step function

A unit step function shown in Fig. 6.10 is defined by the following.

$$\mathcal{U}(t) = \begin{cases} 1 & t \geq 0 \\ 0 & t < 0 \end{cases} \tag{6.26}$$

In general, a unit step function that begins at time t_0 is denoted by $\mathcal{U}(t - t_0)$, where

$$\mathcal{U}(t - t_0) = \begin{cases} 1 & t \geq t_0 \\ 0 & t < t_0 \end{cases} \tag{6.27}$$

A step function of magnitude K is simply given by $f(t - t_0) = K\mathcal{U}(t - t_0)$.

Note that a step function is not the same as a constant function. A constant function by contrast would have been the same value for all time, whereas the step function comes into being at a certain time. It is in fact this *transient* behavior that we are interested in.

Response of a single-degree-of-freedom system to a step force

Let us derive the response of a single-degree-of-freedom (SDOF) damped oscillator to a step force for magnitude F_0.

$$m\ddot{x} + c\dot{x} + kx = F(t)$$
$$= F_0 \mathcal{U}(t) \tag{6.28}$$

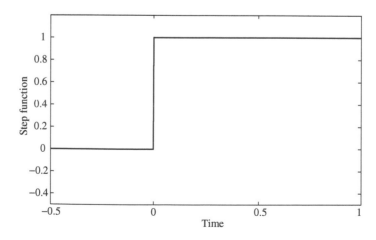

Figure 6.10: *Unit step function*

The technique we are going to use is the one we just derived: the convolution theorem.

$$x(t) = \int_0^t F(\tau) h(t - \tau) \, d\tau \tag{6.29}$$

Recall that the impulse response function is given by

$$h(t) = \frac{1}{m\omega_d} e^{-\zeta\omega_n t} \sin(\omega_d t + \phi) \tag{6.30}$$

where

$$\phi = \tan^{-1} \left[\frac{\sqrt{1 - \zeta^2}}{\zeta} \right] \tag{6.31}$$

So, substituting for the force and the impulse response function $h(t)$,

$$x(t) = \frac{F_0}{m\omega_d} \int_0^t e^{-\zeta\omega_n \tau} \sin(\omega_d \tau + \phi) \, d\tau \tag{6.32}$$

Integration of this is a little tricky, but you have seen many such tricks before in your calculus course. Here, we simply resort to integration tables (integration by parts always works as well; also, refer to Appendix A), and find that it is necessary to separate out three cases: $\zeta < 1$, $\zeta = 1$, and $\zeta > 1$. Hopefully, that will not come as a surprise to you; we surely expect underdamped and overdamped systems to act quite differently; this should be true whether we are talking about free response, impulse response, or forced response to any kind of force.

For $\zeta < 1$, the response turns out to be

$$x(t) = \frac{F_0}{k} \left\{ 1 - \frac{1}{\sqrt{1 - \zeta^2}} e^{-\zeta\omega_n t} \sin(\omega_d t + \phi) \right\} \tag{6.33}$$

For $\zeta = 1$, it is

$$x(t) = \frac{F_0}{k} \left[1 - (1 + \omega_n t) e^{-\omega_n t} \right] \tag{6.34}$$

and for $\zeta > 1$, the response is

$$x(t) = \frac{F_0}{k} \left\{ 1 + \frac{1}{s_1 - s_2} \left[s_2 e^{s_1 t} - s_1 e^{s_2 t} \right] \right\} \tag{6.35}$$

where $s_1 = \zeta\omega_n + \omega_n\sqrt{\zeta^2 - 1}$, and $s_2 = \zeta\omega_n - \omega_n\sqrt{\zeta^2 - 1}$.

Plots of the three responses are shown in Fig. 6.11; they are normalized by the equivalent static displacement F_0/k. Note that their steady state values are all the same, and are in fact given by F_0/k as they should be. After all the transients have died out, after all, the system is in static equilibrium under the action of a static force F_0.

Exercise 6.4

Carry out the integrals for the underdamped, critically damped, and overdamped cases, and derive Eqs. (6.33), (6.34), and (6.35).

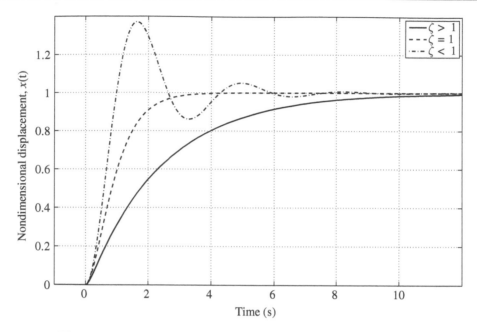

Figure 6.11: *Step response of an SDOF system to a step force*

--- **Example 6.2 Response to a pulse** ---

Consider a mass-spring-damper system subject to a force in the form of a pulse, as shown in Fig. 6.12. We have the equation of motion

$$m\ddot{x} + c\dot{x} + kx = F(t) \tag{6.36}$$

where

$$F(t) = \begin{cases} F_0, & 0 < t \leq t_1 \\ 0, & t > t1 \end{cases} \tag{6.37}$$

In this example, $t_1 = 3$ s and $F_0 = 10$ N. The pulse can be conveniently written as a combination of two-step functions, as shown in Fig. 6.13. Or,

$$F(t) = F_1(t) + F_2(t) \tag{6.38}$$

where

$$F_1(t) = F_0\,\mathcal{U}(t), \quad F_2(t) = -F_0\,\mathcal{U}(t - t_1) \tag{6.39}$$

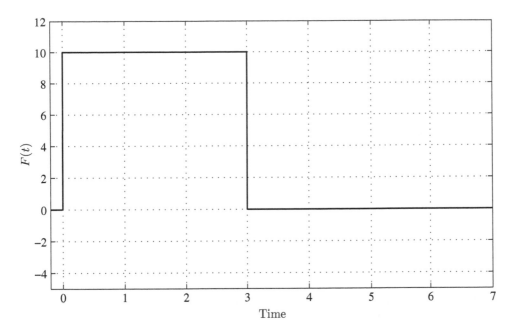

Figure 6.12: *Force in the form of a pulse*

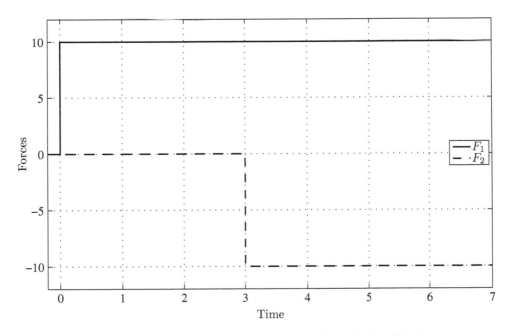

Figure 6.13: *Combination of two-step functions displaced in time*

Hence, the response of the system can now be computed by summing the individual responses to the two "forces."[3]

$$m\ddot{x} + c\dot{x} + kx = F_1(t) + F_2(t) \tag{6.40}$$

So, assuming $\zeta < 1$ for convenience, we compute the response to $F_1(t)$, say $x_1(t)$ with

$$x_1(t) = \frac{F_0}{k}\left\{1 - \frac{1}{\sqrt{1-\zeta^2}}e^{-\zeta\omega_n t}\sin\left(\omega_d t + \phi\right)\right\} \tag{6.41}$$

and $x_2(t)$, the response to $F_2(t)$ from

$$x_2(t) = -\frac{F_0}{k}\left\{1 - \frac{1}{\sqrt{1-\zeta^2}}e^{-\zeta\omega_n(t-t_1)}\sin\left(\omega_d(t - t_1) + \phi\right)\right\} \tag{6.42}$$

where

$$\phi = \tan^{-1}\left[\frac{\sqrt{1-\zeta^2}}{\zeta}\right] \tag{6.43}$$

The two responses are shown together in Fig. 6.14 for an underdamped system with $\zeta = 0.3$. Note how x_2 kicks in at t_1 (here, equal to 3 s). Putting them together, we get the total response, $x(t)$, as shown in Fig. 6.15.

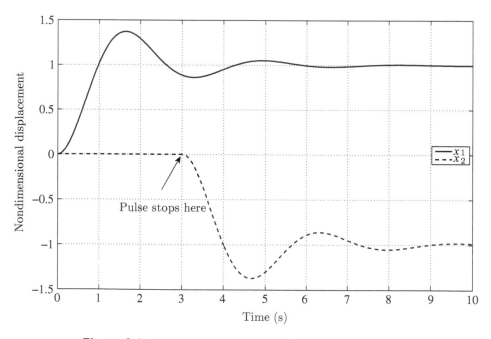

Figure 6.14: *Individual responses to the two-step functions*

---

[3]Whatever would we do without the principle of superposition?!

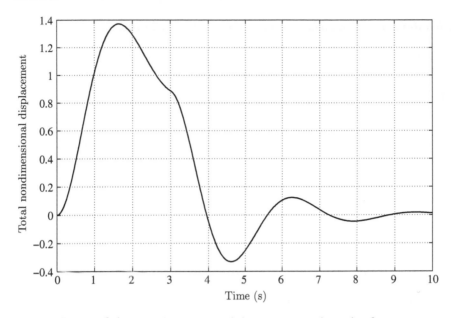

Figure 6.15: *Total response of the system to the pulse force*

Ramp function

A ramp of magnitude K is defined by the following function.

$$r(t - t_0) = \begin{cases} K(t - t_0) & t \geq t_0 \\ 0 & t < t_0 \end{cases} \tag{6.44}$$

A unit ramp would have $K = 1$. Note that the ramp is different from a linear function, since it comes into effect at a certain time, t_0, and is zero before that time (Fig. 6.16).

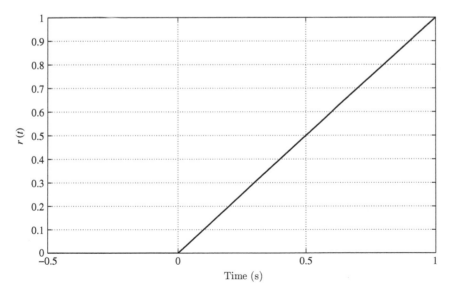

Figure 6.16: *Unit ramp function*

Example 6.3 Response of an undamped oscillator to a triangular function

Consider an undamped mass-spring system subjected to a force shown in Fig. 6.17. Essentially, the force is an ascending ramp followed by a descending ramp, and can be defined by the following.

$$F(t) = \begin{cases} 0, & t < 0 \\ 2F_0 \dfrac{t}{t_1}, & 0 \leq t < t_1/2 \\ 2F_0 \left(1 - \dfrac{t}{t_1}\right), & t_1/2 \leq t < t_1 \\ 0, & t \geq t_1 \end{cases} \tag{6.45}$$

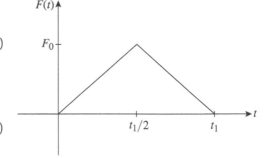

The equation of motion is (of course),

$$m\ddot{x} + kx = F(t) \tag{6.46}$$

We will solve by using the convolution theorem

$$x(t) = \int_0^t F(\tau)h(t - \tau)d\tau \tag{6.47}$$

Figure 6.17: *Triangular periodic force*

where $h(t)$, the unit impulse response function is given by

$$h(t) = \frac{1}{m\omega_n} \sin \omega_n t \tag{6.48}$$

Note that, since the force is discontinuous, we need to consider the four time intervals individually.

$$\boxed{0 \leq t < \frac{t_1}{2}}$$

Substituting the appropriate function into the convolution integral,

$$x(t) = \int_0^t \underbrace{2F_0 \frac{\tau}{t_1}}_{F(\tau)} \underbrace{\frac{1}{m\omega_n} \sin \left[\omega_n (t - \tau)\right]}_{h(t-\tau)} d\tau$$

$$= \frac{2F_0}{m\omega_n t_1} \int_0^t \tau \sin \left[\omega_n (t - \tau)\right] d\tau \tag{6.49}$$

Integrating by parts, and using $\omega_n = \sqrt{k/m}$,

$$x(t) = \frac{2F_0}{k} \left[\frac{t}{t_1} - \frac{1}{\omega_n} \sin \omega_n t\right] \tag{6.50}$$

$$\boxed{\frac{t_1}{2} \leq t < t_1}$$

Note that the convolution integral still starts from $t = 0$, although we are looking at the response in the time interval after $t_1/2$. In other words, the response in this time interval still depends on what happened in the previous time intervals (in theory, all the way back to the beginning of time!). This is somewhat subtle, but is a very important property of all dynamic systems.

$$x(t) = \int_0^t F(\tau)h(t - \tau)d\tau$$

$$= \int_0^{\frac{t_1}{2}} 2F_0 \frac{\tau}{t_1} \frac{1}{m\omega_n} \sin \left[\omega_n (t - \tau)\right] d\tau + \int_{\frac{t_1}{2}}^t 2F_0 \left(1 - \frac{\tau}{t_1}\right) \frac{1}{m\omega_n} \sin \left[\omega_n (t - \tau)\right] d\tau \tag{6.51}$$

Note the limits of integration, and be very careful about not mixing up t and τ. Solution should be easy—surely, this is a piece of cake for you having aced through your mathematics classes. (You may also want to refer to integration tables in Appendix A.)

$$x(t) = \frac{2F_0}{k} \left\{ 1 - \frac{t}{t_1} + \frac{1}{\omega_n t_1} \left[2 \sin\left(\omega_n(t - t_1/2)\right) - \sin \omega_n t \right] \right\} \tag{6.52}$$

$$\boxed{t > t_1}$$

Now, we know the force stops acting at time t_1; however, this does not mean that the oscillations stop. Again, the current displacement of a vibrating system depends on the *entire history* of the force. We just have to make sure that we use the correct limits on the integrals. (Do take a moment to compare the limits here with the previous interval.)

$$\begin{aligned}
x(t) &= \int_0^t F(\tau) h(t - \tau) d\tau \\
&= \int_0^{\frac{t_1}{2}} 2F_0 \frac{\tau}{t_1} \frac{1}{m\omega_n} \sin\left[\omega_n(t - \tau)\right] d\tau \\
&\quad + \int_{\frac{t_1}{2}}^{t_1} 2F_0 \left(1 - \frac{\tau}{t_1}\right) \frac{1}{m\omega_n} \sin\left[\omega_n(t - \tau)\right] d\tau \\
&\quad + \int_{t_1}^t (0) \frac{1}{m\omega_n} \sin\left[\omega_n(t - \tau)\right] d\tau \\
&= \frac{2F_0}{k} \left\{ \frac{1}{\omega_n t_1} \left[2\sin\left(\omega_n(t - t_1/2)\right) - \sin \omega_n(t - t_1) - \sin \omega_n t \right] \right\} \tag{6.53}
\end{aligned}$$

The response is shown in Fig. 6.18. Note that this is a never-decaying response due to the absence of damping in the model.

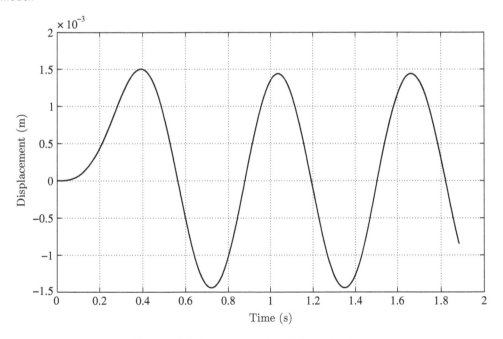

Figure 6.18: *Response to a triangular force*

Sine function

Although we have seen sine functions (ad nauseum, a bored student would groan!), we define it again noting in particular that this function comes into being from time zero, and can hence cause transients that we did not take care of earlier when we analyzed harmonic excitation. We define the sine function (or the harmonic function), by the following.

$$f(t - t_0) = \begin{cases} A \sin\left[\omega(t - t_0)\right] & t \geq t_0 \\ 0 & t < t_0 \end{cases} \tag{6.54}$$

where A is the amplitude of the function, and ω is the frequency. Figure 6.19 shows the function; again, note that it is zero before $t = 0$.

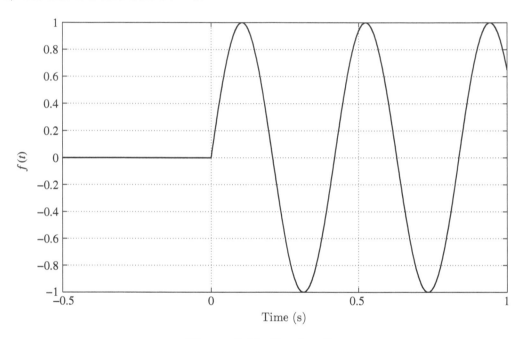

Figure 6.19: *Sine function*

--- **Example 6.4 Car on a bump** ---

Let us consider an automobile that goes over a speed bump. A good model for speed bump is a half-sine function, as shown in Fig. 6.20. It can be described as follows in terms of d, a coordinate to describe distance along the road.

$$u(d) = \begin{cases} u_0 \sin\left(\dfrac{\pi d}{b}\right), & 0 < d \leq b \\ 0 & d > b \end{cases} \tag{6.55}$$

where b is the width of the bump.

We assume that the car moves forward at a constant speed v; then, $d = vt$. Hence, the car, modeled as a particle mass, suffers a base motion $u(t)$ given by

$$u(t) = \begin{cases} u_0 \sin\left(\dfrac{\pi v}{b} t\right), & d \leq b \\ 0 & d > b \end{cases} \tag{6.56}$$

We note that v, the forward speed of the car defines the time period of the excitation (the car would be subject to the base motion longer at a slower speed, and vice versa).

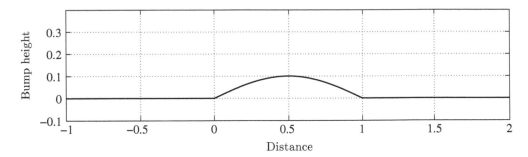

Figure 6.20: *A typical speed bump modeled as a half-sine with $b = 1$ m, $u_0 = 0.1$ m*

Now, we have the same old base excitation problem from Section 5.8.1 discussed on page 144 (Fig. 6.21)

$$m\ddot{x} + c\dot{x} + kx = c\dot{u} + ku \tag{6.57}$$

Hence, we will also need $\dot{u}(t)$

$$\dot{u}(t) = \begin{cases} \dfrac{\pi v}{b} u_0 \cos\left(\dfrac{\pi v}{b} t\right), & t \le \dfrac{b}{v} \\ 0 & t > \dfrac{b}{v} \end{cases} \tag{6.58}$$

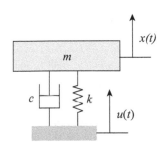

Figure 6.21: *Car with a base excitation*

Next, we use the convolution theorem

$$x(t) = \int_0^t F(\tau) h(t - \tau) d\tau \tag{6.59}$$

where $h(t)$, the unit impulse response function for the damped system is as follows.

$$h(t) = \frac{1}{m\omega_d} e^{-\zeta\omega_n t} \sin(\omega_d t + \phi) \tag{6.60}$$

where

$$\phi = \tan^{-1}\left[\frac{\sqrt{1 - \zeta^2}}{\zeta}\right] \tag{6.61}$$

Here,

$$F(t) = \begin{cases} k u_0 \sin \omega t + \omega c u_0 \cos \omega t & t \le t_1 \\ 0 & t > t_1 \end{cases} \tag{6.62}$$

where we have used

$$\omega = \frac{\pi v}{b}, \quad \text{and} \quad t_1 = \frac{b}{v}$$

for convenience. Again, we will need to consider the two time intervals separately; the first for when the car is on the bump, and the second, when the car has gone past the bump, as shown in Fig. 6.22.

$$\boxed{0 \le t \le b/v}$$

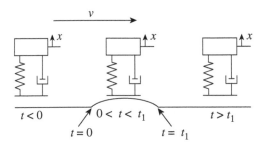

Figure 6.22: *Car going over the bump*

$$x(t) = \int_0^t F(t - \tau)h(\tau)d\tau$$

$$= \frac{1}{m\omega_d} k u_0 \int_0^t \sin[\omega(t - \tau)]\sin(\omega_d\tau + \phi)e^{-\zeta\omega_n\tau}d\tau$$

$$+ \frac{1}{m\omega_d} c\omega u_0 \int_0^t \cos[\omega(t - \tau)]\sin(\omega_d\tau + \phi)e^{-\zeta\omega_n\tau}d\tau \qquad (6.63)$$

$$\boxed{t > b/v}$$

$$x(t) = \int_0^t F(t - \tau)h(\tau)d\tau$$

$$= \frac{1}{m\omega_d} k u_0 \int_0^{t_1} \sin[\omega(t - \tau)]\sin(\omega_d\tau + \phi)e^{-\zeta\omega_n\tau}d\tau$$

$$+ \frac{1}{m\omega_d} c\omega u_0 \int_0^{t_1} \cos[\omega(t - \tau)]\sin(\omega_d\tau + \phi)e^{-\zeta\omega_n\tau}d\tau \qquad (6.64)$$

Note that the contribution from the base excitation is zero for $t > t_1$.

After extensive use of integration tables (some relevant ones are listed in Appendix A.1.2), we get some numerical results. Let us use the following parameter values to get a practical feel: $m = 2000$ kg, $k = 8 \times 10^5$ N/m, $\zeta = 0.5$, $b = 1$ m, $u_0 = 0.1$ m. Let us consider the response for two values of the forward speed. Case 1: 10 km/h (2.78 m/s) and Case 2: 25 km/h (6.9 m/s). We assume of course, that at all speeds, the car remains in contact with the bump! Figures 6.23 and 6.24 show the response for the two cases, respectively. Note the strong undershoot in both the cases. The numbers are not very realistic because the model is too simplistic; still, the response is reasonably correct in a qualitative sense. Note, how the response depends (in a very complex fashion) on the relationship between the excitation period and the natural frequency.

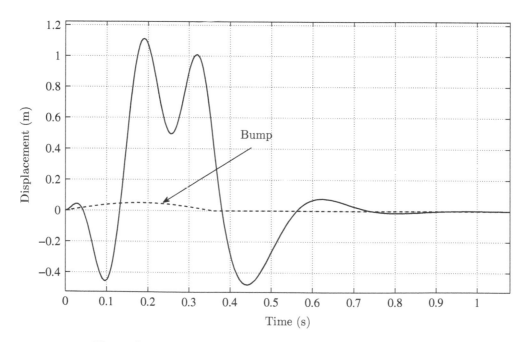

Figure 6.23: *Response of the car moving over a bump; Case 1*

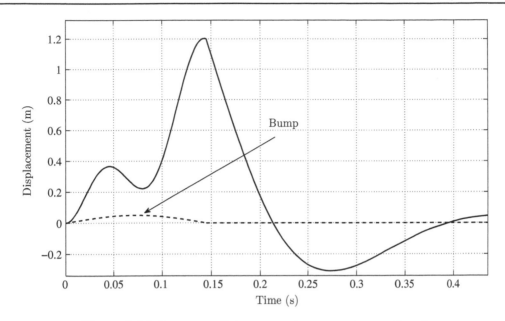

Figure 6.24: *Response of the car moving over a bump; Case 2*

6.2.3 Support Acceleration

Often equipment is required to be able to withstand a support or ground *acceleration*, which can be steady (as in the previous chapter on harmonic and periodic excitation), or transient. This problem is really not very different from anything we have done previously, but is often solved using relative displacement. With a base motion of $u(t)$, we have, for the absolute motion ($x(t)$) of the mass,

$$m\ddot{x} + c\dot{x} + kx = c\dot{u} + ku \tag{6.65}$$

Now, suppose, we are given information only on the support acceleration $u_a(t)$ (often obtained from practical measurement, as discussed later in Chapter 10). Then, we can rewrite the above equation in terms of the relative motion of the motion, $z(t) = x(t) - u(t)$ as

$$m\ddot{z} + m\ddot{u} + c\dot{z} + c\dot{u} + kz + ku = c\dot{u} + ku \tag{6.66}$$

Simplifying,

$$m\ddot{z} + c\dot{z} + kz = -m\ddot{u} = -mu_a \tag{6.67}$$

Now, given the acceleration, we can solve for $z(t)$ using convolution theorem. If we need to solve for the absolute displacement, we would need initial values of the ground displacement and velocity.

— **Example 6.5 Support acceleration response of an instrument** —

A delicate instrument to be mounted on the space shuttle is required to survive a support acceleration experienced during take off. The acceleration is given in the form of the parabolic function

$$u_a(t) = 600\left(1 - \frac{(t-60)^2}{60^2}\right) \text{ m/s}^2 \tag{6.68}$$

for the first 60 s and then flattens out at 600 m/s^2, as shown in Fig. 6.25. Determine the relative response of the instrument. The known parameters are as follows: $m = 5$ kg, $c = 20$ Ns/m, $k = 1$ kN/m.

$$z(t) = -\frac{1}{\omega_n}\int_0^t F(\tau)e^{-\zeta\omega_n(t-\tau)}\sin\omega_d(t-\tau)d\tau \tag{6.69}$$

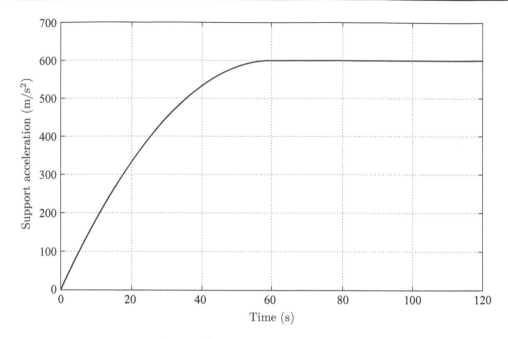

Figure 6.25: *Support acceleration*

where

$$F(\tau) = \begin{cases} -mu_{a0}\left[1 - \frac{(t-t_1)^2}{t_1^2}\right], & t \le t_1 \\ -mu_{a0}, & t > t_1 \end{cases}$$

(6.70)

where $t_1 = 60$ s and $u_{a0} = 600$ m/s^2.

$$\boxed{0 \le t < t_1}$$

$$
\begin{aligned}
z(t) &= \frac{u_{a0}}{\omega_n} \int_0^t \left(\frac{t^2}{t_1^2} - \frac{2t}{t_1}\right) e^{-\zeta\omega_n(t-\tau)} \left(\sin\omega_d t \cos\omega_d\tau - \sin\omega_d\tau \cos\omega_d t\right) d\tau \\
&= \frac{u_{a0}\,e^{-\zeta\omega_n t}}{\omega_n} \left[\sin\omega_d t(I_1 - I_2) - \cos\omega_d t(I_3 - I_4)\right]
\end{aligned}
$$

(6.71)

where we have given the following symbols for the various integrals.

$$I_1 = \int_0^t \frac{t^2 e^{\zeta\omega_n\tau} \cos\omega_d\tau}{t_1^2}\,d\tau$$

$$I_2 = \int_0^t \frac{2t e^{\zeta\omega_n\tau} \cos\omega_d\tau}{t_1}\,d\tau$$

$$I_3 = \int_0^t \frac{t^2 e^{\zeta\omega_n\tau} \sin\omega_d\tau}{t_1^2}\,d\tau$$

$$I_4 = \int_0^t \frac{2t e^{\zeta\omega_n\tau} \sin\omega_d\tau}{t_1}\,d\tau$$

These integrals can be computed from formulas provided in the Appendix.

$$\boxed{t > t_1}$$

Now, we have to consider the fact that the support acceleration changes after t_1 and break up the convolution integral into two parts.

$$z(t) = \frac{u_{a0}}{\omega_n} \int_0^{t_1} \left(\frac{t^2}{t_1^2} - \frac{2t}{t_1} \right) e^{-\zeta\omega_n(t-\tau)} \left(\sin\omega_d t \cos\omega_d \tau - \sin\omega_d \tau \cos\omega_d t \right) d\tau$$

$$+ \frac{u_{a0}}{\omega_n} \int_{t_1}^t e^{-\zeta\omega_n(t-\tau)} \left(\sin\omega_d t \cos\omega_d \tau - \sin\omega_d \tau \cos\omega_d t \right) d\tau \tag{6.72}$$

Again, we rewrite it in terms of manageable integrals.

$$z(t) = \frac{u_{a0} e^{-\zeta\omega_n t}}{\omega_n} \left[\sin\omega_d t(I_{1n} - I_{2n}) - \cos\omega_d t(I_{3n} - I_{4n}) - \sin\omega_d t(I_5) + \cos\omega_d t(I_6) \right] \tag{6.73}$$

where

$$I_{1n} = \int_0^{t_1} \frac{t^2 e^{\zeta\omega_n \tau} \cos\omega_d \tau}{t_1^2} d\tau$$

$$I_{2n} = \int_0^{t_1} \frac{2t e^{\zeta\omega_n \tau} \cos\omega_d \tau}{t_1} d\tau$$

$$I_{3n} = \int_0^{t_1} \frac{t^2 e^{\zeta\omega_n \tau} \sin\omega_d \tau}{t_1^2} d\tau$$

$$I_{4n} = \int_0^{t_1} \frac{2t e^{\zeta\omega_n \tau} \sin\omega_d \tau}{t_1} d\tau$$

$$I_5 = \int_{t_1}^t e^{\zeta\omega_n \tau} \cos\omega_d \tau \, d\tau$$

$$I_6 = \int_{t_1}^t e^{\zeta\omega_n \tau} \sin\omega_d \tau \, d\tau$$

Numerical calculations for the given system yield the response as shown in Fig. 6.26.

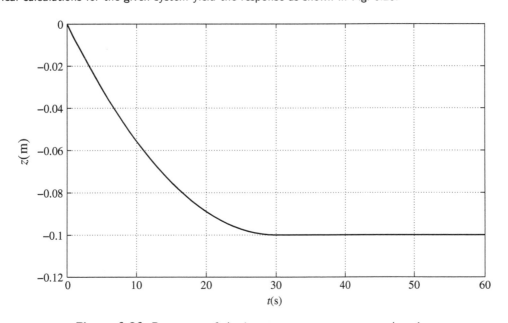

Figure 6.26: *Response of the instrument to support acceleration*

6.3   Laplace Transform Methods of Solution

So far we have been looking at direct solution to determine the response of an SDOF system to several kinds of excitation. The methods we have been using have been mostly familiar to you possibly from your mathematics courses. In this chapter, we will look at alternative methods of obtaining the same solutions, primarily the Laplace transformation method. In addition to adding a new tool to your toolbox giving you a choice of techniques when you are confronting a problem, this will also serve as an introduction to other related areas of study such as system dynamics and control systems. The methods we will study in this chapter are generally useful for any kind of excitation; however, they are particularly suited for transient response.

6.3.1   The Laplace Transform

The Laplace transform of a function of time is defined by the following integral.

$$\mathcal{L} = \int_0^\infty f(t)\, e^{-st}\, dt \tag{6.74}$$

where s, called the Laplace variable, is complex. It is assumed that $f(t)$ is zero for $t < 0$.

Some properties

Laplace transforms have some very useful properties. In the following, a, a_1, etc., are constants (not functions of time). Upper case denotes a Laplace transformed variable; for example,

$$F_1(s) \overset{\text{def}}{=} \mathcal{L}[f_1(t)] \tag{6.75}$$

- Scaling

$$\mathcal{L}[af(t)] = aF(s)$$

- Linearity

$$\mathcal{L}[a_1 f_1(t) + a_2 f_2(t)] = a_1 F_1(s) + a_2 F_2(s)$$

- Shift (in time)

$$\mathcal{L}[f(t-a)] = e^{-as} F(s)$$

- Shift (in Laplace variable)

$$\mathcal{L}[f(t)e^{-at}] = F(s+a)$$

- Initial value theorem

$$f(0+) = \lim_{s \to \infty} [sF(s)]$$

- Final value theorem

$$\lim_{t \to \infty} f(t) = \lim_{s \to 0} [sF(s)]$$

- Differentiation

$$\mathcal{L}\left[\frac{df}{dt}\right] = sF(s) - f(0)$$

$$\mathcal{L}\left[\frac{d^2f}{dt^2}\right] = s^2F(s) - \dot{f}(0) - sf(0)$$

- Integration

$$\mathcal{L}\left[\int_0^t f(\tau)d\tau\right] = \frac{F(s)}{s}$$

Table 6.1 lists some typical functions of time and their Laplace transforms.

$f(t)$	$F(s)$
$\delta(t)$(unit impulse)	1
1 (unit step)	$\dfrac{1}{s}$
t (unit ramp)	$\dfrac{1}{s^2}$
t^2 (parabolic)	$\dfrac{2}{s^3}$
t^n	$\dfrac{n!}{s^{n+1}}$
e^{-at}	$\dfrac{1}{s+a}$
$t^n e^{-at}$	$\dfrac{(n)!}{(s+a)^{n+1}}$
$\sin bt$	$\dfrac{b}{s^2+b^2}$
$\cos bt$	$\dfrac{s}{s^2+b^2}$
$e^{-at}\cos bt$	$\dfrac{s+a}{(s+a)^2+b^2}$
$e^{-at}\sin bt$	$\dfrac{b}{(s+a)^2+b^2}$
$e^{-at}f(t)$	$F(s+a)$
$\dfrac{df}{df}$	$sF(s) - f(0)$
$\dfrac{d^2f}{dt^2}$	$s^2F(s) - sf(0) - \left.\dfrac{df}{dt}\right|_{t=0}$
$\dfrac{d^k f(t)}{dt^k}$	$s^n F(s) - \sum_{k=0}^{n-1} s^{n-k-1}\left[\dfrac{d^k f(t)}{dt^k}\right]\Big|_{t=0}$
$\int_0^t f(t)\,dt$	$\dfrac{F(s)}{s} + \dfrac{1}{s}\left[\int f(t)\,dt\right]\big|_{t=0}$
$f(t) = f(t+T)$, (periodic)	$\dfrac{F(s)}{1-e^{-sT}}$, $f(t)$ written using step functions

Table 6.1: *Laplace transform pairs*

The inverse laplace transform

The inverse Laplace transformation is the mechanism to obtain the time function back from the Laplace domain. It is mathematically defined as follows.

$$f(t) = \mathcal{L}^{-1}\{F(s)\} = \frac{1}{2\pi i} \int_{c-j\infty}^{c+j\infty} F(s)\, e^{st}\, ds \tag{6.76}$$

The above integral is quite complicated and involves the theory of analytic functions. Instead of dealing with the integral, we will use the Laplace tables extensively. For example, we already know the following.

$$\mathcal{L}\{e^{-at}\} = \frac{1}{s+a} \tag{6.77}$$

Hence, it is clear that

$$\mathcal{L}^{-1}\left\{\frac{1}{s+a}\right\} = e^{-at} \tag{6.78}$$

The mass-spring-damper (again!)

Now, let us apply this principle to our SDOF model and see where that will take us. Recall that the equation of motion is

$$m\ddot{x}(t) + c\dot{x}(t) + kx(t) = F(t) \tag{6.79}$$

where we are emphasizing the dependence on time by writing $x(t)$ explicitly. As has been the case so far, we assume that m, c, and k are constant and do not change with time.

Next, we take the Laplace transform of the above equation; in other words, we multiply both sides by $e^{-st}\, dt$, and integrate from 0 to ∞. Or,

$$\mathcal{L}\left[m\ddot{x}(t) + c\dot{x}(t) + kx(t)\right] = \mathcal{L}\left[F(t)\right] \tag{6.80}$$

Note that we do not know (or care at this point in time) what kind of a function $F(t)$ is, or $x(t)$ is. We next use the properties listed earlier to process things further.

$$m\mathcal{L}\left[\ddot{x}(t)\right] + c\mathcal{L}\left[\dot{x}(t)\right] + k\mathcal{L}\left[x(t)\right] = \mathcal{L}\left[F(t)\right] \tag{6.81}$$

$$m\left[s^2 x(s) - \dot{x}(0) - sx(0)\right] + c\left[sx(s) - x(0)\right] + kx(s) = F(s) \tag{6.82}$$

This can be rewritten

$$\left[ms^2 + cs + k\right] x(s) = \left[ms + c\right] x(0) + \left[m\right] \dot{x}(0) + F(s) \tag{6.83}$$

Note that this is an algebraic equation in s. So, by using the Laplace transforms, we have effectively converted an ordinary differential equation into an algebraic equation. This is indeed a very common mathematical technique underlying the principle of many integral transforms such as Fourier, Hankel, etc.

Now, we are interested in the displacement; so let us solve for $x(s)$:

$$x(s) = \frac{ms + c}{ms^2 + cs + k}\, x(0) + \frac{m}{ms^2 + cs + k}\, \dot{x}(0) + \frac{1}{ms^2 + cs + k}\, F(s) \tag{6.84}$$

This looks like a complete solution, but has this strange variable s in it. After all, we live in the real world with time (t), and not s. In other words, we want an answer ultimately in the time domain, not the Laplace domain. So, at least conceptually at this point, let us write down the solution in terms of time.

$$x(t) = \underbrace{\mathcal{L}^{-1}\left\{\frac{ms+c}{ms^2+cs+k}\right\}x(0) + \mathcal{L}^{-1}\left\{\frac{m}{ms^2+cs+k}\right\}\dot{x}(0)}_{\text{Free vibration}} + \underbrace{\mathcal{L}^{-1}\left\{\frac{1}{ms^2+cs+k}F(s)\right\}}_{\text{Forced vibration}} \quad (6.85)$$

Clearly, as shown above, we have the mechanism now to get both the free and the forced responses, the true complete response being the sum of the two. As a side note, observe that the initial conditions are constant, and are hence outside the argument of the inverse Laplace transform, but that for the forced term, we need to be sure to include $F(s)$ inside the argument of the inverse Laplace integral.

Of course, we have the conceptual solution above, which is of limited practical utility. To get the solution that we can actually see and plot however, we have no choice but to carry out the gory details. This requires us to get the actual inverse Laplace transforms for which we have to get familiar with some simple algebraic concepts.

6.3.2 The Transfer Function

Before we get into the details let us define a few terms that are in common usage, which we define in the Laplace domain. The first is the concept of *transfer functions*. Clearly, there are two independent causes for the vibration to occur in this case: the initial conditions and the external excitation. So, if we think of the vibrating oscillator as a system with a block diagram representation, we can think of these causes as *inputs*, and the actual response of the system as the *output*. The two initial conditions and the force act as three inputs, as shown in Fig. 6.27. The system essentially takes each of these inputs, transforms them in some way [as shown by the individual factors in Eq. (6.84)], adds the individual responses, and finally puts it out as the vibrational displacement $x(s)$. This is shown symbolically in Fig. 6.28. Each of these factors that produces the output based on the individual input is called a transfer function. In practice, especially in control systems, the transfer function is only defined for the force input and not for the initial conditions.

Figure 6.27: *The SDOF oscillator with a block diagram representation*

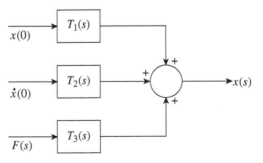

Figure 6.28: *The SDOF oscillator with inputs and output*

To summarize, the definition of the transfer function $T(s)$ is then given by the following.

$$T(s) = \frac{\mathcal{L}\{\text{Output}\}}{\mathcal{L}\{\text{Input}\}} \tag{6.86}$$

with the initial conditions set to zero (this just emphasizes that we are only looking at the third term in Eq. (6.84)). For the SDOF oscillator, this will be as follows.

$$T(s) = \frac{x(s)}{F(s)} \tag{6.87}$$

$$= \frac{1}{ms^2 + cs + k} \tag{6.88}$$

The next concept comes from analytical function theory and concerns the *poles* of the transfer function. Poles are the values of s when the denominator of the transfer function goes to zero. In this case, this would be when

$$ms^2 + cs + k = 0 \tag{6.89}$$

Does this look familiar? It should, because we (both you and I!) spent a lot of time on it in Chapter 4. In fact, it is exactly the same equation as Eq. (4.10), which is the *characteristic equation* of the mass-spring-damper system; hence, the poles are the same as the characteristic roots. So, even though we are talking about integral transforms and such high falutin mathematics, it comes back to the same old thing that you know so well.[4]

6.3.3 Getting the Inverse Laplace Transforms

As mentioned above, the normal procedure is to look to see if the table of Laplace transforms already lists the function that we have, so that we can write the inverse Laplace transform by inspection. In situations where we cannot easily look them up, we will need to use partial fractions before we can use the tables. The procedure depends upon whether the function has simple poles or multiple poles. Simple poles are when the poles are all distinct, and multiple poles are when the poles are repeated.

Simple poles

Given a general function, $G(s)$, that is in the form of a ratio of polynomials, we would follow the procedure given below to separate it into partial fractions.

$$G(s) = \frac{Q(s)}{P(s)} = \frac{Q(s)}{(s - s_1)(s - s_2)\ldots(s - s_n)}$$

$$= \frac{K_1}{s - s_1} + \cdots + \frac{K_n}{s - s_n}$$

where

$$K_i = \left[(s - s_i)\frac{Q(s)}{P(s)}\right]\Bigg|_{(s=s_i)}$$

---

[4]This is often termed—putting old wine in a new bottle!

Multiple poles

If the poles repeat, it is a little more complex algebraically, but is still a straightforward procedure.

$$G(s) = \frac{Q(s)}{P(s)} = \frac{Q(s)}{(s - s_1)(s - s_2)\ldots(s - s_{n-r})(s - s_i)^r}$$

$$= \underbrace{\frac{K_1}{s - s_1} + \cdots + \frac{K_{n-r}}{s - s_{n-r}}}_{\text{simple poles}} + \underbrace{\frac{A_1}{s - s_i} + \cdots + \frac{A_r}{(s - s_i)^r}}_{\text{repeated poles}}$$

where

$$A_r = \left[(s - s_i)^r \frac{Q(s)}{P(s)}\right]\Bigg|_{(s=s_i)}$$

$$A_{r-1} = \frac{d}{ds}\left[(s - s_i)^r \frac{Q(s)}{P(s)}\right]\Bigg|_{(s=s_i)}$$

$$A_{r-2} = \frac{1}{2!}\frac{d^2}{ds^2}\left[(s - s_i)^r \frac{Q(s)}{P(s)}\right]\Bigg|_{(s=s_i)}$$

$$\vdots$$

$$A_1 = \frac{1}{(r-1)!}\frac{d^{r-1}}{ds^{r-1}}\left[(s - s_i)^r \frac{Q(s)}{P(s)}\right]\Bigg|_{(s=s_i)}$$

6.3.4   Free Vibrational Response

Let us now look at the free response of an SDOF system as we have done in the past, but this time using the new tools we have learned. And we hope to get the same answers!

Consider the equation of motion again, but in a standard form this time.

$$\ddot{x} + 2\zeta\omega_n\dot{x} + \omega_n^2 x = 0 \tag{6.90}$$

with initial conditions, $x(0) = x_0$, and $\dot{x}(0) = v_0$. Taking Laplace transforms,

$$\left[s^2 x(s) - sx(0) - \dot{x}(0)\right] + 2\zeta\omega_n\left[sx(s) - x(0)\right] + \omega_n^2 x(s) = 0 \tag{6.91}$$

Solving for $x(s)$,

$$x(s) = \frac{s + 2\zeta\omega_n}{s^2 + 2\zeta\omega_n s + \omega_n^2} x_0 + \frac{1}{s^2 + 2\zeta\omega_n s + \omega_n^2} v_0 \tag{6.92}$$

It is clear that the response due to an initial displacement and an initial velocity can be simply superposed. Hence, for simplicity, we let $v_0 = 0$ for now. Then,

$$x(t) = \mathcal{L}^{-1}\left[\frac{s + 2\zeta\omega_n}{s^2 + 2\zeta\omega_n s + \omega_n^2}\right] x_0 \tag{6.93}$$

As we discovered earlier, the characteristic equation is the denominator of the transfer function

$$s^2 + 2\zeta\omega_n s + \omega_n^2 = 0 \tag{6.94}$$

Solving,

$$s_{1,2} = \omega_n \left(-\zeta \pm \sqrt{\zeta^2 - 1} \right) \tag{6.95}$$

The above solution depends on the nature of the characteristic roots, since they could be real or complex depending on whether $\zeta \geq 1$ or $\zeta < 1$. Hence, we need to look at these cases individually.

Case 1: $\zeta > 1$ (overdamped)

In this case, the roots are real, say s_1 and s_2. Then,

$$s^2 + 2\zeta\omega_n s + \omega_n^2 = (s - s_1)(s - s_2) \tag{6.96}$$

where

$$s_1 = \omega_n \left(-\zeta + \sqrt{\zeta^2 - 1} \right) \tag{6.97}$$

$$s_2 = \omega_n \left(-\zeta - \sqrt{\zeta^2 - 1} \right) \tag{6.98}$$

Hence, the transfer function can be written as

$$
\begin{aligned}
T(s) &= \left[\frac{s + 2\zeta\omega_n}{s^2 + 2\zeta\omega_n s + \omega_n^2} \right] \\
&= \frac{s_1 + 2\zeta\omega_n}{(s_1 - s_2)(s - s_1)} + \frac{s_2 + 2\zeta\omega_n}{(s_2 - s_1)(s - s_2)} \\
&= \frac{1}{2\omega_n\sqrt{\zeta^2 - 1}} \left[\frac{s_1 + 2\zeta\omega_n}{(s - s_1)} - \frac{s_2 + 2\zeta\omega_n}{(s - s_2)} \right]
\end{aligned}
\tag{6.99}
$$

where we have used partial fractions to separate the two terms, and $s_1 - s_2 = 2\omega_n\sqrt{\zeta^2 - 1}$. Substituting into Eq. (6.93), and carrying out the inverse Laplace transform

$$x(t) = \frac{1}{2\omega_n\sqrt{\zeta^2 - 1}} \left[(s_1 + 2\zeta\omega_n)e^{s_1 t} - (s_2 + 2\zeta\omega_n)e^{s_2 t} \right] x_0 \tag{6.100}$$

At this point we can define what are called *time constants* for the system using

$$\tau_1 = -\frac{1}{s_1}, \quad \tau_2 = -\frac{1}{s_2} \tag{6.101}$$

We will see shortly that the time constants are convenient ways of depicting the speed of response. Then, the solution is of the form

$$x(t) = \left[C_1 e^{-t/\tau_1} + C_2 e^{-t/\tau_2} \right] x_0 \tag{6.102}$$

where C_1 and C_2 are constants that depend on the system parameters (ζ and ω_n) and are given below.

$$C_1 = \frac{\zeta + \sqrt{\zeta^2 - 1}}{2} \tag{6.103}$$

$$C_2 = \frac{\zeta - \sqrt{\zeta^2 - 1}}{2} \tag{6.104}$$

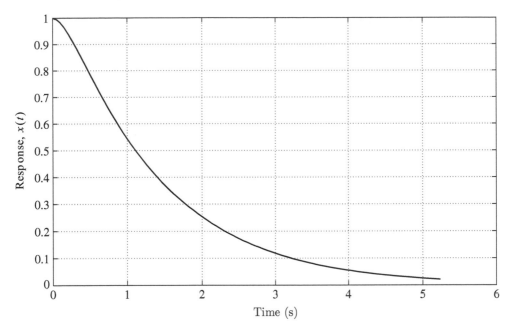

Figure 6.29: *Free response of the overdamped system*

Clearly, the response is a sum of two exponential decays at two different rates corresponding to the time constants and looks as shown in Fig. 6.29. The larger the time constant, the slower the response of that component in the solution (take out that calculator and verify this now with some numbers!).

Case 2: $\zeta < 1$ (underdamped)

In this case, the roots are complex

$$s_{1,2} = -\zeta \omega_n \pm i \omega_n \sqrt{1 - \zeta^2} \tag{6.105}$$

If we define a new parameter

$$\omega_d = \omega_n \sqrt{1 - \zeta^2} \tag{6.106}$$

the roots can be written as

$$s_{1,2} = -\zeta \omega_n \pm i \omega_d \tag{6.107}$$

In this case, the denominator cannot be broken up into two real factors. Hence, we rewrite the transfer function as follows.

$$T(s) = \left[\frac{(s + \omega_n)}{(s + \zeta \omega_n)^2 + \omega_n^2 (1 - \zeta^2)} \right]$$
$$+ \frac{\zeta \omega_n}{\omega_n \sqrt{1 - \zeta^2}} \left[\frac{\omega_n \sqrt{1 - \zeta^2}}{(s + \zeta \omega_n)^2 + \omega_n^2 (1 - \zeta^2)} \right] \tag{6.108}$$

From the Laplace tables, we have

$$\mathcal{L}^{-1}\left[\frac{s+a}{(s+a)^2+b^2}\right] = e^{-at}\cos bt \tag{6.109}$$

$$\mathcal{L}^{-1}\left[\frac{b}{(s+a)^2+b^2}\right] = e^{-at}\sin bt \tag{6.110}$$

Hence,

$$x(t) = e^{-\zeta\omega_n t}\left[\cos\omega_d t + \frac{\zeta}{\sqrt{1-\zeta^2}}\sin\omega_d t\right]x_0 \tag{6.111}$$

Using trigonometry and simplifying

$$x(t) = \frac{1}{\sqrt{1-\zeta^2}}x_0 e^{-\zeta\omega_n t}\sin\left(\omega_d t + \phi\right) \tag{6.112}$$

where

$$\phi = \tan^{-1}\left[\frac{\sqrt{1-\zeta^2}}{\zeta}\right] = \cos^{-1}\left(\zeta\right) \tag{6.113}$$

Sketching the response (Fig. 6.30), we find that the system shows an exponentially damped oscillation whose frequency is ω_d, which is the *damped natural frequency* of the system. The rate

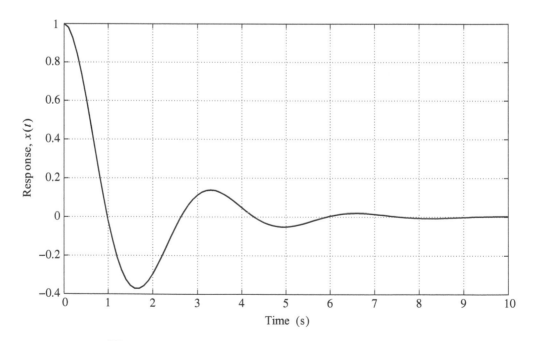

Figure 6.30: *Free response of the underdamped system*

of decay is governed by the exponent $\zeta\omega_n$; we can hence define a time constant (only one this time) as

$$\tau = \frac{1}{\zeta\omega_n} \tag{6.114}$$

Case 3: $\zeta = 1$ (critically damped)

In this case, the roots are real and equal; this is the case of multiple poles.

$$s_{1,2} = -\omega_n \tag{6.115}$$

The transfer function becomes

$$T(s) = \frac{s + 2\omega_n}{(s + \omega_n)^2}$$

$$= \frac{1}{(s + \omega_n)} + \frac{\omega_n}{(s + \omega_n)^2} \tag{6.116}$$

Using the Laplace tables, we have

$$x(t) = x_0 \left[1 + \omega_n t\right] e^{-\omega_n t} \tag{6.117}$$

A sketch of the response is shown in Fig. 6.31 and shows an exponential decay. There is only one time constant for the critically damped system given by

$$\tau = \frac{1}{\omega_n} \tag{6.118}$$

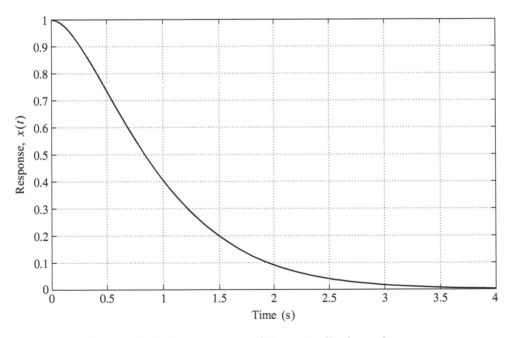

Figure 6.31: *Free response of the critically damped system*

It is interesting to compare the three responses we have discussed so far in one plot (Fig. 6.32). Note how the critically damped system settles the earliest. Not having enough damping keeps the system going on and on; adding too much damping deadens the system in a sense and makes it too sluggish; and, a critically damped system has just the "right" amount of damping if we are interested in minimizing the settling time.

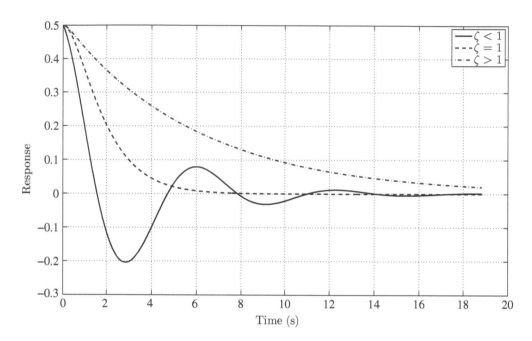

Figure 6.32: *Free response with various damping ratios*

Case 4: $\zeta = 0$ (undamped)

Finally, we look at the undamped system; in this case, the roots are purely imaginary.

$$s_1 = +i\omega_n, \quad s_2 = -i\omega_n \tag{6.119}$$

The transfer function becomes

$$T(s) = \frac{s}{s^2 + \omega_n^2} \tag{6.120}$$

Using the Laplace tables, we have

$$x(t) = x_0 \cos\omega_n t \tag{6.121}$$

In theory, this system oscillates forever, as shown in Fig. 6.33.

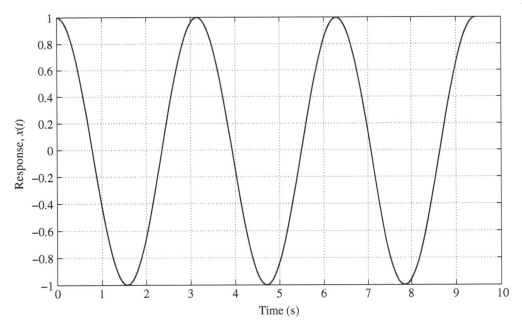

Figure 6.33: *Free response of the undamped system*

Characteristic roots in the complex plane

An alternative way to look at the speed of response as well as stability is to deduce the system behavior from the value of the characteristic roots. This is a more expedient way of analyzing and designing systems rather than looking at their behavior with respect to time.

For the overdamped system, let us place the roots on a complex plane; since the roots do not have imaginary parts, they lie on the real axis (Fig. 6.34). For the underdamped system, we see that the roots are complex, and are mirror images about the real axis, as shown in Fig. 6.35. For the critically damped system, the roots are equal as seen in Fig. 6.36. For the undamped system, we see that the roots are smack on the imaginary axis, as shown in Fig. 6.37.

Consider the simple pendulum with its linearized equations about the two equilibrium positions: 0 and π. The characteristic equation for the hanging position ($\theta_0 = 0$) was

$$s^2 + \frac{g}{\ell} = 0 \tag{6.122}$$

and for the vertically upright position ($\theta_0 = \pi$), it was

$$s^2 - \frac{g}{\ell} = 0 \tag{6.123}$$

In the first case, the roots (say, s_1, s_2) are on the imaginary axis, and in the second case, they are on the real axis (s_3, s_4), with one of them being in the *right half plane (RHP)*. This root in the RHP is the cause of instability. Figure 6.38 shows the roots of the pendulum for the two equilibrium positions. By the definition of stability then, the hanging position (as well as the undamped mass-spring system) is *marginally stable*.

Next, to summarize, we consider systems with different time constants that lead to different characteristic roots, as shown in Fig. 6.39. The following points should be noted.

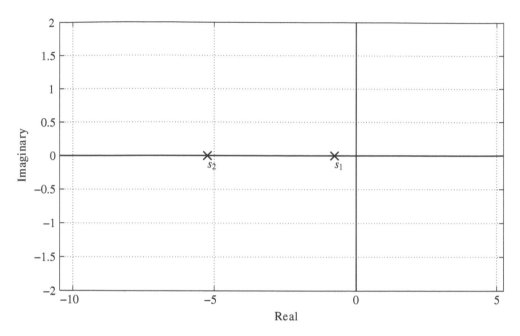

Figure 6.34: *Roots for an overdamped system*

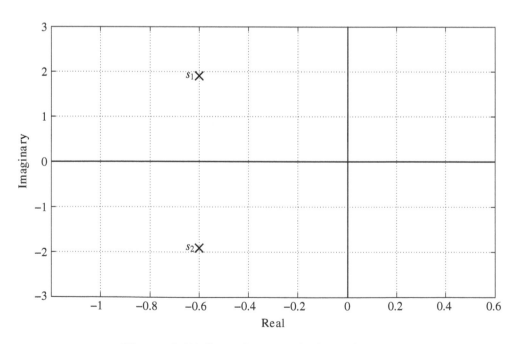

Figure 6.35: *Roots for an underdamped system*

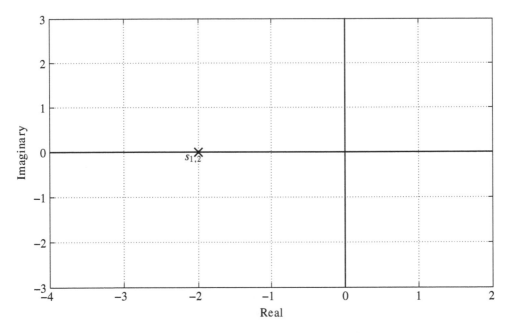

Figure 6.36: *Roots for a critically damped system*

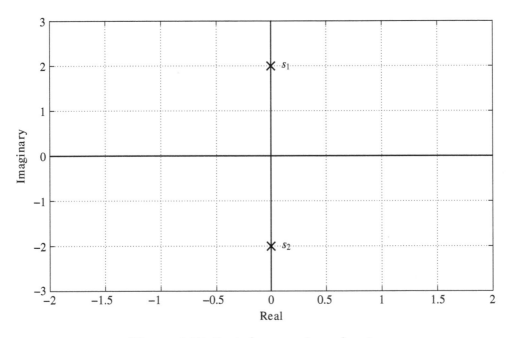

Figure 6.37: *Roots for an undamped system*

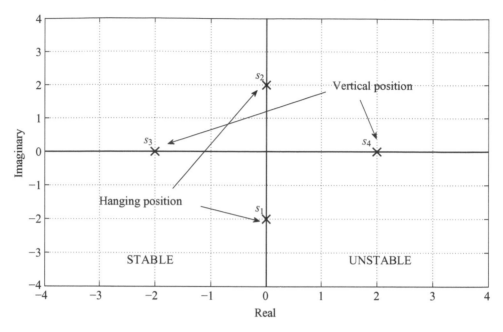

Figure 6.38: *Characteristic roots of the linearized simple pendulum model*

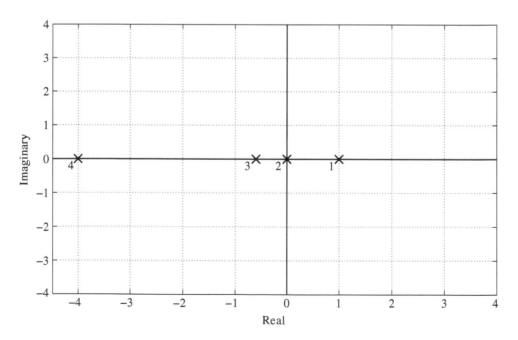

Figure 6.39: *Characteristic roots for different systems*

1. $\tau < 0$; the system is unstable.

2. $\tau = 0$; the system is marginally stable.

3. $\tau > 0$ and large; the system is stable, but has a slow response.

4. $\tau > 0$ and small; the system is stable, and has a quick response.

It is, hence, clear that the following inferences can be made.

- If a characteristic root is in the RHP, the system is unstable.
- For a system to be stable, all the characteristic roots should be in the left half plane (LHP).
- A root close to the imaginary axis leads to a slow response.
- A root far away from the imaginary axis leads to a fast response.

6.3.5   Forced Response of Dynamic Systems

This section will deal with the forced response of the vibrating systems; in other words, we are looking for the response of physical systems in response to external inputs or forces. In the case of linear systems, since the principle of superposition is valid, we can determine the response of the system to each input separately, with the understanding that the responses due to each of these inputs can be added (in addition to the free response) to give us the total response of the system.

We have the equation of motion as usual.

$$\ddot{x} + 2\zeta\omega_n\dot{x} + \omega_n^2 x = \omega_n^2 \frac{1}{k}F(t) \tag{6.124}$$

Taking Laplace transforms with zero initial conditions

$$\left(s^2 + 2\zeta\omega_n s + \omega_n^2\right) x(s) = \omega_n^2 \frac{1}{k}F(s) \tag{6.125}$$

We get the transfer function

$$\frac{x(s)}{F(s)/k} = \frac{\omega_n^2}{s^2 + 2\zeta\omega_n s + \omega_n^2} \tag{6.126}$$

Step response

To demonstrate the Laplace technique, let us determine the step response again. The force is a step function of magnitude F_0.

$$F(t) = \begin{cases} F_0 & t \geq 0 \\ 0 & t < 0 \end{cases} \tag{6.127}$$

whose Laplace transform is

$$F(s) = \frac{F_0}{s} \tag{6.128}$$

Hence, we will need to find the inverse Laplace transform of the following function.

$$x(s) = \frac{F_0}{k} \frac{\omega_n^2}{s(s^2 + 2\zeta\omega_n s + \omega_n^2)} \tag{6.129}$$

The second-order system response again depends on the value of the damping ratio, and we need to consider the various cases.

Overdamped system ($\zeta > 1$)

Since the roots are real, we can factor the denominator into real terms (as we did for the free response).

$$x(s) = \frac{F_0}{k} \frac{\omega_n^2}{s(s - s_1)(s - s_2)} \tag{6.130}$$

where

$$s_1, s_2 = \zeta\omega_n \pm \omega_n\sqrt{\zeta^2 - 1} \tag{6.131}$$

Using partial fractions

$$x(s) = \frac{\omega_n^2 F_0}{ks_1 s_2} \left\{ \frac{1}{s} + \frac{s_2}{s_1 - s_2}\frac{1}{s - s_1} - \frac{s_1}{s_1 - s_2}\frac{1}{s - s_2} \right\} \tag{6.132}$$

Taking inverse Laplace transforms and simplifying

$$x(t) = \frac{F_0}{k} \left\{ 1 + \frac{1}{s_1 - s_2} \left[s_2 e^{s_1 t} - s_1 e^{s_2 t} \right] \right\} \tag{6.133}$$

Clearly, the response to a step input has given us a step (the first term, 1, in the brackets) plus some transient terms which look the same as we got for the free response; i.e., they are exponentials with time constants τ_1 and τ_2. The system approaches a steady state value of F_0/k in an exponential manner as shown in Fig. 6.40.

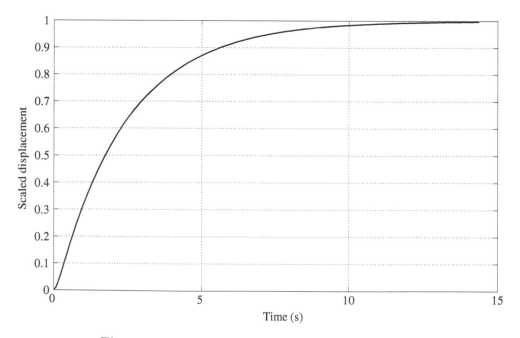

Figure 6.40: *Step response for the overdamped system*

Critically damped system ($\zeta = 1$)

In this case, the roots are real and repeated ($s_1 = s_2 = -\omega_n$) and the partial fractions result in the following.

$$x(s) = \frac{F_0}{k} \frac{\omega_n^2}{s(s-s_1)^2}$$

$$= \frac{F_0}{k} \left[\frac{1}{s} - \frac{1}{s+\omega_n} - \omega_n \frac{1}{(s+\omega_n)^2} \right] \tag{6.134}$$

Taking the inverse Laplace transform

$$x(t) = \frac{F_0}{k} \left[1 - (1+\omega_n t)\, e^{-\omega_n t} \right] \tag{6.135}$$

The response is faster than that for the overdamped system, and looks as shown in Fig. 6.41.

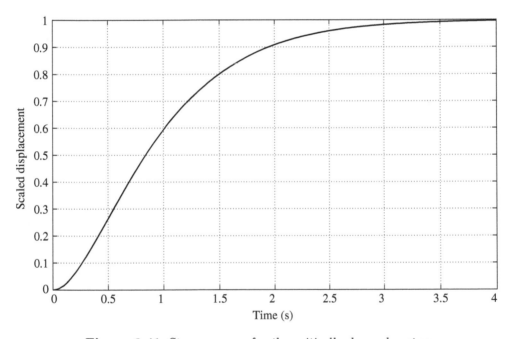

Figure 6.41: *Step response for the critically damped system*

Underdamped system ($\zeta < 1$)

When the system is underdamped, the roots are complex conjugate and the response can be expected to oscillate. To get the inverse Laplace transform we can proceed along the same lines as we did for the overdamped system except that we have to remember that the roots are complex. Hence,

$$x(s) = \frac{F_0}{k} \frac{\omega_n^2}{s_1 s_2} \left\{ \frac{1}{s} + \frac{s_2}{s_1 - s_2} \frac{1}{s - s_1} - \frac{s_1}{s_1 - s_2} \frac{1}{s - s_2} \right\} \tag{6.136}$$

where s_1, $s_2 = -\zeta\omega_n \pm i\omega_d$. Hence, $s_1 s_2 = \omega_n^2$ and $s_1 - s_2 = 2i\omega_d$. Note that $\omega_d = \omega_n\sqrt{1-\zeta^2}$.

$$
\begin{aligned}
x(t) &= \frac{F_0}{k}\left\{1 + \frac{1}{s_1 - s_2}\left[s_2 e^{s_1 t} - s_1 e^{s_2 t}\right]\right\} \\
&= \frac{F_0}{k}\left\{1 - \left[\cos\omega_d t + \frac{\zeta}{\sqrt{1-\zeta^2}}\sin\omega_d t\right]e^{-\zeta\omega_n t}\right\} \\
&= \frac{F_0}{k}\left\{1 - \frac{1}{\sqrt{1-\zeta^2}}e^{-\zeta\omega_n t}\sin\left(\omega_d t + \phi\right)\right\}
\end{aligned}
\tag{6.137}
$$

where

$$
\phi = \tan^{-1}\left[\frac{\sqrt{1-\zeta^2}}{\zeta}\right] = \cos^{-1}\left(\zeta\right)
\tag{6.138}
$$

The response is illustrated in Fig. 6.42.

It is interesting to compare the response of the three cases in a single plot (Fig. 6.43). Note again that the critically damped system settles down in the response in the shortest amount of time.

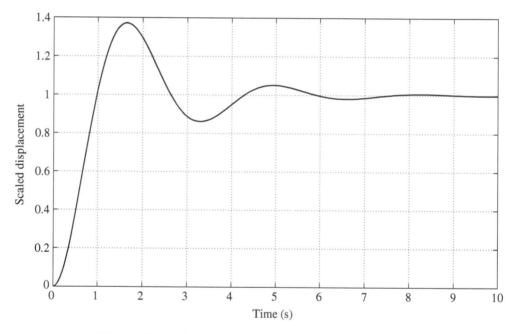

Figure 6.42: *Step response for the underdamped system*

Use the final value theorem for Laplace transforms to derive the *steady-state* value of the displacement with step response. Show that it gives us the same answer as the steady state values obtained from the responses obtained after taking the inverse Laplace transforms [for example, Eq. (6.133)].

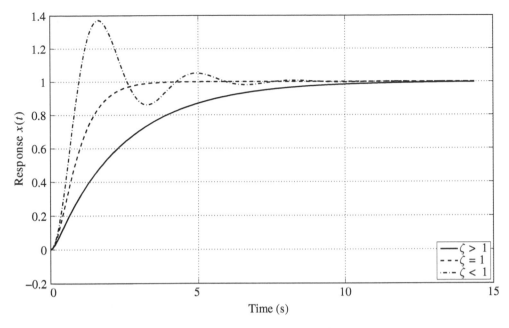

Figure 6.43: *Step Response compared for different damping ratios*

Example 6.6 Step response of a chair

Consider a soft chair modeled as a mass-spring-damper system. A heavy person suddenly sits on the chair. Determine the vibrational response of the chair. Assume that the mass of the seat is 500 g, mass of the person is 120 kg, the spring stiffness is 15,000 N/m, and assume that the designed damping ratio for a nominal mass of 100 kg load is 0.3.

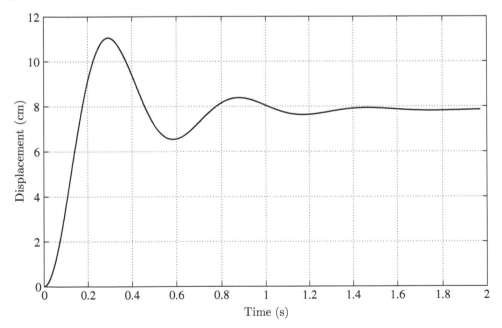

Figure 6.44: *Vibrational response of the chair*

Solution The sudden load can be modeled as a step force of magnitude equal to the weight of the person:

$$F(t) = \begin{cases} F_0, & t > 0 \\ 0, & t \leq 0 \end{cases} \tag{6.139}$$

where $F_0 = 1{,}177$ N. The mass of the system that participates in vibration is $m = 120.5$ kg, and the effective undamped natural frequency is 11.16 rad/s. The damping coefficient can be calculated from the given design ratio.

$$c = 2\zeta\sqrt{km} = (2)(0.6)\sqrt{(15000)(100.5)} = 736.7 \text{ Ns/m} \tag{6.140}$$

The actual damping ratio is then

$$\zeta = \frac{465.9}{2\sqrt{(1500)(120.5)}} = 0.2740 \tag{6.141}$$

and the damped natural frequency is 10.73 rad/s. The system is underdamped, and we have the response from

$$x(t) = \frac{F_0}{k}\left\{1 - \frac{1}{\sqrt{1-\zeta^2}}e^{-\zeta\omega_n t}\sin\left(\omega_d t + \phi\right)\right\}$$

$$= 7.85\left\{1 - \frac{1}{\sqrt{1-(0.2740)^2}}e^{-3.0568t}\sin\left(10.73t + 1.2933\right)\right\} \text{ cm} \tag{6.142}$$

This response is shown in Fig. 6.44. Note that the peak value of the displacement (here, about 11 cm) would be important from the point of view of mechanical design.

— ▨▨▨▨▨▨▨▨▨▨▨▨▨ —

6.3.6   Transient Response Characteristics

Based on the step response of an underdamped second-order system, several transient characteristics can be defined. These are useful to rate and compare different designs for example. Of the many that can be defined, we discuss two that are perhaps the most important.

Percent overshoot

As you must have noticed, the vibrational response of the system shows an overshoot beyond its final steady-state value. This can be an important performance measure depending on the practical application. It is often quantified using a percent overshoot defined as follows

$$\text{P.O.} = 100\left(\frac{M_p - x_{\text{ss}}}{x_{\text{ss}}}\right) = 100\,e^{-\zeta\pi/\sqrt{1-\zeta^2}} \tag{6.143}$$

where M_p is the peak value of the response. For the SDOF system, is clearly just a function of the damping ratio, and its dependence can be plotted easily (Fig. 6.45). Note that, even though we derived an analytical expression for it, it is very often something that can be measured easily in practice without any knowledge of the underlying dynamics. It is hence an extremely useful measure.

Settling time

Yet another measure of the transient response characteristic of a vibrating system is needed to estimate the time taken to settle down. We know that theoretically it takes forever to settle down, but that kind of thinking does not distinguish between a system whose vibration has dropped below a perceptible level in 2 seconds, and another which takes 2 hours! This is where common sense kicks in, and we define "settling down" as the vibration dying down to 2% of the steady-state value. In other words, when the vibration has gone down to that prespecified value, we treat it as if it is done vibrating—practically speaking. (You should also note that there are other prevalent definitions,

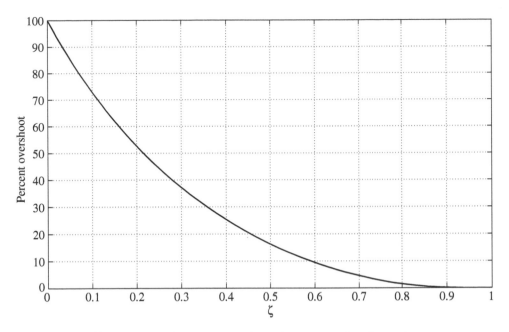

Figure 6.45: *Percent overshoot for a second-order system*

such as a settling time based on 5%.) For the SDOF system we are considering, it is easy to arrive at a closed-form expression for the settling time; it is given by the following formula.

$$t_s = 4\tau = \frac{4}{\zeta\omega_n} \tag{6.144}$$

It should also be noted that often a system with a low overshoot tends to have a large settling time, and vice versa.

Exercise 6.5

Derive the expression given for the settling time. Do it separately for the underdamped and overdamped systems.

Exercise 6.6

Derive an expression given for the 5% settling time. Do it separately for the underdamped and overdamped systems.

Ramp response

Let us consider a force in the form of a ramp. We first define a unit ramp.

$$r(t) = \begin{cases} t & t \geq 0 \\ 0 & t < 0 \end{cases} \tag{6.145}$$

whose Laplace transform is

$$r(s) = \frac{1}{s^2} \tag{6.146}$$

A ramp of magnitude F_0 is given by $F(t) = F_0 r(t)$. Hence, we will need to find the inverse Laplace transform of the following function.

$$x(s) = \frac{F_0}{k} \frac{\omega_n^2}{s^2(s^2 + 2\zeta\omega_n s + \omega_n^2)} \tag{6.147}$$

The system response again depends on the value of the damping ratio, and we need to consider the various cases to find the inverse Laplace transforms.

Overdamped system $(\zeta > 1)$

$$x(s) = \frac{F_0}{k} \omega_n^2 \left\{ \frac{A_1}{s} + \frac{A_2}{s^2} + \frac{K_1}{s - s_1} + \frac{K_2}{s - s_2} \right\} \tag{6.148}$$

Using partial fractions, the following results are obtained.

$$K_1 = \frac{1}{s_1^2(s_1 - s_2)}, \quad K_2 = \frac{1}{s_2^2(s_2 - s_1)}$$

$$A_1 = \frac{s_1 + s_2}{s_1^2 s_2^2}, \quad A_2 = \frac{1}{s_1 s_2} \tag{6.149}$$

Hence, the nondimensional response of the system is

$$\bar{x}(t) = \frac{x(t)}{F_0/k} = A_1 + A_2 t + K_1 e^{s_1 t} + K_2 e^{s_2 t} \tag{6.150}$$

and is shown in Fig. 6.46.

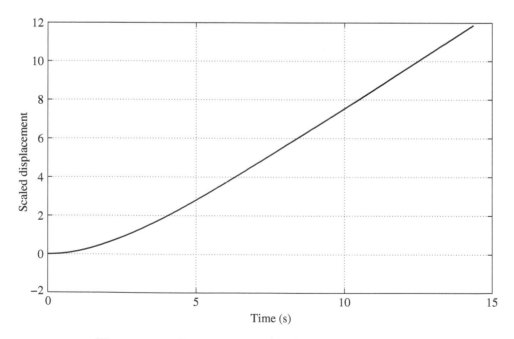

Figure 6.46: *Ramp response for the overdamped system*

Critically damped system ($\zeta = 1$)

$$\bar{x}(s) = \frac{x(s)}{F_0/k} = \frac{1}{s^2(s+\omega_n)^2}$$

$$= \frac{A_1}{s} + \frac{A_2}{s^2} + \frac{B_1}{s+\omega_n} + \frac{B_2}{(s+\omega_n)^2} \tag{6.151}$$

The method of partial fractions yields the following solution.

$$\bar{x}(t) = A_1 + A_2 t + B_1 e^{-\omega_n t} + B_2 t e^{-\omega_n t}$$

$$= k_g \omega_n \left[-2 + \omega_n t + 2e^{-\omega_n t} + \omega_n t e^{-\omega_n t} \right] \tag{6.152}$$

The response is shown in Fig. 6.47.

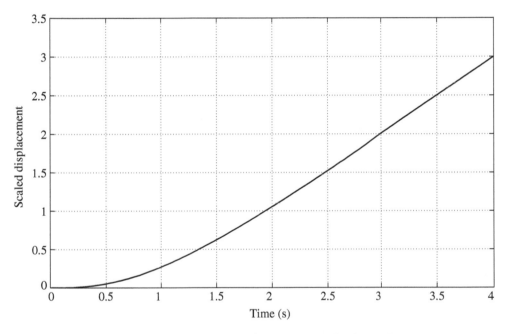

Figure 6.47: *Ramp response for the critically damped system*

Underdamped system ($\zeta < 1$)

The solution we obtained for the overdamped system is valid with the roots having complex conjugate values. Substituting and simplifying

$$\bar{x}(t) = \omega_n^2 \left[-2\zeta + \omega_n t + \frac{1}{\sqrt{1-\zeta^2}} e^{-\zeta\omega_n t} \sin(\omega_d t + \phi) \right] \tag{6.153}$$

where

$$\phi = \tan^{-1} \left[\frac{2\zeta\sqrt{1-\zeta^2}}{2\zeta^2 - 1} \right] \tag{6.154}$$

The response is shown in Fig. 6.48.

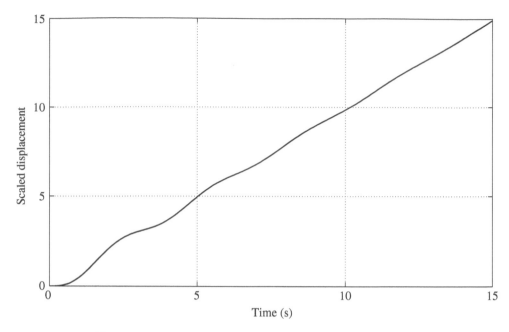

Figure 6.48: *Ramp response for the underdamped system*

6.4 Review Questions

1. What are the units of impulse?
2. Give an example of an impulsive force.
3. Define an arbitrary function in terms of the impulse function.
4. How is the impulse response similar to the initial condition response?
5. How do you interpret the convolution theorem physically?
6. Why is the convolution theorem not quite an explicit solution to the ODE?
7. Can you think of some practical examples of force in the form of the step function?
8. How is the step function different from the constant function?
9. In step response, what does the final value of the response correspond to? Why?
10. Define the Laplace transform. How is it similar to, and different from the Fourier transform?
11. What does the Laplace transformation process do to the ODE?
12. Define the transfer function.
13. How do you obtain the poles of the transfer function?
14. How are the poles of the transfer function related to the characteristic roots? Why?
15. What are simple and multiple poles?
16. Where are the characteristic roots in an overdamped system?
17. Where are the characteristic roots in an underdamped system?
18. Where are the characteristic roots in a critically damped system?
19. Where are the characteristic roots in an undamped system?
20. What is the *qualitative* change when the damping ratio is changed from a value less than 1 to a value greater than 1?

21. For what value of damping ratio do you get the fastest response?
22. How can you determine stability of a system from its characteristic roots plotted on the complex plane?
23. How is the location of the characteristic root on the complex plane related to the speed of response?
24. How is the location of the characteristic root on the complex plane related to the time constant?
25. How do you define settling time? How is it related to the natural frequency of the system and to the damping ratio?
26. How do you define percent overshoot? How is it related to the natural frequency of the system, and to the damping ratio?
27. How do you achieve a mass-spring-damper system with low overshoot and low settling time?

Problems

6.1 Determine the response of an undamped mass-spring system to the rectangular force shown in Fig. 6.49.

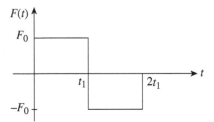

Figure 6.49: *Rectangular force*

6.2 Determine the response of an undamped mass-spring system to the rectangular force shown in Fig. 6.50.

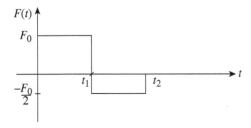

Figure 6.50: *Rectangular force*

6.3 Determine the response of an undamped mass-spring system to the ramp and hold force shown in Fig. 6.51.

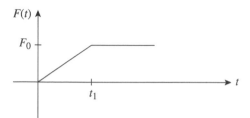

Figure 6.51: *Ramp and hold force*

6.4 Determine the response of an undamped mass-spring system to the triangular force shown in Fig. 6.52.

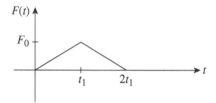

Figure 6.52: *Triangular force*

6.5 Determine the response of an undamped mass-spring system to the triangular force shown in Fig. 6.53.

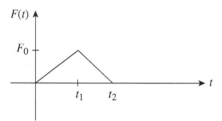

Figure 6.53: *Triangular force*

6.6 Determine the response of an undamped mass-spring system to the falling ramp force shown in Fig. 6.54.

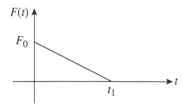

Figure 6.54: *Falling ramp force*

6.7 Determine the response of an undamped mass-spring system to the quad-sinusoidal force, $F(t) = F_0 \sin \pi t/2t_1$, shown in Fig. 6.55 (see Timoshenko et al. [16]).

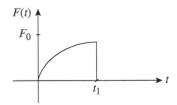

Figure 6.55: *Quad-sinusoidal force*

6.8 Determine the response of an undamped mass-spring system to the quad-sinusoid force, $F(t) = F_0 (1 - \cos \pi t/2t_1)$, shown in Fig. 6.56 (see Timoshenko et al. [16]).

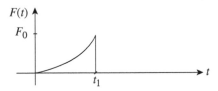

Figure 6.56: *Quad-sinusoidal force*

6.9 Determine the response of an undamped mass-spring system to the parabolic force, $F(t) = F_0 \left(1 - \left(\frac{t}{t_1} \right)^2 \right)$, shown in Fig. 6.57 (see Timoshenko et al. [16]).

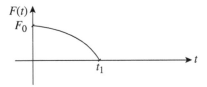

Figure 6.57: *Parabolic force*

6.10 Determine the response of an undamped mass-spring system to the parabolic force, $F(t) = F_0 \frac{(t-t_1)^2}{t_1^2}$, shown in Fig. 6.58.

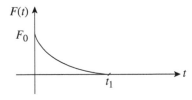

Figure 6.58: *Parabolic force*

6.11 Determine the response of an undamped system to the base motion shown in Fig. 6.59.

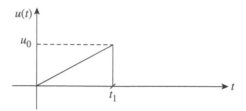

Figure 6.59: *Ramp support motion*

6.12 Determine the response of an undamped system to the base motion, $u(t) = u_0 \left[1 - \frac{(t-t_1)^2}{t_1^2} \right]$, shown in Fig. 6.60 (see Timoshenko et al. [16]).

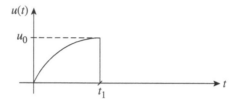

Figure 6.60: *Parabolic support motion*

6.13 Determine the relative displacement of a damped system to the support acceleration shown in Fig. 6.61.

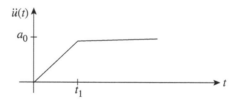

Figure 6.61: *Ramp and hold support acceleration*

6.14 Determine the relative displacement of a damped system to the support acceleration, $\ddot{u}(t) = a_0 \cos \pi t / 2t_1$, shown in Fig. 6.62.

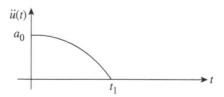

Figure 6.62: *Sinusoidal support acceleration*

6.15 Determine the relative displacement of a damped system to the support acceleration, $\ddot{u}(t) = a_0 \left(1 - \sin \pi t / 2t_1 \right)$, shown in Fig. 6.63.

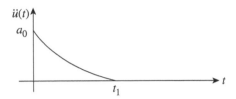

Figure 6.63: *Sinusoidal support acceleration*

6.16 Determine the relative displacement of a damped system to the support acceleration, $\ddot{u}(t)$ $= a_0 t^2/t_1^2$, shown in Fig. 6.64.

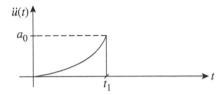

Figure 6.64: *Parabolic support acceleration*

6.17 Determine the response of an underdamped SDOF system that is subjected to the force $F(t)$ $= F_0 \exp(-\alpha t)$. Assume that the system is initially at rest.

6.18 Determine the response of a damped second-order system described to the rectangular pulse. Plot the displacement response for $m = 2\,\text{kg}$, $\zeta = 0.3$, $\omega_n = 5\,\text{rad/s}$, and $F_0 = 10\,\text{N}$. The pulse lasts 2 s.

6.19 Consider the following SDOF system excited by two impulses as follows.

$$\ddot{x} + 2.5\dot{x} + 4x = F(t) + F(t-8) \tag{6.155}$$

Determine the response assuming the system starts from rest.

6.20 An aircraft taxiing on the runway (Fig. 6.65) hits a bump. Assume that the bump is sinusoidal, and analyze the vibrational motion of the wing modeled as a massless cantilever beam, with the engine modeled as a rigid mass located 1/3 the length from the fixed end.

Figure 6.65: *Jet engine on the runway*

6.21 A daredevil motorcyclist hoists his motorcycle in the air (Fig. 6.66) and lands back on the ground from a height of 4 m. Assume that the mass of the vehicle and the rider is 300 kg, and the stiffness of the suspension is 5 kN/m, and assume a damping ratio of 0.5. Determine the subsequent oscillatory motion of the rider.

Figure 6.66: *Motorcycle before coming back to earth*

6.22 Boxing is a controversial sport, enjoyed by many who consider it to be an expression of primal human strength and speed, but frowned upon by some as an uncivilized sport. Nevertheless, an interesting problem to consider is a typical injury that occurs to those who are on the receiving end of a blow on the head. The human brain is suspended in a cushion of meningitic fluid, which nature has designed to be a protective mechanism for exactly this kind of situation. In its simplest form, it can be modeled as a mass-spring-damper system. A blow to the head can be modeled as an impact. Predict the response of the brain-mass if $m = 5\,\text{kg}$, $k = 1.2\,\text{kN/m}$, $\zeta = 0.6$, and the impact magnitude is 2 Ns.

6.23 The vibrational response of seated human subjects is an important problem. One of the simplest models for it is to model the human being as one lumped mass, and the human-seat interface as a spring and a damper, as shown in Fig. 6.67 (see Coermann [18]). (We will look

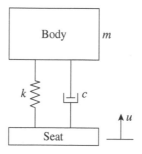

Figure 6.67: *Simplified one-DOF model for a seated human being (see Coermann [18])*

at more complicated models in the forthcoming chapters.) The mass is 56.8 ± 9.4 kg, the spring stiffness is 75.5 ± 28.3 kN/m, and the damping coefficient is 3.840 ± 1.007 kN s/m.

(a) It is subjected to a harmonic base acceleration of amplitude $5\,\text{m/s}^2$, and a frequency range of $0 < \omega < 20$ Hz. Determine the response for the *range* of given parameters.

(b) Determine the response of the system if the base suffers an acceleration in the form shown in Fig. 6.63 with an amplitude of $5\,\text{m/s}^2$.

6.24 Consider a slightly different model for the vibrational response of seated human beings from Problem 6.23. This time, the buttocks and legs of the subject are considered to be rigidly connected to the seat, as shown in Fig. 6.68 (see Wei and Griffin [29]). The body mass, $m_1 = 43.4$ kg, buttocks and legs, $m_0 = 7.8$ kg, $c_1 = 1485$ Ns/m, and $k_1 = 44.13$ kN/m. Determine the response for the same support motions as given in Problem 6.23.

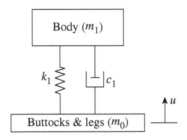

Figure 6.68: *Simplified one-DOF model for a seated human being (see Wei and Griffin [29])*

6.25 Often, during an automobile accident, people suffer what is called *whiplash* in which the upper part of the sitting person moves violently forward and back. We can model this quite well using an inverted pendulum. The head mass sitting on top of the upper body is, in essence, unstable, but is actively stabilized by the neuromuscular system, as shown in Fig. 6.69.[5] Use

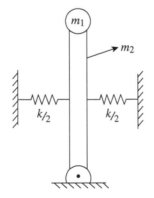

Figure 6.69: *Model for whiplash*

---

[5]This stabilizing neuromuscular system is inactive when we sleep—this is what leads to the "dropping heads" in people sleeping on trains and buses!

the following numerical values: $m_1 = 5.5\,\text{kg}$, $m_2 = 2.5\,\text{kg}$, $\ell = 0.8\,\text{m}$, $k = 2000\,\text{N/m}$, $\zeta = 0.4$. Assume that the spinal column modeled by the rigid rod is uniform, and that the "springs" are attached at the midpoint. Assume that the force of the accident manifests itself by means of an impact force on the head by an amount equal to 10 Ns. Determine the subsequent response of the system.

6.26 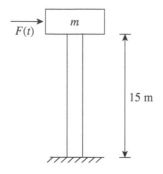 The problem is to obtain the maximum motion of a water tower under the action of a blast wave. Assume that the water tower (shown in Fig. 6.70) can be represented as a concentrated mass of 36,275 kg at the end of a 15 m vertical structure, the weight of which is 47 kg. Calculations show that a 5000 N force at the end of the supporting structure alone will cause a deflection of 3 cm. The force of the air on m is represented as $F(t)$ and may be approximated by a half sine wave reaching a peak of 45,000 N, and dropping to 0 in 0.2 seconds after which the force is zero indefinitely, as shown in Fig. 6.71. Assume negligible damping.

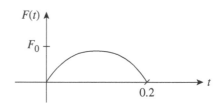

Figure 6.70: *Water tower*   **Figure 6.71:** *Force on the water tower*

(a) Write the equation of motion including all correct numbers for the parameters and defining the variable completely, for the time period from 0 to 0.2 s.

(b) Write the corresponding equation of motion for the time period after 0.2 s.

(c) Obtain the maximum displacement of the mass m.

6.27 Find the free response of the following systems with initial conditions: $x(0) = 0$, and $\dot{x}(0) = 1$ using Laplace transforms. Compare with the solution obtained by straightforward ODE solution.

(a) $\ddot{x} + 25x = 0$

(b) $\ddot{x} + 4\dot{x} + 9x = 0$

(c) $3\ddot{x} + 12\dot{x} + 12x = 0$

(d) $2\ddot{x} + 15\dot{x} + 20x = 0$

6.28 Find the free response of the following systems with initial conditions: $x(0) = 1$, and $\dot{x}(0) = -1$ using Laplace transforms. Compare with the solution obtained by straightforward ODE solution.

(a) $\ddot{x} + 25x = 0$

(b) $\ddot{x} + 4\dot{x} + 9x = 0$

(c) $3\ddot{x} + 12\dot{x} + 12x = 0$

(d) $2\ddot{x} + 15\dot{x} + 20x = 0$

6.29 Plot the characteristic roots of the following systems on a complex plane; state whether each of them is stable, marginally stable, or unstable.

(a) $\ddot{x} + 25x = 0$

(b) $\ddot{x} - 25x = 0$

(c) $\ddot{x} + 4\dot{x} + 9x = 0$

(d) $3\ddot{x} + 12\dot{x} + 12x = 0$

(e) $2\ddot{x} + 15\dot{x} + 20x = 0$

(f) $\ddot{x} + 5\dot{x} - 9x = 0$

(g) $\ddot{x} - 5\dot{x} + 9x = 0$

(h) $-\ddot{x} - 5\dot{x} - 9x = 0$

6.30 A system has a mass 5 kg and a stiffness of 1200 N/m. Its damping changes depending on U, the velocity of air flow in the following manner.

$$c = 20(8 - U) \quad (\mathrm{Ns/m}) \tag{6.156}$$

Plot the change in the characteristic roots (called a *root locus*) as the air velocity changes from 1 to 20 m/s. (Before you scoff at the utility of this made-up problem yielding *negative* damping values, you should know that it is actually a simplified version of the aeroelastic vibration problem that leads to the dangerous phenomenon of flutter in aircraft wings.)

6.31 Find the unit step response of the following systems with zero initial conditions.

(a) $\ddot{x} + 5x = 0$

(b) $4\ddot{x} + 16\dot{x} + 16x = 0$

(c) $5\ddot{x} + 35\dot{x} + 20x = 0$

(d) $5\ddot{x} + 25\dot{x} + 50x = 0$

6.32 Compute the time constant, and frequency of oscillation (if any) for the following systems. Estimate how long it will take for the free response to "disappear."

(a) $\ddot{x} + 5x = 0$

(b) $4\ddot{x} + 16\dot{x} + 16x = 0$

(c) $5\ddot{x} + 35\dot{x} + 20x = 0$

(d) $5\ddot{x} + 25\dot{x} + 50x = 0$

6.33 For each of the following systems, determine the steady state response, and estimate how long it will take for the response to reach steady state. The initial conditions are zero.

(a) $\ddot{x} + 5x = 2\mathcal{U}(t)$

(b) $4\ddot{x} + 16\dot{x} + 16x = 22\mathcal{U}(t)$

(c) $5\ddot{x} + 35\dot{x} + 20x = 40\mathcal{U}(t)$

(d) $5\ddot{x} + 25\dot{x} + 50x = 94\mathcal{U}(t)$

6.34 In each of the following cases, find the forced response of the system $\ddot{x} + 20\dot{x} + 44x = F(t)$, using Laplace transforms and the convolution theorem, and compare.

(a) $F(t) = \sin 3t$

(b) $F(t) = e^{-2t}$

(c) $F(t) = 5t$

(d) $F(t) = 3t^2 + 4$

6.35 Use the method of superposition to determine the response of the system $\ddot{x} + 3\dot{x} + 25x = F(t)$ to $F(t) = 25(1 - 3e^{-5t})$.

6.36 Find the total response of the following systems; in all cases, the initial conditions are $x(0) = 1$, $\dot{x}(0) = 2$.

(a) $\ddot{x} + 25x = 3\mathcal{U}(t)$

(b) $\ddot{x} + 4\dot{x} + 9x = 3\mathcal{U}(t)$

(c) $3\ddot{x} + 12\dot{x} + 12x = 3\mathcal{U}(t)$

(d) $2\ddot{x} + 15\dot{x} + 20x = 3\mathcal{U}(t)$

6.37 Find the response of the following systems using Laplace transforms; assume zero initial conditions.

(a) $\ddot{x} + 5x = 2t$

(b) $4\ddot{x} + 16\dot{x} + 16x = 2t$

(c) $5\ddot{x} + 35\dot{x} + 20x = 2t$

(d) $5\ddot{x} + 25\dot{x} + 50x = 2t$

7 MDOF Systems: Free Response

Most people spend more time and energy going around problems than in trying to solve them.

—Henry Ford

So far we have modeled vibrating systems with one degree of freedom (DOF) only. It turns out that such models can be very accurate and suffice in many practical situations. However, a single DOF simply cannot predict certain kinds of phenomena that we observe in practice.

For example, consider the simple problem of a car on suspensions that we had modeled as a lumped particle mass on springs and dampers. Now, suppose we wish to differentiate between the driver and a rear-seat passenger in the amount of vibration experienced; or, suppose we wish to analyze the pitching motion that we all experience when accelerating, braking, or when going over a bump. Then, we will need at least two DOFs as shown in Fig. 7.1 where the DOFs could for example be the

vertical motion at the two suspension locations. Before we analyze this problem however, let us look at a simpler two-degree-of-freedom model that consists of masses and springs and is an extension of the mass-spring model we have been analyzing. Also, let us drop damping in the beginning to make it easier.

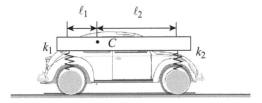

Figure 7.1: *A two-DOF model for the automobile suspension*

7.1 An Undamped Multiple-Degree-of-Freedom Model

Figure 7.2 shows a two-degree-of-freedom (DOF) model with two masses interconnected by springs. They are modeled as two particles, and hence we have two free body diagrams (FBD) to contend with, as shown in Fig. 7.3.

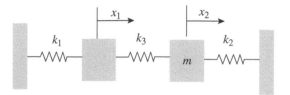

Figure 7.2: *A two-DOF system model*

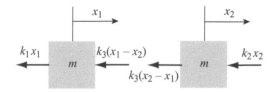

Figure 7.3: *FBDs of the two-DOF system model*

The kinematics is quite straightforward:

$$a_1 = \ddot{x}_1 \tag{7.1}$$
$$a_2 = \ddot{x}_2 \tag{7.2}$$

Writing Newton's second law for both the particle masses

$$\vec{F}_i = m_i \vec{a}_i \tag{7.3}$$

we get

$$m_1 \ddot{x}_1 = -k_1 x_1 - k_3 (x_1 - x_2) \tag{7.4}$$
$$m_2 \ddot{x}_2 = -k_2 x_2 - k_3 (x_2 - x_1) \tag{7.5}$$

These two equations can be written in a matrix form

$$\begin{bmatrix} m_1 & 0 \\ 0 & m_2 \end{bmatrix} \begin{bmatrix} \ddot{x}_1 \\ \ddot{x}_2 \end{bmatrix} + \begin{bmatrix} k_1 + k_3 & -k_3 \\ -k_3 & k_2 + k_3 \end{bmatrix} \begin{bmatrix} x_1 \\ x_2 \end{bmatrix} = \begin{bmatrix} 0 \\ 0 \end{bmatrix} \tag{7.6}$$

They are a set of two linear coupled ordinary differential equations which we will need to solve for two unknown quantities, $x_1(t)$ and $x_2(t)$. They are homogeneous because there are no external forces acting on the system. They are coupled because clearly the motion of each mass influences the other.

The equations can be written more compactly in the form

$$\mathbf{M\ddot{x}} + \mathbf{Kx} = \mathbf{0} \tag{7.7}$$

where $\mathbf{M}$ is called the mass matrix, $\mathbf{K}$ is called the stiffness matrix, and $\mathbf{x}$ is called the displacement vector. $\mathbf{M}$ and $\mathbf{K}$ are 2×2 matrices because we have two DOF. If he had a three-DOF system, they would be square matrices of the order 3; if we had 3298 DOF, they would be of size 3298,... you get the idea.

7.1.1 Solution

To solve this set of homogeneous linear differential equations, we assume an exponential solution (as always). Or,

$$\mathbf{x}(t) = \mathbf{u}e^{st} \tag{7.8}$$

Substituting into the differential equation

$$\left(s^2 \mathbf{M u} + \mathbf{K u} \right) e^{st} = \mathbf{0} \tag{7.9}$$

Then, since $e^{st} \neq 0$, we get the *eigenvalue problem*.[1]

$$\left(s^2\mathbf{M} + \mathbf{K}\right)\mathbf{u} = \mathbf{0} \tag{7.10}$$

This is more often written in the form

$$\mathbf{K}\mathbf{u} = (-s^2)\mathbf{M}\mathbf{u} \tag{7.11}$$

or in the form

$$\boxed{\mathbf{K}\mathbf{u} = \omega^2\mathbf{M}\mathbf{u}} \tag{7.12}$$

where ω has been defined using

$$\omega = is, \quad \omega^2 = -s^2 \tag{7.13}$$

where $i = \sqrt{-1}$. We will use both forms of the eigenvalue problem interchangeably depending on whichever is easier to implement.

It follows then that the following determinant is equal to zero, which is the *characteristic equation*.

$$\left|-\omega^2\mathbf{M} + \mathbf{K}\right| = 0 \tag{7.14}$$

To make the algebra easier for now, let us simplify the problem by letting all the masses and the springs be the same: $k_1 = k_2 = k_3 = k$, $m_1 = m_2 = m$. Then, the stiffness matrix and the mass matrix become

$$\mathbf{M} = \begin{bmatrix} m & 0 \\ 0 & m \end{bmatrix}, \quad \mathbf{K} = \begin{bmatrix} 2k & -k \\ -k & 2k \end{bmatrix} \tag{7.15}$$

Substituting for $\mathbf{M}$ and $\mathbf{K}$ into the characteristic equation, we get a quartic equation (which is however a quadratic in ω^2) and can be easily solved.

$$\begin{vmatrix} -\omega^2 m + 2k & -k \\ -k & -\omega^2 m + 2k \end{vmatrix} = 0 \tag{7.16}$$

$$\Rightarrow \quad m^2\omega^4 - 4km\omega^2 + 3k^2 = 0 \tag{7.17}$$

Defining

$$\lambda = -\frac{\omega^2}{k/m} = \frac{s^2}{k/m} \tag{7.18}$$

the equation becomes

$$\lambda^2 + 4\lambda + 3 = 0 \tag{7.19}$$

The solution yields two roots

$$\lambda = \begin{cases} -1 \\ -3 \end{cases} \tag{7.20}$$

The natural frequencies are then given by

$$\omega_1 = \sqrt{\frac{k}{m}}, \quad \omega_2 = \sqrt{3\frac{k}{m}} \tag{7.21}$$

---

[1] *eigenvalue* is a strange combination word. *eigen* comes from German, and approximately means *characteristic*. The German word for eigenvalue by the way is *Eigenwert*. It is a testament to the contribution of the German scientists to this field that the word, *eigen* has now become an English word.

Eigenvectors

The story does not end with the eigenvalues. Remember that we had a matrix equation, Eq. (7.12). Let us see what happens if we substitute the first eigenvalue ω_1^2 into that equation. Let us also call that particular value of the vector $\mathbf{u}_1$.

$$\left(-\omega_1^2 \mathbf{M} + \mathbf{K}\right) \mathbf{u}_1 = \mathbf{0} \tag{7.22}$$

Or,

$$\begin{bmatrix} -\omega_1^2 m + 2k & -k \\ -k & -\omega_1^2 m + 2k \end{bmatrix} \begin{bmatrix} u_{11} \\ u_{21} \end{bmatrix} = \begin{bmatrix} 0 \\ 0 \end{bmatrix} \tag{7.23}$$

Solving

$$\frac{u_{11}}{u_{21}} = \frac{k}{-m\omega_1^2 + 2k} = 1 \tag{7.24}$$

after using $\omega_1^2 = \frac{k}{m}$. So, we now have a particular vector that satisfies the linear algebraic equation that we called the eigenvalue problem. We call this vector, an *eigenvector*. But you should have noticed something funny here: we did not really get a unique solution for the eigenvector. In other words, the following are all valid eigenvectors here, since they all obey Eq. (7.24).

$$\mathbf{u}_1 = \begin{bmatrix} 1 \\ 1 \end{bmatrix}, \quad \mathbf{u}_1 = \begin{bmatrix} -3.3 \\ -3.3 \end{bmatrix}, \quad \mathbf{u}_1 = \begin{bmatrix} 8i \\ 8i \end{bmatrix} \tag{7.25}$$

Essentially, any constant multiplying an eigenvector also gives us a valid eigenvector. This is a very important property of eigenvectors. Do not get anxious—we will shortly resolve this "arbitrary" issue.

Similarly, the second eigenvector is determined by using the second natural frequency.

$$\begin{bmatrix} -\omega_2^2 m + 2k & -k \\ -k & -\omega_2^2 m + 2k \end{bmatrix} \begin{bmatrix} u_{12} \\ u_{22} \end{bmatrix} = \begin{bmatrix} 0 \\ 0 \end{bmatrix} \tag{7.26}$$

Solving

$$\frac{u_{12}}{u_{22}} = \frac{k}{-m\omega_2^2 + 2k} = -1 \tag{7.27}$$

after using $\omega_2^2 = 3\frac{k}{m}$. Again, this eigenvector is arbitrary to a multiplicative constant, so we could just pick any one as long as it satisfies the relationship given by Eq. (7.27). For example,

$$\mathbf{u}_2 = \begin{bmatrix} 1 \\ -1 \end{bmatrix} \tag{7.28}$$

would be a valid (second) eigenvector for this problem.

The complete solution

Now, let us see if we can make sense of the solution we have arrived at. We started out assuming an exponential solution (as we do with all linear homogeneous ODEs) and ended with four values for the exponent (really, a pair of squares, s_1^2 and s_2^2). In addition to this, we ended up with a constraint on the displacement vectors associated with those exponents. Since this is a linear equation, we obtain the complete solution by simply adding them all up. In addition, since the eigenvectors are arbitrary to a multiplicative constant, we will throw in constants (C_i) and figure out shortly how to determine them.

$$\mathbf{x}(t) = C_1 \mathbf{u}_1 e^{s_1 t} + C_2 \mathbf{u}_1 e^{-s_1 t} + C_3 \mathbf{u}_2 e^{s_2 t} + C_4 \mathbf{u}_2 e^{-s_2 t} \tag{7.29}$$

Note that this method follows along the same lines as we did for the single-DOF system; in that case, we had two exponents, and now we have four. As we did in that case, we now use Euler formula for both the pairs and write them out as trigonometric functions.

$$\mathbf{x}(t) = B_1 \mathbf{u}_1 \cos \omega_1 t + B_2 \mathbf{u}_1 \sin \omega_1 t + B_3 \mathbf{u}_2 \cos \omega_2 t + B_4 \mathbf{u}_2 \sin \omega_2 t \tag{7.30}$$

where we changed the names of the constants to reflect the new form of the solution. This then is the solution to the problem. Clearly, the response will have elements of both the natural frequencies. Exactly what the solution looks like, of course, depends on the initial conditions, something we have conveniently omitted to specify so far. Obviously, we will need to specify four initial conditions: two displacements and two velocities. Let us do that next; that will reveal to us something beautiful about the eigenvectors.

First, let us try a set of initial conditions: $x_1(0) = x_2(0) = x_0$, $\dot{x}_1(0) = \dot{x}_2(0) = 0$. In other words, we will push both the masses by an equal amount (x_0) and let go; we want to know what happens subsequently to the system. Simple algebraic substitution into the solution, Eq. (7.30) will give us the constants: $B_1 = x_0$, $B_2 = B_3 = B_4 = 0$. Or,

$$x_1(t) = x_0 \cos \omega_1 t \tag{7.31}$$
$$x_2(t) = x_0 \cos \omega_1 t \tag{7.32}$$

This means that the motion of the two masses will be identical for all time. They have the same amplitude and frequency and are in phase. (Of course, all this "sameness" is really happening because of the special nature of the eigenvector—take a look at $\mathbf{u}_1$ again.) Now, the system will only exhibit the first natural frequency in its response; there is no hint of the second natural frequency at all. One might almost think that it is a single-DOF system with a natural frequency of ω_1. Figure 7.4 shows the rather obvious response of the two masses. The numerical values used for this figure give us the first natural frequency of $3.26 \, \text{rad/s}$, which corresponds to a natural period of $0.2 \, \text{s}$ (do you see that period in the figure?).

We call it a *mode*, or a modal response, when the two DOF exhibit only one natural frequency and their amplitudes are in the ratio as specified by the corresponding eigenvector. In fact, the motion we showed in Fig. 7.4 is so obvious that we do not normally plot it. Instead, we are more interested in depicting the amplitude ratio; this essentially depicts the shape of vibration, and is hence called the *mode shape*. Figure 7.5 shows the mode shape; note that the actual motion of the masses in this example would be in the horizontal direction, so this is a figurative display of the mode shape.

Now, how about another set of initial conditions? Let us try $x_1(0) = -x_2(0) = x_0$, $\dot{x}_1(0) = \dot{x}_2(0) = 0$. So, this time, we are pushing the two masses away from each other before letting them go. Again, solving for the constants after substituting into Eq. (7.30), we get the following solution.

$$x_1(t) = x_0 \cos \omega_2 t \tag{7.33}$$
$$x_2(t) = -x_0 \cos \omega_2 t \tag{7.34}$$

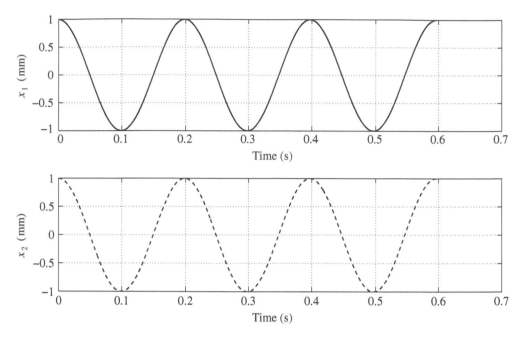

Figure 7.4: *Response to initial conditions; Case 1*

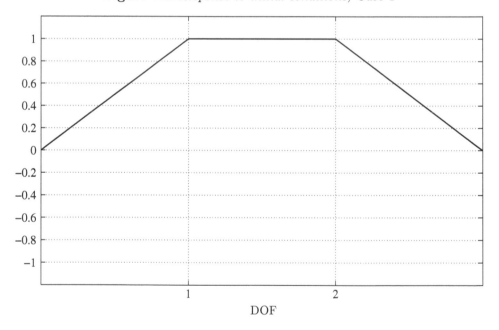

Figure 7.5: *First mode shape*

Now, the two masses will move in exact antiphase. Their amplitudes are the same however, which is a consequence of the amplitude ratio in the second eigenvector $\mathbf{u}_2$. Figure 7.6 plots the two responses. Note that the frequency in evidence is now the second natural frequency; for the numerical values used, it is 54.77 rad/s, or a period of 0.11 s. Figure 7.7 shows the second mode shape and reflects the fact that the two masses are always moving in opposite directions.

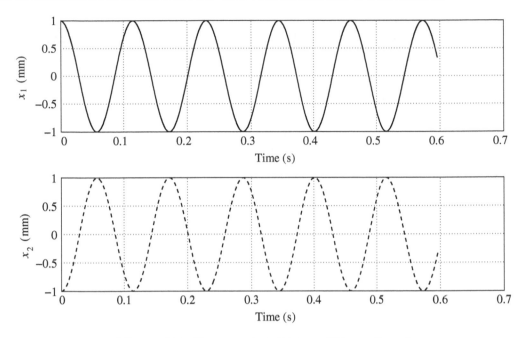

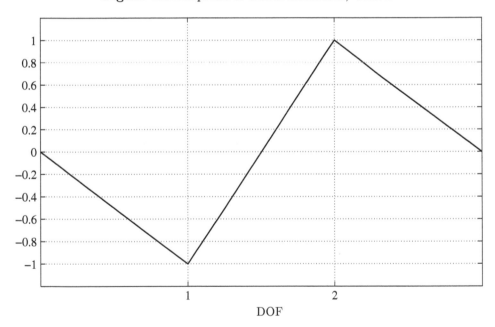

Figure 7.6: *Response to initial conditions; Case 2*

Figure 7.7: *Second mode shape*

It must be clear by now that a mode shape essentially follows from the eigenvector and gives us a nice intuitive way to understand the mathematical concept of eigenvectors. Of course, we already know that the interpretation of eigenvalues is that they correspond to the natural frequencies, the frequencies one would observe in the response of the system in free vibration.

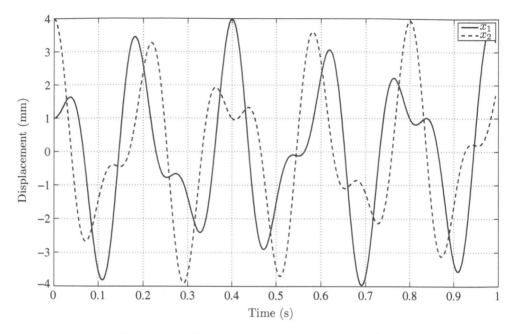

Figure 7.8: *Response to initial conditions; Case 3*

A good way to think about the mode shapes is that it is a snapshot of the system as it is vibrating. In other words, if we choose initial conditions in such a fashion that the system is oscillating in the first mode only, then photographing the system with a flash would reveal a mode shape of that particular mode. Similarly for the second mode. In general, for a general set of initial conditions (as would occur in practice), the response would consist of a combination of the two modes oscillating at the two natural frequencies. An example for such a general set of initial conditions is displayed in Fig. 7.8 and shows how difficult it can be to interpret the motion. Note in particular that the vibration pattern does not repeat because the two natural frequencies are incommensurate. It hence makes enormous sense to look at the individual modes because *any* motion is going to be a simple combination of the the modal responses; so the argument goes—if we can understand the individual modal responses (natural frequencies and mode shapes), we know everything the system can do.[2]

Exercise 7.1

Obtain Eq. (7.30) from Eq. (7.29); also determine the relationships between B_i and C_i.

Exercise 7.2

For both the cases of initial conditions, carry out the algebra to determine the unknown constants B_i. Also, try arbitrary sets of initial conditions and plot the responses.

---

[2]A somewhat unrigorous analogy is the concept of vectors in a plane; if we understand $\hat{\imath}$ and $\hat{\jmath}$, the unit vectors, we understand any vector in the plane because it can be written as a combination of these two.

How did we come up with a set of initial conditions to excite only one of the modes? Was it just dumb luck, or is there a method to our madness? Of course, it is not just luck; we used a property called orthogonality of eigenvectors that you will learn about later in this chapter. Here is a hint, and maybe you can come up with the answer: look carefully at the initial condition vector $[x_1(0) \quad x_2(0)]^T$ and compare it to the eigenvectors for both the cases we considered.

--- **Example 7.1 Coupled pendulums** ---

Two simple pendulums, each of length ℓ, are suspended from a horizontal bar. They are coupled by a spring of stiffness k, a distance a below their points of suspension (Fig. 7.9).

Kinetics

An FBD of each pendulum modeled as a rigid body is shown in Fig. 7.10.

The spring force is given by

$$F_s = ka(\theta_1 - \theta_2) \tag{7.35}$$

where we have assumed small displacements about the vertical equilibrium position.

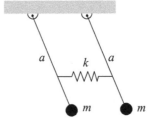

Figure 7.9: *Coupled pendulum*

Kinematics

Since we are not going to use the force equation, we will not determine the linear acceleration of the two centers of mass. The angular accelerations are then easily shown to be

$$\alpha_1 = \ddot{\theta}_1, \quad \alpha_2 = \ddot{\theta}_2 \tag{7.36}$$

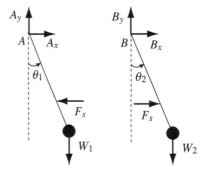

Figure 7.10: *Coupled pendulum FBD*

Newton's laws

Since the two pendulums are modeled as rigid bodies in a plane, Newton's laws are

$$\vec{F} = m\vec{a}_O \tag{7.37}$$
$$M_0 = I_0\alpha \tag{7.38}$$

for both the bodies (where O is a fixed point). Since the force equation will involve the unknown support reactions, we will use the moment equations only (about the respective support points).

$$m\ell^2\ddot{\theta}_1 = -(W_1)(\ell\sin\theta_1) - (F_s)(a\cos\theta_1) \tag{7.39}$$
$$m\ell^2\ddot{\theta}_2 = -(W_2)(\ell\sin\theta_2) + (F_s)(a\cos\theta_2) \tag{7.40}$$

After substituting for F_s and using $W_1 = W_2 = mg$, and linearizing with the assumption of small θ, we get the following matrix equation.

$$\begin{bmatrix} m\ell^2 & 0 \\ 0 & m\ell^2 \end{bmatrix} \begin{bmatrix} \ddot{\theta}_1 \\ \ddot{\theta}_2 \end{bmatrix} + \begin{bmatrix} ka^2 + mg\ell & -ka^2 \\ -ka^2 & ka^2 + mg\ell \end{bmatrix} \begin{bmatrix} \theta_1 \\ \theta_2 \end{bmatrix} = \begin{bmatrix} 0 \\ 0 \end{bmatrix} \tag{7.41}$$

This equation can then be written in the abbreviated form

$$\mathbf{M\ddot{x} + Kx = 0} \tag{7.42}$$

To solve this set of homogeneous linear differential equations, we assume an exponential solution and go through the standard process. This will lead to the eigenvalue problem,

$$\mathbf{Ku} = \omega^2 \mathbf{Mu} \tag{7.43}$$

It follows then that the following determinant is equal to zero, which is the characteristic equation.

$$\left| -\omega^2 \mathbf{M} + \mathbf{K} \right| = 0 \tag{7.44}$$

Substituting for $\mathbf{M}$ and $\mathbf{K}$, we get a quartic equation (which is however a quadratic in ω^2) and can be easily solved.

$$\begin{vmatrix} -\omega^2 m\ell^2 + ka^2 + mg\ell & -ka^2 \\ -ka^2 & -\omega^2 m\ell^2 + ka^2 + mg\ell \end{vmatrix} = 0 \tag{7.45}$$

$$\Rightarrow m^2 \ell^4 \omega^4 - \left(2m\ell^2 (ka^2 + mg\ell) \right) \omega^2 + (ka^2 + mg\ell)^2 - (ka^2)^2 = 0 \tag{7.46}$$

The solution yields two roots

$$\omega^2 = \begin{cases} \dfrac{g}{\ell} \\ \dfrac{g}{\ell} + 2\dfrac{k}{m}\dfrac{a^2}{\ell^2} \end{cases} \tag{7.47}$$

The natural frequencies are then given by

$$\omega_1 = \sqrt{\frac{g}{\ell}}, \quad \omega_2 = \sqrt{\frac{g}{\ell} + 2\frac{k}{m}\frac{a^2}{\ell^2}} \tag{7.48}$$

Mode shapes

The mode shapes are determined by solving for the eigenvectors. To get the first mode shape, we use the first eigenvalue ω_1^2.

$$\left(-\omega_1^2 \mathbf{M} + \mathbf{K} \right) \mathbf{u}_1 = 0 \tag{7.49}$$

Or,

$$\begin{bmatrix} -\omega_1^2 m\ell^2 + ka^2 + mg\ell & -ka^2 \\ -ka^2 & -\omega_1^2 m\ell^2 + ka^2 + mg\ell \end{bmatrix} \begin{bmatrix} u_{11} \\ u_{21} \end{bmatrix} = \begin{bmatrix} 0 \\ 0 \end{bmatrix} \tag{7.50}$$

Solving

$$\frac{u_{11}}{u_{21}} = \frac{ka^2}{-\omega_1^2 m\ell^2 + ka^2 + mg\ell} = 1 \tag{7.51}$$

after using $\omega_1^2 = \frac{g}{\ell}$.

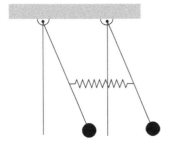

Figure 7.11: *First mode shape*

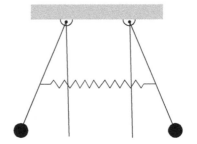

Figure 7.12: *Second mode shape*

Similarly, the second mode is determined from

$$\begin{bmatrix} -\omega_2^2 m\ell^2 + ka^2 + mg\ell & -ka^2 \\ -ka^2 & -\omega_2^2 m\ell^2 + ka^2 + mg\ell \end{bmatrix} \begin{bmatrix} u_{12} \\ u_{22} \end{bmatrix} = \begin{bmatrix} 0 \\ 0 \end{bmatrix} \tag{7.52}$$

Solving

$$\frac{u_{12}}{u_{22}} = \frac{ka^2}{-\omega_2^2 m\ell^2 + ka^2 + mg\ell} = -1 \tag{7.53}$$

after using $\omega_2^2 = \frac{g}{\ell} + 2\frac{k}{m}\frac{a^2}{\ell^2}$.

Figures 7.11 and 7.12 show the two mode shapes.

Note how, in the first mode, the pendulums would move together completely in phase; in the second mode, they move opposite to each other, and are exactly $180°$ out of phase.

7.2   A Two-Degree-of-Freedom Model for an Automobile

Let us get back to the two-DOF model for the automobile on suspension springs that was proposed at the beginning of this chapter (Fig. 7.13). It consists of a rigid rod of length ℓ, supported on springs with stiffness k_1 and k_2. Note how we have a choice in the matter of picking the DOF. We can choose (x, θ), where x is the vertical displacement of the center of mass and θ is the angle of rotation in the plane of the paper. Or, we can choose (x_1, x_2), which are the endpoint displacements of the rigid rod. Clearly, since we have only two DOF, we *have* to choose exactly two independent quantities to describe the dynamics. Hence, the *number* of these DOF is not arbitrary; however, which ones we choose as our DOF seems to be up for grabs. We will see what kind of impact this has on our mathematical model.

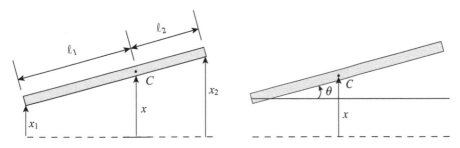

Figure 7.13: *Choices of DOF*

This arbitrariness about the DOF should bother us, don't you think? Clearly, making one or the other of those choices listed above is going to lead us to different mathematical models; hence, we are bound to get different mass and stiffness matrices. This, we would guess, will lead us to different natural frequencies and mode shapes. Wait! That would be really quite ridiculous however, since what we are saying now is that the physical system's natural behavior would depend upon how *we choose* to describe it. If the car we are trying to model has certain natural frequencies, does it make sense that it would suddenly change its behavior simply because we decided to describe its motion using different quantities? Since we have been led to an absurd result by our reasoning, there must be some flaw with our logic somewhere. Can you find that flaw?

First, let us choose (x, θ) as our DOF. Kinematics is then fairly simple.

$$a_C = \ddot{x} \tag{7.54}$$

$$\alpha = \ddot{\theta} \tag{7.55}$$

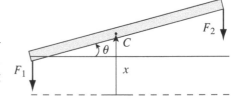

The FBD is as shown in Fig. 7.14. We have dropped gravity since it will not affect the free response that we are interested in. Assuming that the displacements are small, the spring forces can be written as

$$F_1 = k_1(x - \ell_1\theta) \tag{7.56}$$

$$F_2 = k_2(x + \ell_2\theta) \tag{7.57}$$

Figure 7.14: *FBD of the car model*

Newton's second law for a rigid body in a plane is

$$\vec{F} = m\vec{a}_C \tag{7.58}$$

$$M_C = I_C\alpha \quad \text{or} \quad M_O = I_O\alpha \tag{7.59}$$

Here, there is no fixed point on the body and it is quite sensible to use the center of mass. Hence, the equations of motion become

$$\begin{bmatrix} m & 0 \\ 0 & I_C \end{bmatrix} \begin{bmatrix} \ddot{x} \\ \ddot{\theta} \end{bmatrix} + \begin{bmatrix} k_1 + k_2 & -k_1\ell_1 + k_2\ell_2 \\ -k_1\ell_1 + k_2\ell_2 & k_1\ell_1^2 + k_2\ell_2^2 \end{bmatrix} \begin{bmatrix} x \\ \theta \end{bmatrix} = \begin{bmatrix} 0 \\ 0 \end{bmatrix} \tag{7.60}$$

Note that the mass matrix is diagonal, but the stiffness matrix is not diagonal. Hence, the two equations are coupled, but both the second derivatives do not appear in each equation. Such a representation of the system is said to be statically coupled, and dynamically decoupled.

Next, let us consider a new set of coordinates to describe the motion of the system: x_1, x_2, which are the linear displacements at the ends of the automobile, as shown in the first part of Fig. 7.13. This does not mean of course that Newton's laws are any different; we still need the acceleration at the center of mass and the angular acceleration. It is just that we are going to *express* everything in terms of the new set of coordinates. In other words,

$$\theta = \frac{x_2 - x_1}{\ell} \tag{7.61}$$

$$x_C = x_1 + \ell_1\theta = x_1 + \frac{\ell_1}{\ell}(x_2 - x_1) \tag{7.62}$$

Hence, the kinematics is now as follows:

$$a_C = \ddot{x}_1 + \frac{\ell_1}{\ell}(\ddot{x}_2 - \ddot{x}_1) \qquad (7.63)$$

$$\alpha = \frac{\ddot{x}_2 - \ddot{x}_1}{\ell} \qquad (7.64)$$

We draw an FBD as earlier with the new coordinates (Fig. 7.15). The forces are now given by

$$F_1 = k_1 x_1, \quad F_2 = k_2 x_2 \qquad (7.65)$$

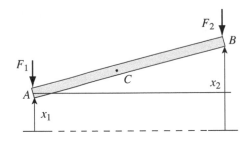

Figure 7.15: *FBD of the car model with the second set of coordinates*

Applying Newton's second law as before, we get the following matrix equation.

$$\begin{bmatrix} m\left(1 - \dfrac{\ell_1}{\ell}\right) & m\dfrac{\ell_1}{\ell} \\ -\dfrac{I_C}{\ell} & \dfrac{I_C}{\ell} \end{bmatrix} \begin{bmatrix} \ddot{x}_1 \\ \ddot{x}_2 \end{bmatrix} + \begin{bmatrix} k_1 & k_2 \\ -k_1\ell_1 & k_2\ell_2 \end{bmatrix} \begin{bmatrix} x_1 \\ x_2 \end{bmatrix} = \begin{bmatrix} 0 \\ 0 \end{bmatrix} \qquad (7.66)$$

Note that now the equations are statically and dynamically coupled, since both the system matrices are nondiagonal.

Finally, let us consider a set of coordinates, x_P and θ, where P is a special point such that $k_1\ell_3 = k_2\ell_4$, as shown in Fig. 7.16. Why we are choosing such a weird point will become clear in the sequel after we derive the equations; so hold on to your horses!

Again, we have to figure what the acceleration at the center of mass is, since Newton's laws are always written for the center of mass, no matter how we choose to describe the system (in other words, no matter which set of coordinates we choose).

$$x_C = x_P + (\ell_3 - \ell_1)\theta \qquad (7.67)$$

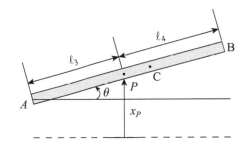

Figure 7.16: *Car model with x_P and θ*

Hence, the kinematics expressions are as follows

$$a_C = \ddot{x}_P + (\ell_3 - \ell_1)\ddot{\theta} \qquad (7.68)$$

$$\alpha = \ddot{\theta} \qquad (7.69)$$

The FBD is as shown in Fig. 7.17, where the forces are now given by

$$F_1 = k_1(x_P - \ell_3\theta), \quad F_2 = k_2(x_P + \ell_4\theta) \qquad (7.70)$$

Next, we apply the same Newton's second laws as earlier; this leads to the matrix equations as follows.

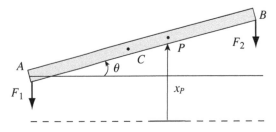

Figure 7.17: *FBD of the car model with the third set of coordinates*

$$\begin{bmatrix} m & m(\ell_3 - \ell_1) \\ m(\ell_3 - \ell_1) & I_P \end{bmatrix} \begin{bmatrix} \ddot{x}_P \\ \ddot{\theta} \end{bmatrix} + \begin{bmatrix} k_1 + k_2 & 0 \\ 0 & k_1\ell_3^2 + k_2\ell_4^2 \end{bmatrix} \begin{bmatrix} x_P \\ \theta \end{bmatrix} = \begin{bmatrix} 0 \\ 0 \end{bmatrix} \qquad (7.71)$$

Now, the equations are dynamically coupled, but statically decoupled. This means, that if a *static* force were applied at P, pure translation would result because the first term, $\mathbf{M\ddot{x}}$, would be set to zero; hence, there would be no rotation of the rod. Now, you see that the point P was cleverly chosen so that the system would become statically uncoupled.

This is what has happened so far: we modeled the *same* system with *same* assumptions but with different coordinates (or different ways of describing the motion), and found that the equations of motion are very different. In particular, the mass matrix and the stiffness matrix would be very different, depending on the choice of coordinates. The next question to ask is: if the system matrices are different, would not the natural frequencies be different? However, we realize that the system is the same, no matter how we describe it. Hence, the natural frequencies should not be different; or, the natural frequencies should be independent of the choice of coordinates. This is in fact true and can also be shown to be always true using principles of linear algebra. In the present case, we will skip the proof, but we will verify that the results are independent of the choice of coordinates by solving this problem numerically.

Suppose, we consider a car with the following parameters in consistent SI units: $m = 1500$, $\ell = 2.5$, $\ell_1 = 1.35$, $I_C = 2233$, $k_1 = k_2 = 50{,}000$. The first set of coordinates, (x_c, θ), results in the following matrices:

$$\mathbf{M} = \begin{bmatrix} 1500 & 0 \\ 0 & 2233 \end{bmatrix}; \quad \mathbf{K} = \begin{bmatrix} 1.0000 & -0.1000 \\ -0.1000 & 1.5725 \end{bmatrix} \times 10^5 \tag{7.72}$$

For the second set of coordinates, (x_1, x_2), the matrices are as follows:

$$\mathbf{M} = \begin{bmatrix} 690 & 810 \\ -893 & 893 \end{bmatrix}; \quad \mathbf{K} = \begin{bmatrix} 5.0000 & 5.0000 \\ -6.7500 & 5.7500 \end{bmatrix} \times 10^4 \tag{7.73}$$

For the third set of coordinates, (x_P, θ), the matrices are as follows:

$$\mathbf{M} = \begin{bmatrix} 1.5000 & 0.1500 \\ 0.1500 & 2.2476 \end{bmatrix} \times 10^3; \quad \mathbf{K} = \begin{bmatrix} 1.0000 & 0 \\ 0 & 1.5625 \end{bmatrix} \times 10^5 \tag{7.74}$$

Solving the eigenvalue problem, in all the three cases, the natural frequencies can be shown to be 7.9228 and 8.6215 rad/s. The mode shapes are shown in Fig. 7.18; note, how there is increased pitching motion in the second mode.

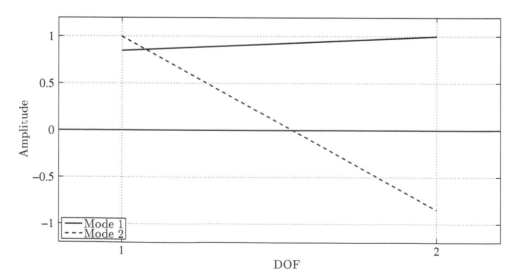

Figure 7.18: *Mode shapes of the two-DOF automobile model*

So, we have verified our assertion (at least for this problem) that the natural frequencies of the system do not depend upon our choice of coordinates. The next question that arises is: Is there a choice of coordinates that would give us a statically *and* dynamically uncoupled model? That question will be answered (in the affirmative) when we establish some general properties of the eigenvalue problem later in this chapter.

Exercise 7.3

Solve the car model problem we analyzed for the three sets of coordinates and verify the results. Also, derive the mode shapes for each set of coordinates. Do you get different eigenvectors? You should! If you did not, you need to check your work. Plot each of them and you will see that they will give us the same picture. In fact, you can transform one set of eigenvectors into another by the same kind of relations we used to transform the coordinates, for example, Eqs. (7.61) and (7.62).

The fact that the eigenvalues do not depend on the choice of coordinates is an important one. The system matrices that we obtain when we transform the coordinates are said to be *similar*, and it can be shown that the similar matrices have identical eigenvalues (see Theorem A.3.3 on page 422; also see Applied Linear Algebra [30]).

7.3   The Zero Eigenvalue

Consider the problem of a car or a truck pulling a trailer (Fig. 7.19). Suppose, we can model each of these two bodies as two particle masses; they are coupled by a linear spring of stiffness k, which approximates the connection mechanism. The model is shown pictorially in Fig. 7.20, and its FBD is shown in Fig. 7.21.

The EOM can be very easily derived to be

$$\mathbf{M\ddot{x} + Kx} = 0 \qquad (7.75)$$

where

$$\mathbf{M} = \begin{bmatrix} m_1 & 0 \\ 0 & m_2 \end{bmatrix}, \quad \mathbf{K} = \begin{bmatrix} k & -k \\ -k & k \end{bmatrix} \qquad (7.76)$$

The characteristic equation is

$$m_1 m_2 \omega^4 - (km_1 + km_2)\,\omega^2 = 0 \qquad (7.77)$$

Figure 7.19: *A truck pulling a trailer*

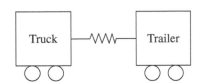

Figure 7.20: *A truck pulling a trailer*

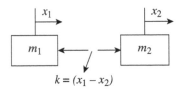

Figure 7.21: *A truck pulling a trailer*

Solution yields the natural frequencies (surprise!) to be

$$\omega_1 = 0, \quad \omega_2 = \sqrt{\frac{k\,(m_1 + m_2)}{m_1 m_2}} = \sqrt{k\left(\frac{1}{m_1} + \frac{1}{m_2}\right)} \tag{7.78}$$

The second one is OK, but it is the first one that is worrisome. What the deuce does a zero natural frequency mean? Is it right? Does it mean that we made an untenable modeling approximation that is leading to a nonsensical result?

Of course, the assumption of an oscillatory solution that led us to this is no longer valid (with a zero frequency), so we are going to have to revert to the first step. These are all the questions we (actually, you, the student!) should indeed be asking. However, let us look at the mode shapes first because that might give us some clue as to what is happening.

We solve for the eigenvectors the same way we have been doing, which leads us to the following.

$$\mathbf{u}_1 = \begin{bmatrix} 1 \\ 1 \end{bmatrix}, \quad \mathbf{u}_2 = \begin{bmatrix} 1 \\ -m_2/m_1 \end{bmatrix} \tag{7.79}$$

This does seem to tell us something. The first mode result in fact tells us that the system does not oscillate, and it moves in such a way that the two masses have the same displacement. Come to think of it, that is what we expect the car-trailer combination to do! That is, we expect them to *move as a rigid body*, because they are not constrained to any fixed point like the other systems we have been looking at. So, that is the root of the problem: we have an incompletely constrained system, which led us to what we call a *rigid body mode*.

In addition to the rigid body mode, we also have a regular vibrational mode, often called a *flexural mode*, which is associated with the second natural frequency. This of course involves the two masses moving in such a fashion that the spring between them gets compressed and extended. Figures 7.22 and 7.23 show the mode shapes conceptually. The actual "vibration" of the system would of course involve a linear combination of the two modes.

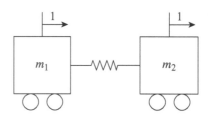

Figure 7.22: *Rigid body mode of truck-trailer*

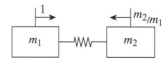

Figure 7.23: *Flexible mode of truck-trailer*

Example 7.2 Numerical example of a truck-trailer

Consider a truck-trailer combination modeled with the following parametric values. $m_1 = 1500$ kg, $m_2 = 6000$ kg, $k = 12$ kN/m. Determine the natural frequencies and mode shapes.

Substituting the values into the characteristic equation solution, Eq. (7.78),

$$\omega_1 = 0, \quad \omega_2 = 3.16\,\text{rad/s} \tag{7.80}$$

The eigenvectors are

$$\mathbf{u}_1 = \begin{bmatrix} 1 \\ 1 \end{bmatrix}, \quad \mathbf{u}_2 = \begin{bmatrix} 1 \\ -4 \end{bmatrix} \tag{7.81}$$

The mode shapes are displayed symbolically in Figs. 7.24 and 7.25. The first is the rigid body mode, and the second is the first flexural mode.

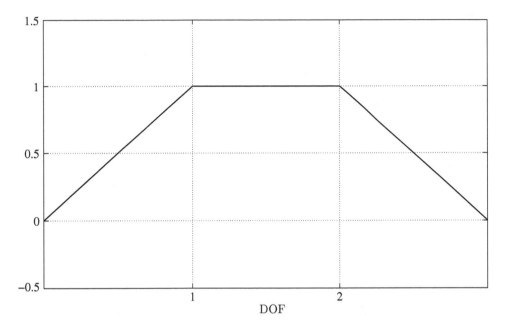

Figure 7.24: *Rigid body mode for truck-trailer system*

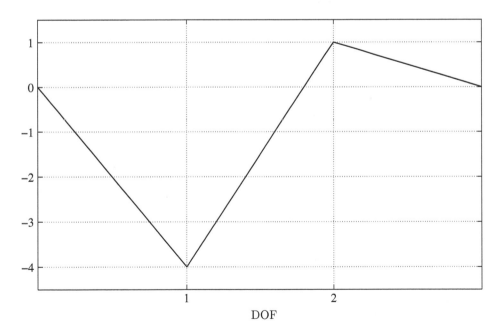

Figure 7.25: *Second mode for truck-trailer system*

Let us look at the problem of zero eigenvalues mathematically. Careful examination of the stiffness matrix reveals a curious property; its determinant is zero! In other words, the stiffness matrix is *singular*. Review of some linear algebra reveals that a singular matrix will indeed lead to a zero eigenvalue, and ergo, a zero natural frequency. In fact, this is pretty neat. We have a real live system which is a nice example of the theoretical concept of a singular matrix. This, hence, gives us a satisfying practical interpretation of zero eigenvalues and singular matrices! Whenever a vibrating system is not constrained completely, we can expect it to have a rigid body motion, and hence can expect the stiffness matrix to be singular. Examples are numerous and include rotating shafts, robotic arms, and moving vehicles such as automobiles, aircraft, rockets, missiles, ships, etc.

Running problem: string vibration

Two- and three-DOF models

We continue the string vibration problem we looked at with a single-DOF model in Chapter 3 on page 64. This time, let us model the string with two lumped masses. Again, the lumping process is not obvious, but we could try a common sense approach, and lump the total mass of the string in equal amounts at positions $\ell/3$ from each end (Fig. 7.26). Now, $m_1 = m_2 = m = \rho A \ell / 2$.

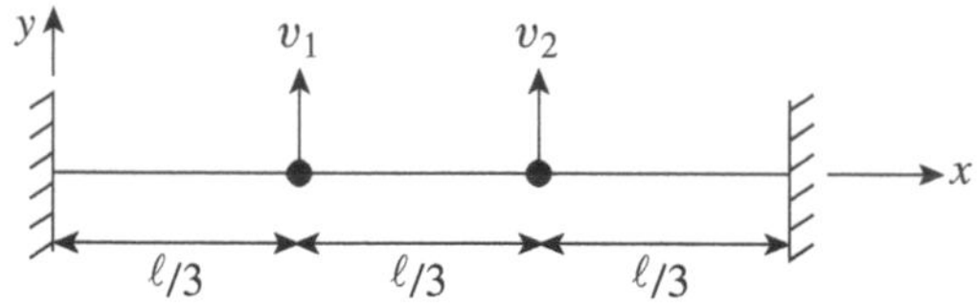

Figure 7.27 shows the FBD; we will continue to use $v(t)$ to notate the displacement, this time with two indices. The kinematics is quite simple, with $a_1 = \ddot{v}_1$ and $a_2 = \ddot{v}_2$. Hence, the equations of motion follow easily.

Figure 7.26: *String with two lumped masses*

$$m_1 \ddot{v}_1 = -T\left(\frac{v_1}{\ell/3}\right) + T\left(\frac{v_2 - v_1}{\ell/3}\right) \qquad (7.82)$$

$$m_2 \ddot{v}_1 = -T\left(\frac{v_2 - v_1}{\ell/3}\right) - T\left(\frac{v_2}{\ell/3}\right) \qquad (7.83)$$

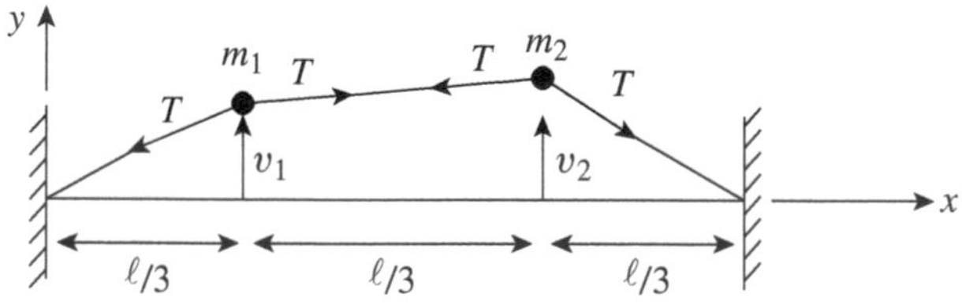

In matrix form

Figure 7.27: *FBD of string with two lumped masses*

$$\mathbf{M}\ddot{\mathbf{x}} + \mathbf{K}\mathbf{x} = 0 \qquad (7.84)$$

where

$$\mathbf{M} = \frac{\rho A \ell}{2}\begin{bmatrix} 1 & 0 \\ 0 & 1 \end{bmatrix}, \quad \mathbf{K} = \frac{3T}{\ell}\begin{bmatrix} 2 & -1 \\ -1 & 2 \end{bmatrix} \qquad (7.85)$$

Solving the eigenvalue problem, we get the following natural frequencies.

$$\omega_1 = \frac{\sqrt{6}}{\ell}\sqrt{\frac{T}{\rho A}}, \quad \omega_2 = \frac{3\sqrt{2}}{\ell}\sqrt{\frac{T}{\rho A}} \qquad (7.86)$$

This is a good time to go back and review the one-DOF treatment (Come on, do it! It is on page 64.). Leaving out the parameters (which are common), the natural frequency was 2 in that case; now we have two natural frequencies. it is reasonable to assume that the one-DOF model was an approximate way to get the fundamental, or the first natural frequency, which is $\sqrt{6}$, or 2.45; this is about 22% different than the first approximation. Since we now have a more complex model, presumably we have a more accurate answer. Alas, that is not always the case! Also, there is no way to tell at this point which one is a better approximation. However, as promised, we will continue to look into it, and it will hopefully be obvious by the time we are done with the course.

Next, let us look at the eigenvectors or mode shapes. After substituting the eigenvalues, we can determine the eigenvectors to be as follows:

$$\mathbf{u}_1 = \begin{bmatrix} 1 \\ 1 \end{bmatrix}, \quad \mathbf{u}_2 = \begin{bmatrix} 1 \\ -1 \end{bmatrix} \quad (7.87)$$

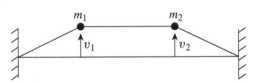

Figure 7.28: *First mode for string with two lumped masses*

Figures 7.28 and 7.29 picture the two modes. Clearly, the first mode is the same motion as was modeled by the one-DOF model, and hence, the first natural frequency predicted by the two-DOF model should be the one that is close to the natural frequency predicted by the one-DOF model.

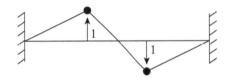

Figure 7.29: *Second mode for string with two lumped masses*

We continue the saga of the string vibration, this time with a three-DOF model. As you can assume, the process is a straight forward extension of the two-DOF model discussion. This time, we divide the mass into three lumped particle masses, at $\ell/4$, $\ell/2$, and $3\ell/4$ from the left end, as shown in Fig. 7.30. We would like to remind you again that this kind of lumping is quite arbitrary (but not crazy; it does follow somewhat of a common

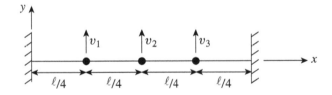

Figure 7.30: *String with three lumped masses*

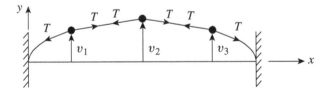

Figure 7.31: *FBD of string with three lumped masses*

sense approach). There are indeed better ways of converting a distributed parameter system into a lumped one, and we will get to it in the sequel, but for now, this is fine. Each of those lumped masses is $\rho A\ell/3$.

Figure 7.31 shows the FBD of the system. Again, linearizing the components of the tension, the equations of motion can be easily derived.

$$\mathbf{M}\ddot{\mathbf{x}} + \mathbf{K}\mathbf{x} = 0 \tag{7.88}$$

where

$$\mathbf{M} = \frac{\rho A\ell}{3} \begin{bmatrix} 1 & 0 & 0 \\ 0 & 1 & 0 \\ 0 & 0 & 1 \end{bmatrix}, \quad \mathbf{K} = \frac{4T}{\ell} \begin{bmatrix} 2 & -1 & 0 \\ -1 & 2 & -1 \\ 0 & -1 & 2 \end{bmatrix} \tag{7.89}$$

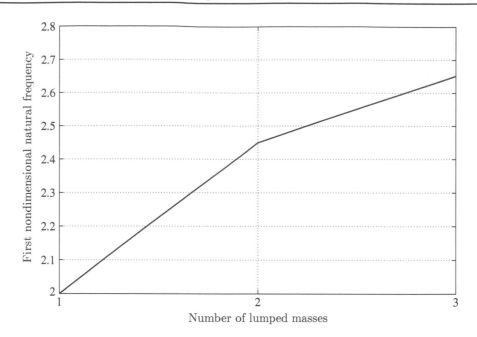

Figure 7.32: *Computed first natural frequency versus lumped masses*

Solving, using MATLAB, the natural frequencies turn out to be

$$\omega_1 = 2.6513\alpha, \quad \omega_2 = 4.8890\alpha, \quad \omega_3 = 6.4008\alpha \tag{7.90}$$

where

$$\alpha = \frac{1}{\ell}\sqrt{\frac{T}{\rho A}}$$

Note, how the first natural frequency compares to 2.45 we obtained with the two-DOF model, and 2.00 we obtained with the one-DOF model. It looks like we are converging to some answer! This convergence is shown graphically in Fig. 7.32. Do Problem 7.41 if you really want to discover what it will converge to when you use more DOF.

The mode shapes can be determined from MATLAB and are as follows:

$$\mathbf{u}_1 = \begin{bmatrix} 1 \\ \sqrt{2} \\ 1 \end{bmatrix}, \quad \mathbf{u}_2 = \begin{bmatrix} 1 \\ 0 \\ -1 \end{bmatrix}, \quad \mathbf{u}_3 = \begin{bmatrix} -1 \\ \sqrt{2} \\ -1 \end{bmatrix} \tag{7.91}$$

Figures 7.33, 7.34, and 7.35 display the three modes. Note, how the first mode is now a lot more intuitive in how we think a string might vibrate (actually, can you guess what kind of mathematical shape our model is trying to approximate?). The second mode displays a node at the center, and the third displays two nodes.

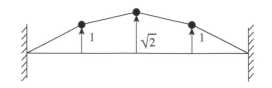

Figure 7.33: *First mode of string with three lumped masses*

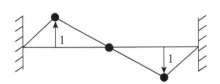

Figure 7.34: *Second mode of string with three lumped masses*

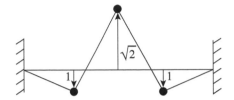

Figure 7.35: *Third mode of string with three lumped masses*

Example 7.3 A three-DOF system

Consider a three-DOF model for a vibrating system shown schematically using masses and springs in Fig. 7.36. Although this problem can also be carried out by hand, let us use MATLAB to get an exposure to computational procedures for MDOF systems.

The FBDs are shown in Fig. 7.37. The EOM follows from applying Newton's second law for each of the masses modeled as particles.

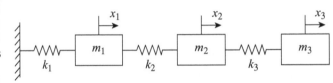

Figure 7.36: *Example problem with three DOF*

$$m_1\ddot{x}_1 = -k_1x_1 - k_2(x_1 - x_2) \tag{7.92}$$
$$m_2\ddot{x}_2 = -k_2(x_2 - x_1) - k_3(x_2 - x_3) \tag{7.93}$$
$$m_3\ddot{x}_3 = -k_3(x_3 - x_2) \tag{7.94}$$

Rewriting in matrix form

$$\mathbf{M}\ddot{\mathbf{x}} + \mathbf{K}\mathbf{x} = 0 \tag{7.95}$$

where

$$\mathbf{M} = \begin{bmatrix} m_1 & 0 & 0 \\ 0 & m_2 & 0 \\ 0 & 0 & m_3 \end{bmatrix} \tag{7.96}$$

and

$$\mathbf{K} = \begin{bmatrix} k_1 + k_2 & -k_2 & 0 \\ -k_2 & k_2 + k_3 & -k_3 \\ 0 & -k_3 & k_3 \end{bmatrix} \tag{7.97}$$

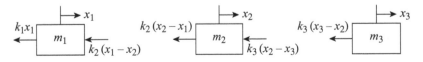

Figure 7.37: *FBD for example problem with three DOF*

Suppose $m_1 = 1$ kg, $m_2 = m_3 = 2$ kg, and $k_1 = 1000$ N/m, $k_2 = 1500$ N/m, $k_3 = 600$ N/m. Then, the matrices are as follows:

$$\mathbf{M} = \begin{bmatrix} 1 & 0 & 0 \\ 0 & 2 & 0 \\ 0 & 0 & 2 \end{bmatrix}, \quad \mathbf{K} = \begin{bmatrix} 2500 & -1500 & 0 \\ -1500 & 2100 & -600 \\ 0 & -600 & 600 \end{bmatrix} \tag{7.98}$$

Listing 7.1 shows the MATLAB implementation to specify the matrices, and to solve the eigenvalue problem. The natural frequencies turn out to be 10.4, 26.0, and 55.4 rad/s, and the eigenvectors are as follows:

$$\mathbf{u}_1 \begin{bmatrix} 0.23 \\ 0.37 \\ 0.58 \end{bmatrix}, \quad \mathbf{u}_2 \begin{bmatrix} -0.41 \\ -0.50 \\ 0.40 \end{bmatrix}, \quad \mathbf{u}_3 \begin{bmatrix} -0.88 \\ 0.33 \\ -0.04 \end{bmatrix} \tag{7.99}$$

Figure 7.38 illustrates these mode shapes.

Listing 7.1: *Natural frequencies and mode shapes of an MDOF model*

```
% mdof_three_example.m
% 3 DOF example
% in consistent SI units

% system matrices
  m1 = 1; m2 = 2; m3 = 2;
  k1 = 1000; k2 = 1500; k3 = 600;
  mass = [m1 0 0; ...
          0 m2 0; ...
          0 0 m3];
  stiff= [ k1+k2 -k2      0;  ...
          -k2     k2+k3 -k3;  ...
           0     -k3      k3];

% solve for the natural frequencies and modeshapes
  [modmat, eigval] = eig(stiff,mass);
% sort the eigenvalues and vectors
  [lambda,index] = sort(diag(eigval));
  modmat = modmat(:,index);
% natural frequencies are the square roots of eigenvalues
% columns of modmat are the eigenvectors
  natural_frequencies = sqrt(lambda);

  xvar=[1 2 3 ];
  plot(xvar,modmat(:,1),'-', ...
       xvar,modmat(:,2),'-.', ...
       xvar,modmat(:,3),'--', ...
       'LineWidth',2.0);
  grid on;
  xlabel('Degree_of_Freedom');
% notice how we can place the legend at a specific place
  legend('Mode_1','Mode_2','Mode_3','Location','SouthEast');
% change where the 'ticks' go
  set(gca,'XTick',[1  2  3 ]);
% notice how this stretches out the plot
  set(gca,'PlotBoxAspectRatio',[2 1 1]);
```

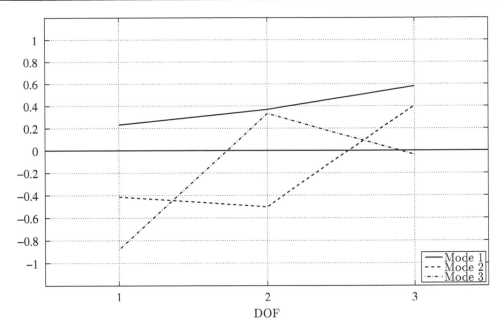

Figure 7.38: *Mode shapes for the three-DOF example*

7.4 General Theory

Now that we have looked at a relatively simple example of a two- (and three-) DOF oscillator and have been exposed to the fascinating phenomena that occur, we are ready (or should be!) for a general theory. We will also explore, in a more rigorous fashion, the strange properties we just witnessed with the car model. First, we summarize what we have already learnt.

7.4.1 Summary

Clearly, free undamped vibration of an MDOF system (of order N) leads to the following set of coupled differential equations. We assume that M and K are symmetric.

$$\mathbf{M\ddot{x}} + \mathbf{Kx} = \mathbf{0} \tag{7.100}$$

Seeking an exponential solution, $\mathbf{x}(t) = \mathbf{u}e^{st}$, we get the eigenvalue problem

$$\left(s^2\mathbf{M} + \mathbf{K}\right)\mathbf{u} = \mathbf{0} \tag{7.101}$$

Since we know that it will lead to an oscillatory solution, it might be more intuitive in this case to let $s = i\omega$; then, the eigenvalue problem can be rewritten

$$\mathbf{Ku} = \omega^2\mathbf{Mu} \tag{7.102}$$

The solution of this problem now reduces to solving for the eigenvalues using the characteristic equation

$$\left|-\omega^2\mathbf{M} + \mathbf{K}\right| = 0 \tag{7.103}$$

This will be an algebraic equation of order $2N$ in ω, with only even powers; hence, it can be solved as an algebraic equation of order N in ω^2. The solution will lead to the eigenvalues, square roots of which are the natural frequencies of the system. Clearly, there will be N natural frequencies. We assume that the eigenvalues are all distinct and nonzero—we will not handle the special cases of repeated eigenvalues in this book.

We normally arrange the natural frequencies in numerically increasing order

$$\omega_1 < \omega_2 < \ldots \omega_N \tag{7.104}$$

The first of these (ω_1) is often called the *fundamental* natural frequency, and is a very important property of any vibratory system.

The next step is to solve for the eigenvectors corresponding to each eigenvalue.

$$\mathbf{K}\mathbf{u}_r = \omega_r^2 \mathbf{M}\mathbf{u}_r \quad r = 1, 2, \ldots N \tag{7.105}$$

where ω_r is the rth natural frequency, and $\mathbf{u}_r$ is the rth eigenvector. We will have N of these eigenvectors, which are often called the *natural modes* of the system. The eigenvectors are not unique in the sense that they are arbitrary to a multiplicative constant. In other words, if $\mathbf{u}_r$ is an eigenvector, so is $c\mathbf{u}_r$, where c is any constant. This implies that the *shape* of the eigenvector is unique and important, but not its amplitude. It is for this reason that they are also called *mode shapes*.

Since the eigenvector has an arbitrariness associated with it (this really upsets us engineers; we really have to standardize everything!), we would like to normalize them in some way. Although there are numerous schemes for normalization (the process of normalization is—you guessed it—arbitrary), the most common approach in vibration is to use the following procedure. We multiply each eigenvector by a constant such that

$$\mathbf{u}_r^T \mathbf{M}\mathbf{u}_r = 1 \quad r = 1, 2, \ldots N \tag{7.106}$$

Premultiplying Eq. (7.105) by $\mathbf{u}_r^T$, we get

$$\mathbf{u}_r^T \mathbf{K}\mathbf{u}_r = \omega_r^2 \quad r = 1, 2, \ldots N \tag{7.107}$$

The response of the system in the rth mode is called the *natural motion* and is given by

$$\mathbf{x}_r(t) = C_r \mathbf{u}_r \cos(\omega_r t - \phi_r) \tag{7.108}$$

The complete solution is given by a superposition of all the natural motions:

$$\mathbf{x}(t) = \sum_{r=1}^{N} C_r \mathbf{u}_r \cos(\omega_r t - \phi_r) \tag{7.109}$$

There are $2N$ unknown constants in the complete solution, specified here as C_r and ϕ_r. These are determined from the $2N$ initial conditions (displacements and velocities at time zero).

Sometimes, we like to adjoin the eigenvectors together as follows:

$$\mathbf{U} = \begin{bmatrix} \mathbf{u}_1 & \mathbf{u}_2 & \ldots & \mathbf{u}_N \end{bmatrix} \tag{7.110}$$

This $N \times N$ matrix is called the *modal matrix*. Remember that each eigenvector is normalized individually (using possibly different constants) before assembly into the modal matrix. We can also assemble the eigenvalues into the diagonals of a matrix (the way MATLAB returns the values) as follows:

$$\Lambda = \begin{bmatrix} \omega_1^2 & 0 & \ldots & 0 \\ 0 & \omega_2^2 & \ldots & 0 \\ \ldots & \ldots & \ldots & \ldots \\ 0 & 0 & \ldots & \omega_N^2 \end{bmatrix} \tag{7.111}$$

7.4.2 Orthogonality

We next discuss a very important property of eigenvectors (so pay attention!) that is the underlying principle of most vibration analysis techniques. Although it appears to be quite theoretical, this property has a far-reaching practical value. With the assumptions we have made (symmetric mass and stiffness matrices, and distinct natural frequencies), it is quite easy to prove.

Consider two distinct eigensolutions, say r and s.

$$\mathbf{K}\mathbf{u}_r = \omega_r^2 \mathbf{M}\mathbf{u}_r \tag{7.112}$$

$$\mathbf{K}\mathbf{u}_s = \omega_s^2 \mathbf{M}\mathbf{u}_s \tag{7.113}$$

Let us premultiply the first of these by $\mathbf{u}_s^T$, and the second by $\mathbf{u}_r^T$. Then,

$$\mathbf{u}_s^T \mathbf{K}\mathbf{u}_r = \omega_r^2 \mathbf{u}_s^T \mathbf{M}\mathbf{u}_r \tag{7.114}$$

$$\mathbf{u}_r^T \mathbf{K}\mathbf{u}_s = \omega_s^2 \mathbf{u}_r^T \mathbf{M}\mathbf{u}_s \tag{7.115}$$

We then transpose the second of these, and use the symmetry properties of K and M.

$$\mathbf{u}_s^T \mathbf{K}\mathbf{u}_r = \omega_s^2 \mathbf{u}_s^T \mathbf{M}\mathbf{u}_r \tag{7.116}$$

We now subtract this from Eq. (7.114).

$$\left(\omega_r^2 - \omega_s^2\right) \mathbf{u}_s^T \mathbf{M}\mathbf{u}_r = 0 \tag{7.117}$$

Since $\omega_r \neq \omega_s$, we have

$$\mathbf{u}_s^T \mathbf{M}\mathbf{u}_r = 0 \quad \forall\, r \neq s \tag{7.118}$$

Also, using Eq. (7.114),

$$\mathbf{u}_s^T \mathbf{K}\mathbf{u}_r = 0 \quad \forall\, r \neq s \tag{7.119}$$

If we normalized them by the procedure outlined earlier,

$$\mathbf{u}_s^T \mathbf{M}\mathbf{u}_r = \delta_{rs} \tag{7.120}$$

and

$$\mathbf{u}_s^T \mathbf{K}\mathbf{u}_r = \delta_{rs}\omega_r^2 \tag{7.121}$$

where δ_{rs} is called the *Kronecker delta* and is defined by

$$\delta_{rs} = \left\{ \begin{array}{ll} 1, & r = s \\ 0, & r \neq s \end{array} \right. \tag{7.122}$$

This is the *orthogonality* property of eigenvectors. In terms of the modal matrix, it can be written as

$$\mathbf{U}^T \mathbf{M}\mathbf{U} = \mathbf{I} \tag{7.123}$$

$$\mathbf{U}^T \mathbf{K}\mathbf{U} = \Lambda \tag{7.124}$$

7.4.3 Linear Independence

The eigenvectors $\mathbf{u}_r$ form a linearly independent set of vectors. In other words, any $N \times 1$ vector can be constructed as a linear combination of these vectors.

Proof of linear independence

We will prove this by *reductio ad absurdum*. First, we assume that the eigenvectors *are* linearly independent. Then, it follows that

$$c_1 \mathbf{u}_1 + c_2 \mathbf{u}_2 + \cdots + c_N \mathbf{u}_N = \sum_{r=1}^{N} c_r \mathbf{u}_r = 0 \tag{7.125}$$

where c_r are constants with at least one being nonzero. Premultiplying by $\mathbf{u}_s^T \mathbf{M}$

$$\sum_{r=1}^{N} c_r \mathbf{u}_s^T \mathbf{M} \mathbf{u}_r = 0 \tag{7.126}$$

Expanding the above sum, we find that each of the terms does go to zero for $r \neq s$ because of orthogonality. However, the term when $r = s$, $\mathbf{u}_s^T \mathbf{M} \mathbf{u}_s$ does not go to zero. Hence, the above is true only if $c_s = 0$. Repeating the above process for $s = 1, 2, \ldots N$, it follows that it can only be true if $c_1 = c_2 = \ldots c_N = 0$. In other words, the initial supposition is wrong, and the eigenvectors are linearly independent.

Given a set of linearly independent vectors, it follows from linear algebra that they form a basis for the N-dimensional vector space, and hence any vector $\mathbf{u}$ can be expressed as a combination of the eigenvectors as follows:

$$\mathbf{u} = c_1 \mathbf{u}_1 + c_2 \mathbf{u}_2 + \cdots + c_N \mathbf{u}_N = \sum_{r=1}^{N} c_r \mathbf{u}_r \tag{7.127}$$

In terms of vibration of a dynamic system, this means that any possible motion of the system can be described by a linear combination of the eigenvectors. This profound result is called the *expansion theorem*.

To determine the coefficients c_r of the above expansion we use the now-familiar procedure of premultiplying by appropriate quantities and using orthogonality property of the eigenvectors. Premultiply Eq. (7.127) by $\mathbf{u}_s^T \mathbf{M}$. Then,

$$\mathbf{u}_s^T \mathbf{M} \mathbf{u} = \sum_{r=1}^{N} c_r \mathbf{u}_s^T \mathbf{M} \mathbf{u}_r \tag{7.128}$$

Orthogonality implies that every term on the right-hand side is zero except for one: $\mathbf{u}_s^T \mathbf{M} \mathbf{u}_s$, which is unity because of normalization. Hence, it follows that

$$c_r = \mathbf{u}_r^T \mathbf{M} \mathbf{u} \tag{7.129}$$

7.4.4 Modal Analysis

Using the expansion theorem as a basis, let us derive a general procedure to determine the response of an MDOF system to initial conditions. Consider again the undamped free vibration model

$$\mathbf{M}\ddot{\mathbf{x}} + \mathbf{K}\mathbf{x} = 0 \tag{7.130}$$

with initial conditions

$$\mathbf{x}(0) = \mathbf{x}_0 \tag{7.131}$$
$$\dot{\mathbf{x}}(0) = \mathbf{v}_0 \tag{7.132}$$

We wish to determine $x(t)$ for all time.

First, we solve the eigenvalue problem, and determine the natural frequencies ω_r, and the eigenvectors $\mathbf{u}_r$. Then, we use the expansion theorem to express the solution as a linear combination of the eigenvectors.

$$\mathbf{x}(t) = \mathbf{u}_1 \eta_1(t) + \mathbf{u}_2 \eta_2(t) + \cdots + \mathbf{u}_N \eta_N(t) = \sum_{r=1}^{N} \mathbf{u}_r \eta_r(t) = \mathbf{U}\eta(t) \tag{7.133}$$

where

$$\eta(t) = \begin{bmatrix} \eta_1(t) & \eta_2(t) & \cdots & \eta_N(t) \end{bmatrix}^T \tag{7.134}$$

Using the concepts developed in the previous section, it follows that

$$\eta_r(t) = \mathbf{u}_r^T \mathbf{M}\mathbf{x}(t) \quad r = 1,2,\ldots N \tag{7.135}$$

Substituting into the EOM,

$$\mathbf{M}\mathbf{U}\ddot{\eta} + \mathbf{K}\mathbf{U}\eta = 0 \tag{7.136}$$

Premultiplying (here we go again!) by $\mathbf{U}^T$

$$\mathbf{U}^T\mathbf{M}\mathbf{U}\ddot{\eta} + \mathbf{U}^T\mathbf{K}\mathbf{U}\eta = 0 \tag{7.137}$$

Using orthogonality [Eqs. (7.123) and (7.124)],

$$\mathbf{I}\ddot{\eta} + \Lambda\eta = 0 \tag{7.138}$$

Since both $\mathbf{I}$ and Λ are diagonal matrices, this amounts to N uncoupled second order ODEs in η_r.

$$\ddot{\eta}_r(t) + \omega_r^2 \eta(t) = 0 \tag{7.139}$$

The solution is of course laughably easy (remember when the solution of a single second-order ODE was not so trivial? Go ahead and pat yourself on the back for the progress you have made—you deserve it!), and is given by

$$\eta_r(t) = C_r \cos(\omega_r t - \phi_r) \tag{7.140}$$

$\eta_r(t)$ are called the *normal coordinates* of the system.

Now, $x(t)$, the *physical coordinates* of the system, and $\eta(t)$, the normal coordinates of the system can be considered to be two different ways of describing the motion of the system, with the modal matrix $\mathbf{U}$ serving as the linear transformation matrix between them as expressed in Eq. (7.133). In fact, as we discovered earlier with the car vibration example, there are many (actually, infinite) ways of picking these coordinates to describe what is happening in the physical system. In particular, $\eta_r(t)$ happen to be very special and unique because they lead to completely (statically and dynamically) decoupled system, as in Eq. (7.138). Hence, conceptually, we can think of the normal coordinates as N *independent* mass-spring oscillators that together give us a system that is completely equivalent to the original system. Of course, they may not make as much sense in a real physical system as using a physical coordinate such as displacement of a particular mass, for example. The normal coordinates are nevertheless very important for analysis, experimental characterization, and control. So, you will encounter them again and again.

Now, back to determining the *real* displacement coordinate, $\mathbf{x}(t)$.

$$\mathbf{x}(t) = \sum_r \mathbf{u}_r \eta_r(t)$$

$$= \sum_r C_r \mathbf{u}_r \cos(\omega_r t - \phi_r) \tag{7.141}$$

The question is, how do we easily determine the many unknown constants C_r and ϕ_r from the initial conditions. You should have guessed the answer by now; yes, the answer is again orthogonality! From the given initial conditions

$$\mathbf{x}(0) = \mathbf{x}_0 = \sum_r C_r \mathbf{u}_r \cos\phi_r \tag{7.142}$$

$$\dot{\mathbf{x}}(0) = \mathbf{v}_0 = \sum_r C_r \omega_r \mathbf{u}_r \sin\phi_r \tag{7.143}$$

premultiplying by $\mathbf{u}_s^T M$, and using orthogonality (you should work out the algebraic details here)

$$C_r \cos\phi_r = \mathbf{u}_r^T \mathbf{M} \mathbf{x}_0 \tag{7.144}$$

$$C_r \sin\phi_r = \frac{1}{\omega_r} \mathbf{u}_r^T \mathbf{M} \mathbf{v}_0 \tag{7.145}$$

Substituting this in Eq. (7.141), we have the complete solution

$$\mathbf{x}(t) = \sum_{r=1}^N \left\{ \mathbf{u}_r^T \mathbf{M} \mathbf{x}_0 \cos\omega_r t + \frac{1}{\omega_r} \mathbf{u}_r^T \mathbf{M} \mathbf{v}_0 \sin\omega_r t \right\} \mathbf{u}_r \tag{7.146}$$

Note how the response is a superposition of the normal modes (each vibrating at its natural frequency). The weightage of each of the modes is determined by the term $\mathbf{u}_r^T \mathbf{M}$ multiplying the initial condition vectors. Hence, the contribution of each of the modes to the real response of the system is not identical, and is essentially determined by the *shape* of the initial condition vector and how much it resembles an eigenvector (mode shape), or if it is "far" from it.

To elucidate this, consider that the system has an initial displacement that is identical to the second mode; in other words, let $\mathbf{x}_0 = x_0 \mathbf{u}_2$, where x_0 is a constant, and with the initial velocity being zero: $\mathbf{v}_0 = 0$. Then, every term in the summation in Eq. (7.146) is zero except for only one, namely $\mathbf{u}_2^T \mathbf{M} \mathbf{u}_2$, by virtue of orthogonality. Hence, the response reduces to the following.

$$\mathbf{x}(t) = x_0 \mathbf{u}_2 \cos\omega_2 t \tag{7.147}$$

Now, the response in *every* DOF is simply a simple harmonic oscillation at the second natural frequency, and moreover, the system response resembles the second mode shape exactly. In other words, we started with an initial condition that resembled the second mode shape, and this resulted in a response that is exactly like the second mode. There is no contribution to the final response from any of the other modes. Of course, instead of the second, if we had chosen third, first, or sth mode, we would have had a response in only that mode. In general, of course, the response is hardly so simple; it is often a confusing collection of all of the simple harmonic responses occurring at the individual natural frequencies, with varying contributions from each of the modes.

— **Example 7.4 General response of the coupled pendulums** —

Consider the coupled pendulums we looked at earlier. The mass and stiffness matrices are reproduced here for convenience.

$$\mathbf{M} = \begin{bmatrix} m\ell^2 & 0 \\ 0 & m\ell^2 \end{bmatrix}, \quad \mathbf{K} = \begin{bmatrix} ka^2 + mg\ell & -ka^2 \\ -ka^2 & ka^2 + mg\ell \end{bmatrix} \tag{7.148}$$

Recall that the natural frequencies are given by

$$\omega_1 = \sqrt{\frac{g}{\ell}}, \quad \omega_2 = \sqrt{\frac{g}{\ell} + 2\frac{k}{m}\frac{a^2}{\ell^2}} \tag{7.149}$$

The *unnormalized* eigenvectors are as follows:

$$\mathbf{u}_1 = \begin{bmatrix} 1 \\ 1 \end{bmatrix}, \quad \mathbf{u}_2 = \begin{bmatrix} 1 \\ -1 \end{bmatrix} \tag{7.150}$$

Let us go ahead and normalize them now, using the convention we just studied. Recognizing that we can multiply the eigenvector by any constant, we let

$$\mathbf{u}_1 = C_1 \begin{bmatrix} 1 \\ 1 \end{bmatrix}, \quad \mathbf{u}_2 = C_2 \begin{bmatrix} 1 \\ -1 \end{bmatrix} \tag{7.151}$$

Next, we let $\mathbf{u}_1^T \mathbf{M} \mathbf{u}_1 = 1$. Or,

$$C_1 \begin{bmatrix} 1 & 1 \end{bmatrix} \begin{bmatrix} m\ell^2 & 0 \\ 0 & m\ell^2 \end{bmatrix} C_1 \begin{bmatrix} 1 \\ 1 \end{bmatrix} = 1 \tag{7.152}$$

This simplifies to

$$2m\ell^2 C_1^2 = 1 \Rightarrow C_1 = \frac{1}{\sqrt{2m\ell^2}}$$

Hence, the normalized first eigenvector is

$$\mathbf{u}_1 = \frac{1}{\sqrt{2m\ell^2}} \begin{bmatrix} 1 \\ 1 \end{bmatrix} \tag{7.153}$$

Similarly, we can obtain the normalized second eigenvector to be

$$\mathbf{u}_2 = \frac{1}{\sqrt{2m\ell^2}} \begin{bmatrix} 1 \\ -1 \end{bmatrix} \tag{7.154}$$

The complete solution (reproduced from earlier, with $N = 2$) is

$$\mathbf{x}(t) = \sum_{r=1}^{2} \left\{ \mathbf{u}_r^T \mathbf{M} \mathbf{x}_0 \cos \omega_r t + \frac{1}{\omega_r} \mathbf{u}_r^T \mathbf{M} \mathbf{v}_0 \sin \omega_r t \right\} \mathbf{u}_r \tag{7.155}$$

We will determine the response of the system with various initial conditions and follow up with a discussion about the meaning of all this.

Suppose we push both pendulums by an equal amount in the same direction, and let go ("letting go" is important if we did not intend to give it an initial velocity). Then,

$$\theta_1(0) = \theta_2(0) = \theta_0, \quad \dot{\theta}_1(0) = \dot{\theta}_2(0) = 0 \tag{7.156}$$

Or,

$$\mathbf{x}_0 = \theta_0 \begin{bmatrix} 1 \\ 1 \end{bmatrix}^T$$

and $\mathbf{v}_0 = \mathbf{0}$. Then, simple algebraic verification will show that $\mathbf{u}_2^T \mathbf{M} \mathbf{x}_0 = 0$. Even though we know this from the general theory, you should really verify this by yourself. The summation over the two modes then reduces to one term, and the resulting solution is as follows:

$$\mathbf{x}(t) = \frac{1}{\sqrt{2m\ell^2}} (2m\ell^2)\theta_0 \frac{1}{\sqrt{2m\ell^2}} \begin{bmatrix} 1 \\ 1 \end{bmatrix} \cos \omega_1 t = \begin{bmatrix} \theta_0 \\ \theta_0 \end{bmatrix} \cos \omega_1 t \tag{7.157}$$

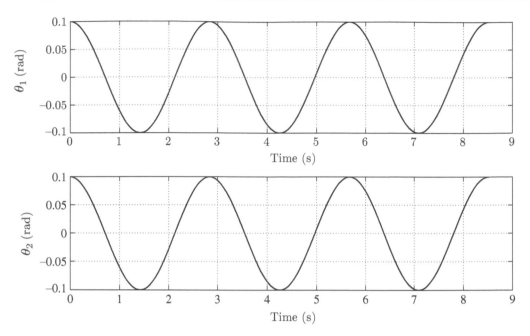

Figure 7.39: *Initial condition response of coupled pendulums to moving both pendulums together*

Let us write out both degrees of freedom for clarity.

$$\theta_1(t) = \theta_0 \cos \omega_1 t \tag{7.158}$$

$$\theta_2(t) = \theta_0 \cos \omega_1 t \tag{7.159}$$

Hence, the two pendulums will move in sync with exactly the same displacement at any instant of time. The spring coupling them sees no action whatsoever (note that this spring does not appear in the first natural frequency—does this have a bearing on the mode shape?). In other words, the system is now exhibiting purely the first mode of vibration. Figure 7.39 shows the two responses with time.

Now, suppose we push them apart by the same amount and let go; i.e.,

$$\theta_1(0) = -\theta_2(0) = \theta_0, \quad \dot{\theta}_1(0) = \dot{\theta}_2(0) = 0 \tag{7.160}$$

Or,

$$\mathbf{x}_0 = \theta_0 \begin{bmatrix} 1 \\ -1 \end{bmatrix}^T$$

and $\mathbf{v}_0 = \mathbf{0}$. This time, $\mathbf{u}_1^T \mathbf{M} \mathbf{x}_0 = 0$. Again, verify! The summation over the two modes then reduces to the second term, and the resulting solution is as follows:

$$\mathbf{x}(t) = \frac{1}{\sqrt{2m\ell^2}}(2m\ell^2)\theta_0 \frac{1}{\sqrt{2m\ell^2}} \begin{bmatrix} 1 \\ -1 \end{bmatrix} \cos \omega_2 t == \begin{bmatrix} \theta_0 \\ -\theta_0 \end{bmatrix} \cos \omega_2 t \tag{7.161}$$

Let us again write out both degrees of freedom for clarity.

$$\theta_1(t) = \theta_0 \cos \omega_2 t \tag{7.162}$$

$$\theta_2(t) = -\theta_0 \cos \omega_2 t \tag{7.163}$$

This time, the two pendulums are moving out of sync with exactly the opposite displacement at any instant of time. The spring coupling them sees a lot of action (note that the spring does appear in the second natural

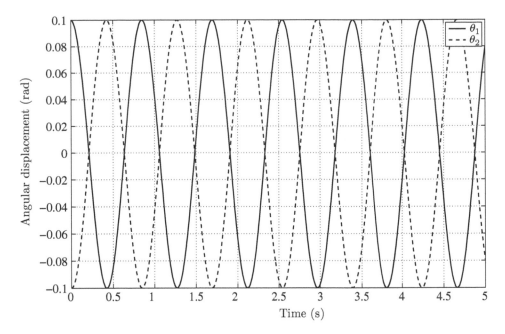

Figure 7.40: *Initial condition response of coupled pendulums to opposite displacements*

frequency—comments?). The system is now exhibiting purely the second mode of vibration. Figure 7.40 shows the two responses with time.

Next, let us give them an initial condition that is not like either of the two modes. Let us say, we just push the first pendulum, and watch the fun. Then,

$$\theta_1(0) = \theta_0, \quad \theta_2(0) = 0, \quad \dot{\theta}_1(0) = \dot{\theta}_2(0) = 0 \tag{7.164}$$

Or,

$$\mathbf{x}_0 = \theta_0 \begin{bmatrix} 1 \\ 0 \end{bmatrix}^T$$

and $\mathbf{v}_0 = \mathbf{0}$. The summation over the two modes does not simplify, since the initial condition vector is not orthogonal to either of the two modes, and the resulting solution is as follows:

$$\theta_1(t) = \frac{\theta_0}{2} [\cos \omega_1 t + \cos \omega_2 t] \tag{7.165}$$

$$\theta_2(t) = \frac{\theta_0}{2} [\cos \omega_1 t - \cos \omega_2 t] \tag{7.166}$$

Figure 7.41 shows the response of both the pendulums. Note that both the modes (and the natural frequencies) are evident in the response of both the masses.

The last initial condition we will consider is an impulse delivered to the second mass. This time, $\mathbf{x}_0 = \mathbf{0}$, and

$$\mathbf{v}_0 = v_0 \begin{bmatrix} 0 \\ 1 \end{bmatrix}^T$$

The algebra is left to you for exercise. The results are shown in Fig. 7.42. Note, how the response can be quite confusing to interpret; fortunately we are smart, and figured out the natural frequencies and mode shapes *first* from which we gained tremendous insight into the dynamics of the system.

To summarize, the final responses obtained in the above example would be difficult to interpret (say if we had obtained them from experiment or numerical simulation *without* the benefit of the eigenvalue problem); but

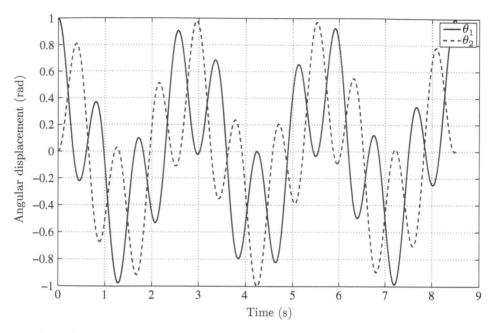

Figure 7.41: *Initial condition response of coupled pendulums to moving the first pendulum*

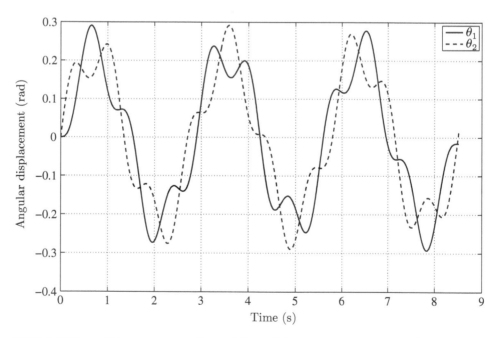

Figure 7.42: *Initial condition response of coupled pendulums to an impulse on the second pendulum*

they can be clearly seen as a relatively simple superposition of two harmonic modal oscillators only because we solved the eigenvalue problem first, and then obtained the initial condition response by modal analysis. In fact, if our system had more DOF resulting in a large number of natural frequencies, many of which could appear in the transient response, we would be completely lost in an apparently chaotic response. Our only survival technique from this hopeless situation is the method we have followed—that of first determining the natural frequencies and mode shapes. In fact, if the response to the initial conditions is going to be so confusing, in practice, is it even necessary to determine it?

This brings us to the final discussion on the importance (or lack of it) of initial conditions. When we determine the response of a model to a set of initial conditions, we are essentially seeking to represent the real behavior of a physical system in realistic conditions. However, there are three problems with this. The first is something we have already noted: the response obtained is often very confusing because it is a combination of incommensurate frequencies. Moreover, different sets of initial conditions reveal very different responses that tell us close to nothing insightful about the system; and, there are infinite possibilities for the initial conditions, and we will run out of time before we run out of initial conditions (look at the "think-box" for an interesting problem about time). The third problem is that it is pretty much impossible to guess what initial conditions we would get in practice in order to be able to simulate them (next time you are riding in a car, try to guess the initial conditions for the bouncing response of the car!). So, it is really not very useful to simulate the response of a vibrating system to initial conditions; at any rate, it does not give us anything more than the understanding that we have already gleaned from the natural frequencies and the mode shapes.

Here is a completely irrelevant problem that was sparked from the comment above about the time it takes to evaluate response to *all possible* initial conditions. This is a very old puzzle possibly dating from ancient times, but also possibly re-invented by the French mathematician Edouard Lucas in the nineteenth century [31]. In any event, the puzzle, called "Tower of Brahma" (also known as "Tower of Hanoi") is as follows. In the holy city of Varanasi (India), there is a dome that marks the center of the world. Below that dome is a brass plate in which there are three diamond needles. Lord Brahma placed 64 disks of pure gold on one of these needles at the time that he created the world. Each disk is of a different size, and each is placed so that it rests on top of another disk of greater size, with the largest resting on the plate at the bottom, and the smallest at the top. The job of the temple priests is to transfer all the gold disks from their original needle to one of the others, without ever moving more than one disk at a time. No priest can ever place any disk on top of a smaller one, or anywhere else except on one of the needles. When the task is done, and all the disks have been successfully transferred to another needle, the world will vanish!

Do not panic, because it turns out that the number of steps required to transfer all the disks is $2^{64} - 1$, which is approximately 1.8×10^{19}. Assuming one second per move, this would take about five times longer than the current age of the universe! Can you prove this?

Notwithstanding the above argument there are times when it does make sense to carry out an initial condition response of vibrating systems. An example is what we just did (with four sets of initial conditions) that enhanced our understanding of the concept of the modes. The second is, when we need to compare different systems in their response to some standard transient excitations (such as an impulse); this is often useful to come up with design standards, or to come up with standardized performance measures.

7.5   MDOF Systems: Free Damped Response

Let us now look at what will happen to the modes if we had damping. Perhaps it is best to start with an example. We introduce damping in the two-DOF model we considered earlier in the beginning

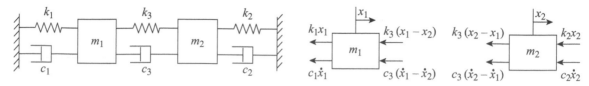

Figure 7.43: *A damped MDOF system*   **Figure 7.44:** *FBD of the damped MDOF system*

of the chapter, shown in Fig. 7.43 with linear viscous dampers. Then, the FBDs would be as shown in Fig. 7.44.

The equations of motion become

$$\begin{bmatrix} m_1 & 0 \\ 0 & m_2 \end{bmatrix}\begin{bmatrix} \ddot{x}_1 \\ \ddot{x}_2 \end{bmatrix} + \begin{bmatrix} c_1 + c_3 & -c_3 \\ -c_3 & c_2 + c_3 \end{bmatrix}\begin{bmatrix} \dot{x}_1 \\ \dot{x}_2 \end{bmatrix} + \begin{bmatrix} k_1 + k_3 & -k_3 \\ -k_3 & k_2 + k_3 \end{bmatrix}\begin{bmatrix} x_1 \\ x_2 \end{bmatrix} = \begin{bmatrix} 0 \\ 0 \end{bmatrix} \quad (7.167)$$

Or,

$$\mathbf{M\ddot{x}} + \mathbf{C\dot{x}} + \mathbf{Kx} = \mathbf{0} \quad (7.168)$$

where $\mathbf{C}$ is called the damping matrix. Hence, it is clear that, in general, we will end up with the additional term $(\mathbf{C\dot{x}})$ in the matrix EOM.

7.5.1   Rayleigh Damping

There are a couple of approaches at this point. One is to solve the undamped problem the usual way and to obtain eigenvalues and eigenvectors. We will sometimes call them *undamped eigenvalues* and *undamped eigenvectors* to emphasize the point. So, we now have ω_r and $\mathbf{u}_r$ for $r = 1, 2, \ldots N$. Now, like we did with the undamped problem, suppose we try the linear transformation to modal coordinates with

$$\mathbf{x}(t) = \sum_{r=1}^{N} \mathbf{u}_r \eta_r(t) = \mathbf{U}\eta(t) \quad (7.169)$$

Substituting into the EOM, Eq. (7.168), we get

$$\mathbf{MU}\ddot{\eta}(t) + \mathbf{CU}\dot{\eta}(t) + \mathbf{KU}\eta(t) = \mathbf{0} \quad (7.170)$$

Premultiplying by $\mathbf{U}^T$,

$$\mathbf{U}^T\mathbf{MU}\ddot{\eta}(t) + \mathbf{U}^T\mathbf{CU}\dot{\eta}(t) + \mathbf{U}^T\mathbf{KU}\eta(t) = \mathbf{0} \quad (7.171)$$

It is time to invoke that magical orthogonality property now

$$\mathbf{U}^T\mathbf{MU} = \mathbf{I}, \quad \mathbf{U}^T\mathbf{KU} = \Lambda \quad (7.172)$$

Applying the property,

$$\mathbf{I}\ddot{\eta}(t) + \mathbf{U}^T\mathbf{CU}\dot{\eta}(t) + \Lambda\eta(t) = \mathbf{0} \quad (7.173)$$

Clearly, the equations are no longer decoupled like they were in the undamped case unless somehow the middle term becomes diagonal. Well, there is a special kind of damping model when that could happen. It is called proportional damping, or more often, *Rayleigh damping*. That is when the damping matrix has the form

$$\mathbf{C} = \alpha\mathbf{M} + \beta\mathbf{K} \tag{7.174}$$

where α and β are some proportionality constants. If that is so, then

$$\mathbf{U}^T\mathbf{C}\mathbf{U} = \mathbf{U}^T\alpha\mathbf{M}\mathbf{U} + \mathbf{U}^T\beta\mathbf{K}\mathbf{U} = \alpha\mathbf{I} + \beta\Lambda \tag{7.175}$$

which is a diagonal matrix. Hence, now the EOM in terms of the modal coordinates would be decoupled. Or, we would have

$$\mathbf{I}\ddot{\eta}(t) + (\alpha\mathbf{I} + \beta\Lambda)\,\dot{\eta}(t) + \Lambda\eta(t) = \mathbf{0} \tag{7.176}$$

If we define

$$\alpha\mathbf{I} + \beta\Lambda = \operatorname{diag}(2\zeta_r\omega_r) \tag{7.177}$$

we get N modal equations in the following form.

$$\ddot{\eta}_r(t) + 2\zeta_r\omega_r\dot{\eta}_r(t) + \omega_r^2\eta_r(t) = 0 \tag{7.178}$$

This equation resembles the single-DOF damped oscillator equation, with ζ_r being called the *modal viscous damping ratio*, and is an important quantity.

The solution of this equation is trivial (since you have done this a hundred times by now) and is given by

$$\eta_r(t) = \left[\frac{1}{\sqrt{1-\zeta_r^2}}\,\eta_r(0)\cos\left(\omega_{dr}t - \psi_r\right) + \frac{1}{\omega_{dr}}\,\dot{\eta}_r(0)\sin\left(\omega_{dr}t\right)\right]e^{-\zeta_r\omega_r t} \tag{7.179}$$

where ω_{dr} is the rth modal damped natural frequency given by

$$\omega_{dr} = \omega_r\sqrt{1-\zeta_r^2} \tag{7.180}$$

and,

$$\psi_r = \tan^{-1}\left[\frac{\zeta_r}{\sqrt{1-\zeta_r^2}}\right] \tag{7.181}$$

Now, the displacement $\mathbf{x}$ can be computed from Eq. (7.169). So, if we assume Rayleigh damping, the mathematics works out very well and we get a convenient answer. There is of course a fly in this ointment; that is, how realistic is this Rayleigh damping? Probably not very! However, the truth is that it is just a model, or just a way to explain away the dissipation mechanism in MDOF models, and if it explains what is being observed fairly well, then it *is* a good model. And, of course, it does simplify the mathematics quite a bit. So, we will end up using it many times.

A better approach in MDOF models might be to think *from the modal space up*. That means, no matter where the damping comes from, it is clearly going to be visible in the modal responses. So, let us assume that there is some damping in each of the modes specified by ζ_r. This means that we do not bother to put the damping in at the system level at all. To clarify, we start with undamped models, carry out the eigenanalysis, and then, introduce appropriate amounts of damping at the modal level. It turns out that modal damping is relatively easy to estimate experimentally anyway, so this approach is actually in good consonance with real practice. Of course, an important exception to this approach is when sufficient information is available from the physics of the problem that would enable us to construct a reliable damping matrix, $\mathbf{C}$, in which case that would be the preferred approach.

7.5.2 Polynomial Eigenvalue Problem

There is yet another approach especially if we know enough about the physics so that we have the damping matrix with a sufficient degree of confidence. Then, we start with the damped EOM and proceed as usual; i.e., we assume an exponential solution:

$$\mathbf{x}(t) = \mathbf{A}e^{st} \tag{7.182}$$

Substituting into the differential EOM, Eq. (7.168),

$$\left(s^2\mathbf{M}\mathbf{A} + s\mathbf{C}\mathbf{A} + \mathbf{K}\mathbf{A}\right)e^{st} = 0 \tag{7.183}$$

Then, since $e^{st} \neq 0$, we get what is called the *polynomial* eigenvalue problem,

$$\left(s^2\mathbf{M} + s\mathbf{C} + \mathbf{K}\right)\mathbf{A} = 0 \tag{7.184}$$

Assuming for illustration that we have two DOF, this will now lead to a fourth-order characteristic equation in s that *cannot* be written as a quadratic equation in s^2 (as we did with the undamped EOM), since it has odd powers of s in it. This equation is easily solved numerically (although an analytical solution does exist for general quartic equations). In general, the solution of this equation yields complex roots that are two pairs of complex conjugates. Let us assume that both the modes are of the form

$$s_{1,1}, \; s_{1,2} = a_1 \pm ib_1, \quad s_{2,1}, \; s_{2,2} = a_2 \pm ib_2 \tag{7.185}$$

where $i = \sqrt{-1}$. From analogy with the single-DOF damped system, we can now rewrite them in terms of damping ratios and natural frequencies as follows.

$$s_{1,1}, \; s_{1,2} = -\zeta_1\omega_1 \pm i\omega_{d1}, \quad s_{2,1}, \; s_{2,2} = -\zeta_2\omega_2 \pm i\omega_{d2} \tag{7.186}$$

where ω_{dr} has the same meaning as before. We will also get the eigenvectors from the numerical analysis, which are now *damped mode shapes*. Interpretation of the damped modes is quite difficult, but we can still use them in our solution in a straightforward manner. Also, note that the eigenvectors also appear in complex conjugate pairs (as required by a theorem from linear algebra).

We will be able to write the solution for the system displacement by simply combining the four solutions.

$$\mathbf{x}(t) = \sum_{i=1}^{4} \mathbf{A}_i e^{s_i t}$$

$$= [C_1 \sin\omega_{d1}t + C_2 \cos\omega_{d1}t]\,\mathbf{u}_1 e^{-\zeta_1\omega_1 t} + [C_3 \sin\omega_{d2}t + C_4 \cos\omega_{d2}t]\,\mathbf{u}_2 e^{-\zeta_2\omega_2 t} \tag{7.187}$$

C_i would be computed from the four initial conditions. This would hence be a simple combination of two single-DOF damped systems.

Suppose, we let $k_1 = k_2 = k_3 = k$, $m_1 = m_2 = m$, and $c_1 = c_2 = c_3 = c$. As we will be employing a numerical solution, we will also use numerical values as follows: $k = 100\,\text{N/m}$, $m = 1\,\text{kg}$, $c = 5\,\text{Ns/m}$. A MATLAB solution (see Listing 7.2) yields the damped eigenvalues as follows:

$$\begin{bmatrix} -2.5000 - 9.6825i \\ -2.5000 + 9.6825i \\ -7.5000 - 15.6125i \\ -7.5000 + 15.6125i \end{bmatrix} \tag{7.188}$$

The damped natural frequencies are the imaginary parts of the eigenvalues, namely 9.68 and 15.61 rad/s. Solving for the modal damping ratios, we get $\zeta_1 = 0.250$, and $\zeta_2 = 0.433$. We can also calculate the undamped natural frequencies, and they turn out to be 10 and 17.32 rad/s, which agree with what you would get by solving the undamped eigenvalue problem (Do it!).

Listing 7.2: *The eigenvalue problem for the damped MDOF system*

```
% solve eigenvalue problem of a two dof damped system

% parameter values in consistent units
  m1=1; m2=1;
  k1=100; k2=100; k3=100;
  c1=5; c2=5; c3=5;

% system matrices
  mass=[m1 0; 0 m2];
  stiff=[k1+k3 -k3; -k3 k2+k3];
  damp=[c1+c3 -c3; -c3 c2+c3];

% undamped system for comparison
% solve for the natural frequencies and modeshapes
  [modmat_undamped, eigval] = eig(stiff,mass);
% sort the eigenvalues and vectors
  [lambda_undamped,index] = sort(diag(eigval));
  modmat_undamped = modmat_undamped(:,index);
% natural frequencies are the square roots of eigenvalues
% columns of modmat are the eigenvectors (already normalized)
  undamped_natural_frequency = sqrt(lambda_undamped);

% damped system
% solve for the natural frequencies and modeshapes
  [modmat_damped, eigval] = polyeig(stiff,damp,mass);
% sort the eigenvalues and vectors
  imagval=abs(imag(eigval));
  [sorted_imgaval,index]=sort(imagval);
  lambda = eigval(index);
  modmat_damped = modmat_damped(:,index);
```

The damped modal matrix turns out to be

$$\begin{bmatrix} 1.0000 + 3.5958i & 1.0000 - 3.5958i & 1.0000 - 0.0034i & 1.0000 + 0.0034i \\ 1.0000 + 3.5958i & 1.0000 - 3.5958i & -1.0000 + 0.0034i & -1.0000 - 0.0034i \end{bmatrix} \qquad (7.189)$$

Note that there are two pairs of complex conjugate eigenvectors. For comparison, the two undamped eigenvectors (which are real) are

$$\begin{bmatrix} 1 & 1 \\ 1 & -1 \end{bmatrix} \qquad (7.190)$$

Let us also evaluate the response to a couple of sets of initial conditions. Say, we need the response of the system to $\mathbf{x}(0) = [1 \quad 0]^T$, and $\dot{\mathbf{x}}(0) = [0 \quad 0]^T$. Then, the response is calculated by substituting into Eq. (7.187) and solving for the unknown constants. Figure 7.45 shows the response of the two masses. Notice that, in this case, the first mode with a natural frequency of 9.68 rad/s dominates the response.

Let us try the set of initial conditions: $\mathbf{x}(0) = [1 \quad 1]^T$, and $\dot{\mathbf{x}}(0) = [5 \quad -10]^T$. Now, the response (Fig. 7.46) shows evidence of both the modes.

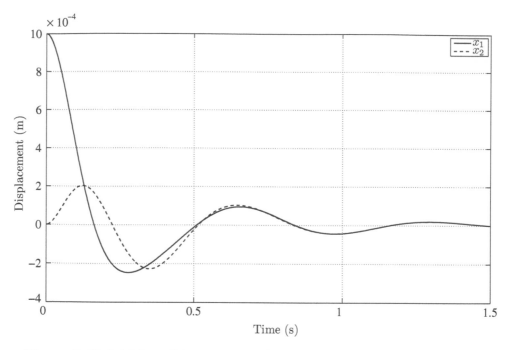

Figure 7.45: *Initial condition response of the damped two-DOF system; Case 1*

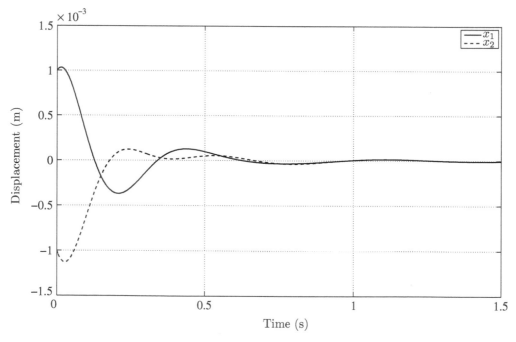

Figure 7.46: *Initial condition response of the damped two-DOF system; Case 2*

7.6 Review Questions

1. Referring to the car example, what is the necessity of including more than one DOF in the analysis?

2. What is the physical meaning of the eigenvalues?

3. What is the physical meaning of the eigenvectors?

4. In exactly what way are the eigenvectors arbitrary?

5. Why is that the fact that the eigenvectors are arbitrary *not* create a problem for us?

6. How many natural frequencies does a two-DOF model have?

7. How many modes does a two-DOF model have?

8. Define *fundamental* natural frequency.

9. What is the order of the stiffness matrix in a three-DOF model?

10. How many natural frequencies (and modes) does a 10,266-DOF model have?

11. Is the choice of coordinates unique? Why or why not?

12. Is the number of independent coordinates arbitrary? Why or why not?

13. How does the choice of coordinates influence the structure of the system matrices?

14. What is a statically coupled system?

15. What is a dynamically coupled system?

16. Do the natural frequencies depend on the choice of coordinates? Why or why not?

17. Do the eigenvectors depend on the choice of coordinates? Why or why not?

18. What is a *node* in MDOF vibration?

19. What is the physical interpretation of a zero eigenvalue? When does it occur?

20. What kind of stiffness matrix leads to a zero eigenvalue?

21. What is a modal matrix?

22. State the principle of orthogonality of eigenvectors.

23. What is the common way of normalizing the eigenvectors?

24. What is the expansion theorem?

25. What is modal analysis, and which theorem is it based on?

26. What are normal coordinates?

27. If the initial conditions resemble a particular mode, why is only that mode represented in the final motion?

28. What is the best way to interpret the motion of an MDOF system with arbitrary initial conditions?

29. What is Rayleigh damping, and how does it simplify the analysis?

30. What is modal damping?

31. When is the polynomial eigenvalue problem solution useful?

32. In case of damped modes, are the eigenvectors real or complex? Do pairs of them have a special property?

Problems

7.1 Two identical masses m are coupled by springs of stiffness k, as shown in Fig. 7.47. Derive the equations of motion, solve the eigenvalue problem, determine the natural frequencies, and sketch the modes of vibration.

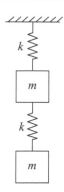

Figure 7.47: *Two-DOF system*

7.2 Analyze the system shown in Fig. 7.48 as a function of n. Derive the equations of motion, solve the eigenvalue problem, determine the natural frequencies, and sketch the modes of vibration [32].

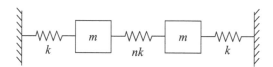

Figure 7.48: *Two-DOF system*

7.3 Two masses m and $3m$ are coupled by three springs, as shown in Fig. 7.49. Derive the equations of motion, solve the eigenvalue problem, determine the natural frequencies, and sketch the modes of vibration.

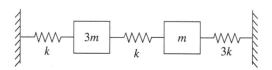

Figure 7.49: *Two-DOF system*

7.4 For each of the following shown in general form in Fig. 7.50, derive the equations of motion, solve the eigenvalue problem, determine the natural frequencies, and sketch the modes of vibration.

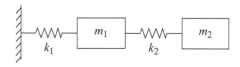

Figure 7.50: *Two-DOF system*

(a) $m_1 = m$, $m_2 = 2m$, $k_1 = k$, $k_2 = 2k$
(b) $m_1 = m_2 = m$, $k_1 = 20k$, $k_2 = k$
(c) $m_1 = 5m$, $m_2 = m$, $k_1 = k_2 = k$
(d) $m_1 = 5\,\mathrm{kg}$, $m_2 = 8\,\mathrm{kg}$, $k_1 = 10\,\mathrm{kN/m}$, $k_2 = 6\,\mathrm{kN/m}$
(e) $m_1 = 200\,\mathrm{kg}$, $m_2 = 50\,\mathrm{kg}$, $k_1 = 220\,\mathrm{kN/m}$, $k_2 = 600\,\mathrm{kN/m}$
(f) $m_1 = 20\,\mathrm{mg}$, $m_2 = 4\,\mathrm{mg}$, $k_1 = 100\,\mathrm{kN/m}$, $k_2 = 60\,\mathrm{kN/m}$

7.5 For each of the following shown in general form in Fig. 7.51, derive the equations of motion, solve the eigenvalue problem, determine the natural frequencies, and sketch the modes of vibration.

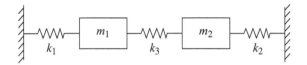

Figure 7.51: *Two-DOF system*

(a) $m_1 = m$, $m_2 = 2m$, $k_1 = k_2 = k_3 = k$
(b) $m_1 = m_2 = m$, $k_1 = 200k$, $k_2 = 200k$, $k_3 = k$
(c) $m_1 = 5m$, $m_2 = m$, $k_1 = k_2 = k$, $k_3 = 100k$
(d) $m_1 = 5\,\mathrm{kg}$, $m_2 = 8\,\mathrm{kg}$, $k_1 = 10\,\mathrm{kN/m}$, $k_2 = 6\,\mathrm{kN/m}$, $k_3 = 20\,\mathrm{kN/m}$
(e) $m_1 = 200\,\mathrm{kg}$, $m_2 = 50\,\mathrm{kg}$, $k_1 = 220\,\mathrm{kN/m}$, $k_2 = 600\,\mathrm{kN/m}$, $k_3 = 440\,\mathrm{kN/m}$
(f) $m_1 = 20\,\mathrm{mg}$, $m_2 = 4\,\mathrm{mg}$, $k_1 = 100\,\mathrm{kN/m}$, $k_2 = 60\,\mathrm{kN/m}$, $k_3 = 100\,\mathrm{kN/m}$

7.6 Assuming small amplitudes, analyze the double pendulum shown in Fig. 7.52; the two masses are identical (m), and the lengths are also the same (ℓ). Derive the equations of motion, solve the eigenvalue problem, determine the natural frequencies, and sketch the modes of vibration. Note that there are many choices for coordinates. In particular, you can exercise the following two choices; verify that the two choices lead to identical natural frequencies.

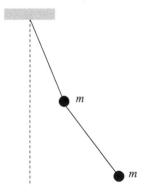

Figure 7.52: *Double pendulum*

(a) Use angles from the vertical as coordinates to describe both the pendulums.
(b) Use the relative angle for the second pendulum from the line of the first pendulum.

7.7 Assuming small amplitudes, analyze the double pendulum shown in Fig. 7.53; the two rods are identical with mass m and length ℓ. Derive the equations of motion, solve the eigenvalue problem, determine the natural frequencies, and sketch the modes of vibration. Note that there are many choices for coordinates. In particular, you can exercise the following two choices; verify that the two choices lead to identical natural frequencies.

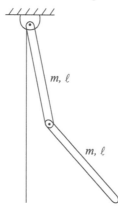

Figure 7.53: *Double rod pendulum*

(a) Use angles from the vertical as coordinates to describe both the pendulums.

(b) Use the relative angle for the second pendulum from the line of the first pendulum.

7.8 Two masses m_1 and m_2 are attached to a light string with tension T, as shown in Fig. 7.54. Assume that the displacement is small and that the tension remains unchanged when the masses are displaced normal to the string. We have already analyzed this problem in the text with $m_1 = m_2 = m$. Now, assume $m_1 = m$, and $m_2 = 2m$ and carry out the analysis. Derive the equations of motion, solve the eigenvalue problem, determine the natural frequencies and sketch the modes of vibration.

Figure 7.54: *Masses on a string*

7.9 Two slender rigid rods of length ℓ hanging from a firm support are coupled by a spring of stiffness k, a distance a below their points of suspension (Fig. 7.55). Each rod has a mass m. Derive

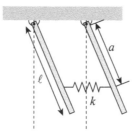

Figure 7.55: *Coupled rod pendulum*

the equations of motion, solve the eigenvalue problem, determine the natural frequencies, and sketch the modes of vibration.

7.10 The model shown in Fig. 7.56 simulates the effect of earthquakes on buildings. The building is assumed to be rigid and its base is connected to the ground through two springs: k_h, a translational spring, and k_r, a torsional spring (not shown). Note that the ground motion during an earthquake is a function of time. The mass of the building is m, and its center of mass C is at a distance ℓ from the bottom. Derive the equations of motion, solve the eigenvalue problem, determine the natural frequencies, and sketch the modes of vibration [32].

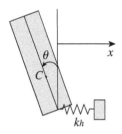

Figure 7.56: *A two-DOF model of a building*

7.11 An automobile can be modeled as a two-DOF system, as shown in Fig. 7.57. Note that the rigid rod shown is not uniform; its center of mass C is located a distance ℓ_1 from the left end, and a distance ℓ_2 from the right end. Use the following numerical values: mass $m = 4400$ kg, radius of gyration $r = 1.2$ m, $\ell_1 = 1.3$ m, $\ell_2 = 1.6$ m, $k_1 = k_2 = 35$ kN/m.

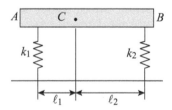

Figure 7.57: *Two-DOF model of an automobile*

7.12 Two identical springs support a rigid rod and two identical masses as shown in Fig. 7.58. Determine the natural frequencies and mode shapes.

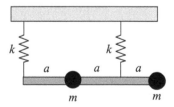

Figure 7.58: *Two-DOF system*

7.13 Consider the torsional vibration of the system shown in Fig. 7.59 that consists of a massless shaft that carries two identical rigid moments of inertia equally spaced. The mass moments of inertia are I_0, the length of the shaft is ℓ, and the radius of the shaft is r. The shear modulus of the material is G. Determine the natural frequencies and mode shapes.

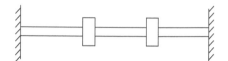

Figure 7.59: *Torsional vibration*

7.14 Consider the torsional vibration of the system shown in Fig. 7.60 that consists of a massless shaft that carries two rigid moments of inertia spaced as shown. The mass moments of inertia are I_1 and I_2, the length of the shaft is ℓ, and the radius of the shaft is r. The shear modulus of the material is G. Determine the natural frequencies and mode shapes if $I_1 = 2I_2 = I_0$.

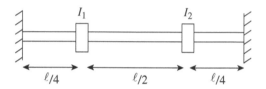

Figure 7.60: *Torsional vibration*

7.15 Consider the torsional vibration of the system shown in Fig. 7.61 that consists of a massless shaft that carries two rigid moments of inertia spaced as shown. The mass moments of inertia are I_1 and I_2, the length of the shaft is ℓ, and the radius of the shaft is r. The shear modulus of the material is G. Determine the natural frequencies and mode shapes if $I_1 = I_0$, and $I_2 = I_0/3$.

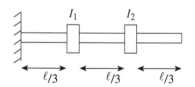

Figure 7.61: *Torsional vibration*

7.16 Consider the torsional vibration of the system shown in Fig. 7.62 that consists of a massless shaft that carries three rigid moments of inertia spaced as shown. The mass moments of inertia are I_1, I_2, and I_3, the length of the shaft is ℓ, and the radius of the shaft is r. The shear modulus of the material is G. Determine the natural frequencies and mode shapes if $I_1 = I_3 = I_0$ and $I_2 = 2I_0$.

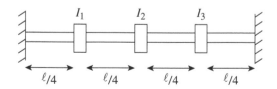

Figure 7.62: *Torsional vibration*

7.17 Consider the system shown in Fig. 7.63 that models a motor-driven shaft. The motor has a mass moment of inertia I_1, and the load has a mass moment of inertia I_2. The length of the shaft is ℓ, the radius is r, and its material has shear modulus G. Determine the natural frequencies and mode shapes.

Motor   Load

Figure 7.63: *Motor-driven shaft*

7.18 A rigid frame is free to move horizontally and supports a pendulum as shown in Fig. 7.64. The pendulum swings free in the same plane in which the frame moves. Derive the equations of motion and determine the natural frequencies and mode shapes. Explain the meaning of the first natural frequency.

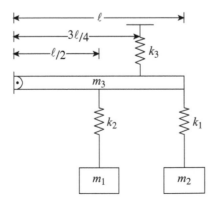

Figure 7.64: *Pendulum in a cart*

7.19 If the system of Problem 7.1 is started with the initial conditions, $x_1(0) = 0$, $x_2(0) = 1.0$, $\dot{x}_1(0) = 0$, $\dot{x}_2(0) = 0$, determine the displacements as functions of time.

7.20 If the system of Problem 7.8 is started with the initial conditions, $x_1(0) = 0.065$ m, $x_2(0) = 0.0$, $\dot{x}_1(0) = 0$, and $\dot{x}_2(0) = 0$, determine the displacements as functions of time. Use $m_1 = 5$ kg, $m_2 = 1$ kg, $\ell = 0.25$ m.

7.21 Two sets of mass and spring mechanisms (m_1, k_1, m_2, k_2) are suspended at a distance $\ell/2$ and ℓ on a hinged bar of mass m_3, as shown in Fig. 7.65. A third spring, k_3, is attached at a distance $3\ell/4$ from the left end as shown. Derive the EOM of the system.

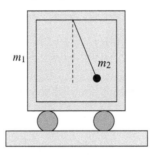

Figure 7.65: *Two-DOF system for Problem 7.21*

7.22 A crane is lifting a mass m_1, as shown in Fig. 7.66. Find the linearized equations of motion for small oscillations, and determine the natural frequencies and mode shapes. The crane may be assumed to be a uniform rod of length ℓ. The mass of the crane, $m_2 = 1200$ kg, load, $m_1 = 100$ kg, $k_1 = 2$ kN/m, $k_2 = 100$ kN/m, and $\ell = 12$ m. Assume that a holding torque, $T = 20$ Nm, is applied at the base of the arm of the crane.

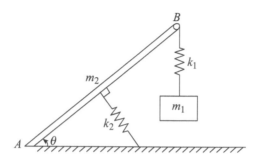

Figure 7.66: *Crane lifting a load*

7.23 Figure 7.67 shows an inverted pendulum coupled to a moving mass.

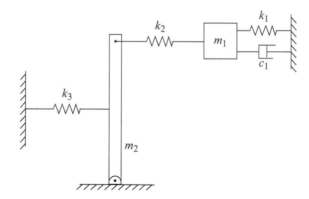

Figure 7.67: *Inverted pendulum with spring and mass mechanisms*

(a) Find the nonlinear equations of motion.
(b) Assume small motions about the equilibrium position, and determine the linearized equations.
(c) Find the natural frequencies and mode shapes.

7.24 A hinged inverted pendulum of length ℓ, and mass m, is supported by a spring and is connected to a particle mass m_2 at the other end through a massless, frictionless pulley and a spring k_2, as shown in Fig. 7.68. The spring k_1 is at a height $2\ell/3$. Derive the linearized equation for small motions about the vertical equilibrium position and determine the natural frequencies and mode shapes. You may assume the following numerical values: $m_1 = 5$ kg, $m_2 = 250$ g, $k_1 = 500$ N/m, $k_2 = 125$ N/m, and $\ell = 1.5$ m.

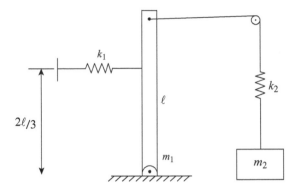

Figure 7.68: *Inverted pendulum mechanism in Problem 7.24*

7.25 A mass m_1 is placed at a distance $3\ell/4$ from the left end of a rod of mass m_2, as shown in Fig. 7.69. At the far end, a mass m_3 is suspended by a spring k_2. Find the natural frequencies and mode shapes of the system if $m_1=2$ kg, $m_2=3$ kg, $m_3=1.5$ kg, $k_1 = 2.5$ kN/m, $k_2 = 4$ kN/m, and $\ell = 0.8$ m.

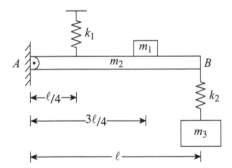

Figure 7.69: *System for Problem 7.25*

7.26 The lower part of the human body can be modeled as a double inverted pendulum with a lumped mass m_1 at the top, and another lumped mass m_2 at the knee, as shown in Fig. 7.70. Assume that the neuromuscular system provides stabilizing springs k_1 and k_2

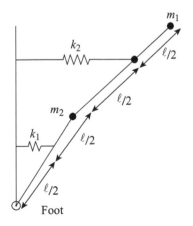

Figure 7.70: *Double pendulum model for the lower part of the body*

as shown. Assume that the springs are attached at the midpoints of the pendulums. Derive a mathematical model for the system. Assume the following numerical values and determine the natural frequencies and mode shapes. $m_1 = 20\,\text{kg}$, $m_2 = 2\,\text{kg}$, $k_1 = 2000\,\text{N/m}$, $k_2 = 3000\,\text{N/m}$, $\ell = 0.7\,\text{m}$.

7.27 Consider the three-DOF system shown in Fig. 7.71. For each of the following cases, determine the natural frequencies of motion and sketch the mode shapes.

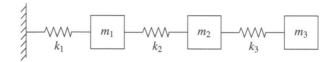

Figure 7.71: *Three-DOF system*

(a) $m_1 = m_2 = m_3 = m$, and $k_1 = k_2 = k_3 = k$
(b) $m_1 = 5m$, $m_2 = m_3 = m$, and $k_1 = k_2 = k_3 = k$
(c) $m_1 = 2000\,\text{kg}$, $m_2 = 500\,\text{kg}$, $m_3 = 350\,\text{kg}$, $k_1 = 250\,\text{kN/m}$, $k_2 = 175\,\text{kN/m}$, $k_3 = 150\,\text{kN/m}$
(d) $m_1 = 200\,\text{kg}$, $m_2 = 100\,\text{kg}$, $m_3 = 100\,\text{kg}$, $k_1 = 150\,\text{kN/m}$, $k_2 = 150\,\text{kN/m}$, $k_3 = 150\,\text{kN/m}$

7.28 Consider the three-DOF system shown in Fig. 7.72. Assume $m_1 = 5m$, $m_2 = m$, $m_3 = m$, $k_1 = k_2 = k$. Determine the natural frequencies of motion and sketch the mode shapes. Verify orthogonality, and present the eigenvectors in normalized form. What kind of a practical system could this be a model for?

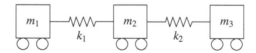

Figure 7.72: *Three-DOF system*

7.29 Consider the three-DOF system shown in Fig. 7.73. Assume $m_1 = 3m$, $m_2 = m$, $m_3 = 0.5m$, $k_1 = k$, $k_2 = 2k$, $k_3 = k$, $k_4 = 1.5k$. Determine the natural frequencies of motion and sketch the mode shapes. Verify orthogonality, and present the eigenvectors in normalized form.

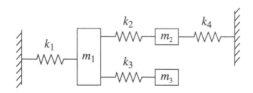

Figure 7.73: *Three-DOF system*

7.30 Consider the three-DOF system shown in Fig. 7.74. Assume $m_1 = m$, $m_2 = 2m$, $m_3 = m$, $k_1 = k$, $k_2 = 0.5k$, $k_3 = 0.5k$, $k_4 = k$. Determine the natural frequencies of motion and sketch the mode shapes. Verify orthogonality, and present the eigenvectors in normalized form.

Figure 7.74: *Three-DOF system*

7.31 Consider the three-DOF system shown in Fig. 7.75. Assume $m_1 = m$, $m_2 = 4m$, $m_3 = 0.5m$, $k_1 = 2k$, $k_2 = 2k$, $k_3 = k$, $k_4 = 1.5k$, $k_5 = 2k$, $k_6 = 3k$. Determine the natural frequencies of motion and sketch the mode shapes. Verify orthogonality, and present the eigenvectors in normalized form.

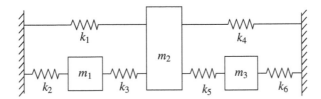

Figure 7.75: *Three-DOF system*

7.32 Consider the four-DOF model for an automobile shown in Fig. 7.76. Determine the natural frequencies of motion and sketch the mode shapes. Present the eigenvectors in normalized form. $m_1 = 3000\,\text{kg}$, $m_2 = 300\,\text{kg}$, $k_1 = 100\,\text{kN/m}$, $k_2 = 120\,\text{kN/m}$, $\ell = 5\,\text{m}$, $r_g = 1.8\,\text{m}$. Assume that m_1 is a uniform rod, and note that the moment of inertia is $I_C = m_1 r_g^2$, where r_g is the radius of gyration.

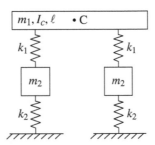

Figure 7.76: *Four-DOF system*

7.33 Consider the five-DOF system shown in Fig. 7.77. Determine the natural frequencies of motion and sketch the mode shapes. Present the eigenvectors in normalized form. $m_1 = 50\,\text{kg}$, $m_2 =$

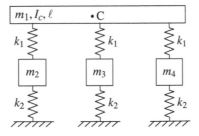

Figure 7.77: *Five-DOF system*

$m_3 = m_4 = 25$ kg, $k_1 = 28$ kN/m, $k_2 = 15$ kN/m, $\ell = 0.5$ m, $r_g = 0.05$ m. Assume that m_1 is a uniform rod, and note that the moment of inertia is $I_C = m_1 r_g^2$, where r_g is the radius of gyration.

7.34 Consider the system shown in Fig. 7.78. The two rigid rods are connected by means of a torsional spring that has no force when the rods are horizontal. Determine the natural frequencies of motion and sketch the mode shapes. What practical system could this be a model for? $m_1 = 1200$ kg, $m_2 = 400$ kg, $k_1 = 60$ kN/m, $k_2 = 40$ kN/m, $\ell_1 = 2$ m, $\ell_2 = 1.5$ m, $r_{g1} = 0.5$ m, $r_{g2} = 0.1$ m, $k_3 = 8$ kNm/rad. Assume that m_1 and m_2 are uniform rods, and note that the moment of inertia is $I_C = mr_g^2$, where r_g is the radius of gyration of each rod.

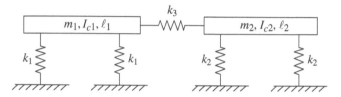

Figure 7.78: *Connected rods*

7.35 Consider the model of an automobile as considered earlier. But this time we will add a driver to the model, as shown in Fig. 7.79, with the assumption that the body is a uniform rod. Determine the natural frequencies of motion and sketch the mode shapes. For the numerical calculations you may use the following numerical values: $m_1 = 1250$ kg, $m_2 = 80$ kg, $k_1 = k_2 = 45$ kN/m, $\ell_1 = 1.575$ m, $\ell_2 = 1.125$ m, $\ell_3 = 1.8$ m, $k_3 = 15$ kN/m, $r_g = 0.7$ m. Assume that m_1 is a uniform rod, and note that the moment of inertia is $I_C = m_1 r_g^2$, where r_g is the radius of gyration.

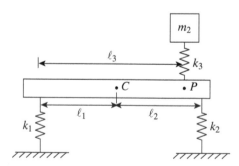

Figure 7.79: *Model for a car with a driver*

7.36 Consider the system shown in Fig. 7.80. The two rigid rods are connected by means of a torsional spring that has no force when the rods are horizontal. Determine the natural frequencies of motion and sketch the mode shapes. What practical system could this be a model for? $m_1 = 1400$ kg, $m_2 = 1200$ kg, $m_3 = m_4 = m_5 = m_6 = 225$ kg, $k_1 = 50$ kN/m, $k_2 = 120$ kN/m, $k_3 = 40$ kN/m, $k_4 = 100$ kN/m, $\ell_1 = 3.2$ m, $\ell_2 = 2.5$ m, $r_{g1} = 0.8$ m, $r_{g2} = 0.3$ m, $k_5 = 12$ kNm/rad. Assume that m_1 and m_2 are uniform rods, and note that the moment of inertia is $I_C = mr_g^2$, where r_g is the radius of gyration of each rod.

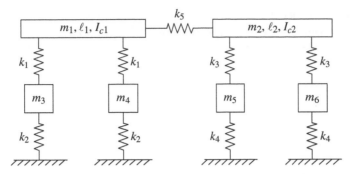

Figure 7.80: *Connected rods with additional masses*

7.37 The atomic force microscope (AFM) is a nifty tool for measuring properties at the nanometer scale; its invention has galvanized research in many areas including mechanics and biology. The essential element in an AFM is a probe that is held against a surface. It can be modeled as a multi-DOF system consisting of a rod coupled to a simple pendulum, as shown in Fig. 7.81. Note that there is also a torsional spring, k_3, coupled to the rod at the end A. Derive the EOM of the system; assume small motions about the equilibrium point.

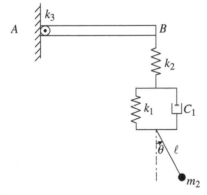

Figure 7.81: *Atomic force microscope*

7.38 Derive the EOM of the model shown in Fig. 7.82. Derive the equation of motion and solve for the undamped natural frequencies and mode shapes.

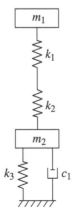

Figure 7.82: *Two-DOF model for Problem 7.38*

7.39 A building frame is modeled as shown in Fig. 7.83. Assume that the masses m_1 and m_2 are rigid and that the columns have flexural rigidities EI_1 and EI_2. Find the natural frequencies and mode shapes of the system when $m_1 = m_2$, $h_1 = h_2$, and $EI_1 = EI_2$.

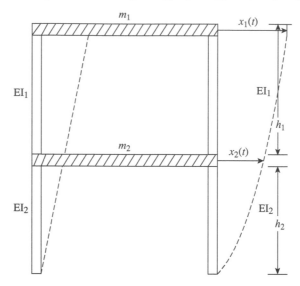

Figure 7.83: *Building frame model*

7.40 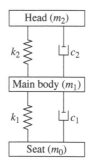 We analyzed the vibration of a string with a single mass, and two and three masses earlier in the text. Develop a *general* model to represent a string with n masses that are uniformly distributed. Solve for the natural frequencies and mode shapes using a large n. Then, plot the first three natural frequencies versus the number of lumped masses; in other words, extend the plot, Fig. 7.32 to say, $n = 15$. Show that the answers converge slowly to some value. (We will find out later that this is indeed the correct value as derived by using a continuous systems model in Chapter 9.) You will certainly need to solve this problem using a computer program unless you want to spend the better part of the year working on just this problem!

7.41 The vibrational response of seated human subjects is an important problem. Earlier, we looked at one-DOF models for this problem.

(a) Let us now consider a slightly more realistic two-DOF model, as shown in Fig. 7.84 from the article by Allen [33]. The parameters are known to be in the following ranges:

Figure 7.84: *Two-DOF model for a seated human being (see Allen [33])*

$m_1 = 51.3 \pm 8.5\,\text{kg}$, $m_2 = 5.5 \pm 0.9\,\text{kg}$, $k_1 = 74.3 \pm 17.4\,\text{kN/m}$, $c_1 = 2.807 \pm 1.007\,\text{kNs/m}$, $k_2 = 41.0 \pm 24.1\,\text{kN/m}$, and $c_2 = 318 \pm 161\,\text{Ns/m}$. Determine the undamped and damped natural frequencies and mode shapes for the parameter ranges given. Also determine the response to various initial conditions.

(b) Here is another two-DOF model. This model is shown in Fig. 7.85 and is taken from the article by Wei and Griffin [29]. The buttock and legs are assumed to be rigidly attached to the seat. The parameter values are as follows: $m_0 = 6.7\,\text{kg}$, $m_1 = 33.4\,\text{kg}$, $m_2 = 10.7\,\text{kg}$, $k_1 = 35.776\,\text{kN/m}$, $c_1 = 761\,\text{Ns/m}$, $k_2 = 38.374\,\text{kN/m}$, and $c_2 = 458\,\text{Ns/m}$. Determine the undamped and damped natural frequencies and mode shapes for the parameter ranges given. Also, determine the response to various initial conditions.

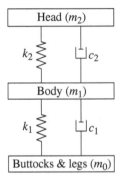

Figure 7.85: *Two-DOF model for a seated human being (see Wei and Griffin [29])*

(c) Yet another model was proposed by Suggs et al. [34], as shown in Fig. 7.86. In the model, the upper torso and lower torso are assumed to be rigidly connected from which it follows that $k_2 = \infty$, $c_2 = \infty$. The parameter ranges are as follows: $m_1 = 15.3 \pm 2.5\,\text{kg}$, $m_2 = 36.0 \pm 6.0\,\text{kg}$, $m_3 = 5.5 \pm 0.9\,\text{kg}$, $k_1 = 40.9 \pm 22.7\,\text{kN/m}$, $c_1 = 2.806 \pm 1.000\,\text{kNs/m}$, $k_3 = 74.3 \pm 17.4\,\text{kN/m}$, and $c_3 = 318 \pm 42\,\text{Ns/m}$. Determine the undamped and damped natural frequencies and mode shapes for the parameter ranges given. Also, determine the response to various initial conditions.

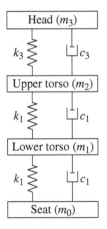

Figure 7.86: *Two-DOF model for a seated human being from (see Suggs et al. [34])*

7.42 A cylinder rolls without slipping and has a simple pendulum hinged to its center, as shown in Fig. 7.87. Assume small displacements about the vertical equilibrium position. Find the natural frequencies and mode shapes of the system.

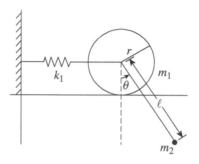

Figure 7.87: *Pendulum coupled to a cylinder*

7.43 Consider the system shown in Fig. 7.88 in which three equal masses ($m_1 = m_2 = m_3 = m$) are located on a uniform cantilever beam with flexural rigidity EI. The flexibility matrix (which is the inverse of the stiffness matrix) is given by:

$$\mathbf{A} = \frac{\ell^3}{3EI} \begin{bmatrix} 27 & 14 & 4 \\ 14 & 8 & 2.5 \\ 4 & 2.5 & 1 \end{bmatrix}$$

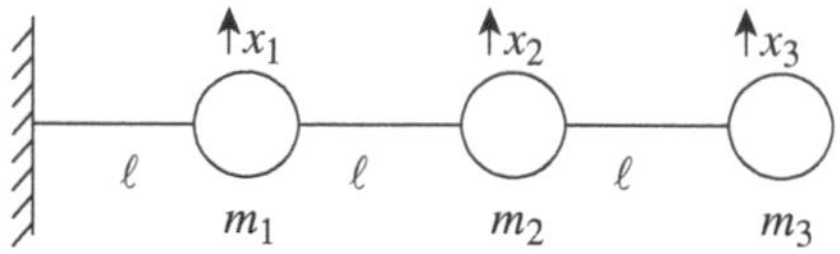

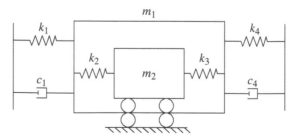

Figure 7.88: *Masses on cantilever beam*

The mass matrix would simply be a diagonal matrix with the masses m ($m_1 = m_2 = m_3 = m$) along the diagonal.

(a) Determine the stiffness matrix.

(b) Derive the governing equations of motion.

(c) Determine the natural frequencies and mode shapes.

7.44 Derive the equations of motion for the model of a train shown in Fig. 7.89.

Figure 7.89: *Model of a train*

7.45 For the free vibration of the mechanism shown in Fig. 7.90, derive the equations of motion.

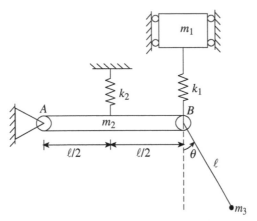

Figure 7.90: *Mechanism for Problem 7.45*

7.46 For the simple two-DOF model shown in Fig. 7.91, derive the equations of motion. Solve for the undamped natural frequencies and mode shapes.

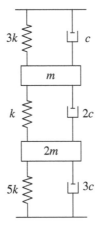

Figure 7.91: *System for Problem 7.46*

7.47 A dynamic vibration absorber, shown in Fig. 7.92, works by *adding* a mass-spring system to the original vibrating system in order to reduce its vibration. This is somewhat of a counter-intuitive technique and can be accomplished without adding damping! We will discuss more of this in detail including design aspects in the next chapter. Derive the equation of motion of the system and determine the natural frequencies and mode shapes assuming $m_1 = 3m_2$ and $k_1 = 4k_2$. The article by Khazanov [35] provides some interesting further reading. Compare this system to the one without the absorber.

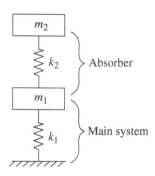

Figure 7.92: *Dynamic vibration absorber*

7.48 Tie-rods are used to hold machinery in place on ships. These structures are attached to the machinery, or machinery raft systems, and then attached to a bulk head or hull of a ship. Since the tie-rods are very rigid, they easily transmit vibration from machinery to the hull of the ship. A tie-rod structure is shown in Fig. 7.93. Derive the equation of motion and find the natural frequencies. More details can be found in the article by Brennan et al. [36]. Use the following numerical values: $m_1 = 3500$ kg, $m_2 = 4600$ kg, $m_3 = 7600$ kg, $m_4 = 5500$ kg, $k_1 = 25$ kN/m, $k_2 = 35$ kN/m, $k_3 = 42$ kN/m, $k_4 = 38$ kN/m.

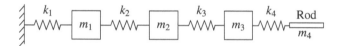

Figure 7.93: *A model for tie-rods*

7.49 An overhead crane has a beam with $I = 1.2 \times 10^{-3}$ m^4, $E = 2.1 \times 10^{11}$ Pa, $\ell = 50$ m. The hoist has a mass $m_2 = 1200$ kg, and it is lifting a mass $m_1 = 5000$ kg through a cable of stiffness $k = 200$ kN/m as shown in Fig. 7.94. Determine the natural frequencies and mode shapes of the system.

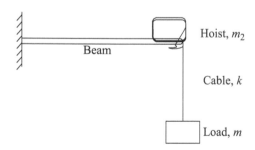

Figure 7.94: *Overhead crane*

7.50 A rod of length ℓ is hinged at one end and has a damper at the other end, as shown in Fig. 7.95. A mass-spring oscillator is attached at the midlength of the rod. Derive the equation of motion of the system.

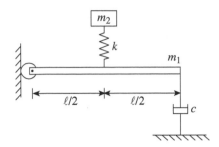

Figure 7.95: *System for Problem 7.50*

7.51 Derive the equation of motion of the system shown in Fig. 7.96.

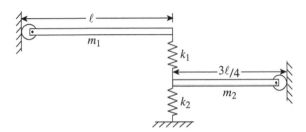

Figure 7.96: *Connected rods*

7.52 The human body is a very complex dynamic system whose mechanical properties vary from moment to moment and from individual to individual. Experimental evidence has shown that the humans are most sensitive to the whole body vibration under low frequency excitation in the seated posture. Research on the vibration effects on seated workers has indicated that the side effects could be very harmful and in some cases lead to permanent injuries. In the measurement of kinetic responses of human subjects, such as assessment of human tolerance to vibration exposure, live humans are usually used as test subjects. Many mathematical models have been developed to describe the biodynamic responses of human beings, the simplest of them being lumped-parameter models. These models consider the human body as several concentrated masses interconnected by springs and dampers.

In this problem, a lumped parameter model with three-DOF is proposed and is shown in Fig. 7.97. The head of mass m_3 is connected to the body of mass m_2 with the help of a spring and a damper (k_2, c_2). The body and the legs of mass m_1 are also connected with the help of a

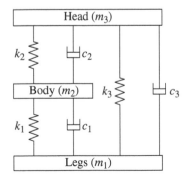

Figure 7.97: *A three-DOF model for the human body*

spring and damper(k_1, c_1) and the head and legs are again connected with the help of another spring and damper (k_3, c_3). Derive the equation of motion of the system. Also, determine the natural frequencies and mode shapes. Use the following parameter values. $m_1 = 15$ kg, $m_2 = 25$ kg, $m_3 = 18$ kg, $k_1 = 12$ N/m, $k_2 = 15$ N/m, $k_3 = 20$ N/m, $c_1 = 4$ Ns/m, $c_2 = 6$ Ns/m, $c_3 = 8$ Ns/m.

7.53 Consider again the human body, but with a simpler model. Now assume that the mass m_1 and m_2 of Fig. 7.97 are clamped together. The new model is shown in Fig. 7.98. Consider the total mass to be m_1 and assume that it is connected to the head m_3 through a spring and a damper (k_2, c_2). Derive the equation of motion. Solve for the natural frequencies and mode shapes using parameter values from Problem 7.52.

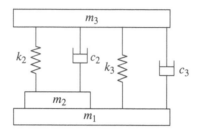

Figure 7.98: *A simpler model for the human body, Problem 7.53*

7.54 Brake squeal is one of the main issues in the overall context of noise, vibration, and harshness in the development of modern passenger cars and also lowers the subjective quality of the vehicle. This squeal is due to friction induced, self-excited vibrations of the brake system. A two-DOF model is presented in the article by Hochlenert et al. [37], which consists of an elastically hinged rigid wobbling disk that represents the actual flexible brake disk. The disk is in frictional contact with idealized brake pads m_1 and m_2. Each brake pad is elastically supported by two springs (stiffness k_2 and k_3). In the present problem, the disk is replaced with a linear viscous damper. So, we now have a two-DOF model shown in Fig. 7.99. Derive the

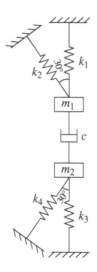

Figure 7.99: *Brake squeal model, Problem 7.54*

equation of the motion of the system and determine the natural frequencies and mode shapes. Use the following parameter values. $m_1 = 750$ kg, $m_2 = 860$ kg, $k_1 = 120$ N/m, $k_2 = 75$ N/m, $k_3 = 135$ N/m, $k_4 = 83$ N/m, $c = 49$ Ns/m.

7.55 Vehicle suspensions are capable of providing a more comfortable ride by serving the basic function of isolating passengers from the roughness of the road. The most important role of suspension systems is ride quality improvement, a problem that is drawing increasing attention because of the impact of vibration on driver fatigue, health, and discomfort. A control synthesis problem is investigated in the article by Zhao et al. [38] for a suspension system. In this problem, however, we are just concerned with the dynamics for which a mathematical model is shown in Fig. 7.100. m_0 is the mass of the cabin floor, m_1 is the mass of the seat frame, m_{21} and m_{22} are the masses of human thighs and the seat cushion, respectively, and $m_2 = m_{21} + m_{22}$; m_3 is the mass of the upper body of the seated human being. The mass of lower legs and feet is neglected because of their small contribution to the biodynamic response of the seated body. c_1, c_2 and k_1, k_2 are the damping and stiffnesses of the passive suspension system, respectively; c_3 and k_3 stand for the damping and stiffness of the components inside the human body such as spinal column, muscles, etc. Derive the equations of motion of the system.

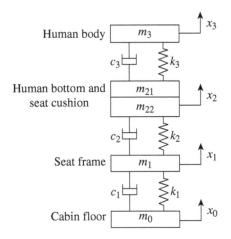

Figure 7.100: *Vehicle suspension model*

7.56 During machining, undesirable vibrations between the tool and the workpiece can lead to an unacceptable surface finish. Perhaps, the most troublesome cause of such vibrations is regenerative chatter, where unstable self-excited oscillations lead to a wavy surface finish that can result in tool wear, tool failure, or damage to the machine. Now, assume the turning tool is cutting a workpiece that had a wavy surface left behind by the previous tool revolution. Removal of the chip by the cutting tool results in a cutting force exerted upon the tool. This cutting force results in a vibration of the tool relative to the workpiece. In our present problem, we model the force experienced by the cutting tool as a mass and spring mechanism, as shown in Fig. 7.101. Derive the EOM of the model. You may want to read more about this important problem of cutting tool vibrations in an article by Taylor et al. [39].

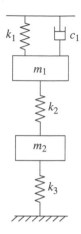

Figure 7.101: *Cutting tool vibration*

7.57 Vehicle suspension systems can be broadly classified as passive, semiactive, and active suspension systems. In the passive suspension system, the stiffness and damping parameters are fixed and are effective over a certain range of frequencies. To overcome this problem, the use of active suspension systems, which have the capability of adapting to changing road conditions by the use of an actuator, has been considered. In this problem, the control of the response of a half-car vehicle model to random road excitation is considered with optimal semiactive magnetorheological (MR) dampers.

The schematic of the half-car model with semiactive suspension provided by MR dampers is shown in Fig. 7.102. The MR damper is modeled by the so-called modified Bouc-Wen hysteretic model, as shown by the simplified spring-damper models on the right side in Fig. 7.103. m is the sprung mass of the vehicle body, I is the mass moment of inertia of the vehicle body with respect to the center of gravity, m_f and m_r are the unsprung masses, y_c is the absolute displacement of the center of gravity of the vehicle body (sprung mass), y_f and y_r are the unsprung mass displacements, θ is the pitch angle, k_f and k_r are the tire stiffnesses, and u_f and u_r are the road excitations. $(U_f)_{\mathrm{MR}}$ and $(U_r)_{\mathrm{MR}}$ are the control forces generated by the MR dampers. In all these quantities, the subscripts "f" and "r" refer to the front and rear end of the vehicle, respectively, and a and b are the distances of the front and rear ends from the

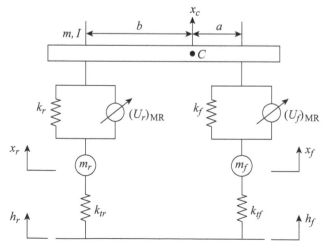

Figure 7.102: *Half-car model with MR dampers*

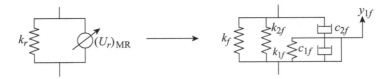

Figure 7.103: *MR damper model*

center of gravity of the vehicle. Derive the equations of motion. If this problem piques your interest, you may want to read a recent article on MR dampers by Prabakar et al. [40].

7.58 Complex robotic systems have recently garnered increased attention in many fields. An example of such a system is a dexterous manipulator mounted on a flexible base. A simplified model of such a robot is discussed in the present problem. Figure 7.104 shows the model of a manipulator mounted on an oscillating base. This robot has a manipulator that can be modeled as a double-rod pendulum with a loaded mass at its end; in addition, the dynamics of the base is modeled as a lumped mass with a spring and damper.

Figure 7.104: *Robotic manipulator*

Derive the equations of motion of the system. Assume that the two manipulator links have identical lengths, ℓ, and masses, m_1. The load mass is m_2, and the base mass is m_3. The base flexibility is modeled with k and c for stiffness and damping, respectively. Also assume that the base mass can only move up and down in the vertical direction.

7.59 Derive the equations of motion for the system shown in Fig. 7.105; see the article by Dong et al. [41] for an interesting discussion of this problem.

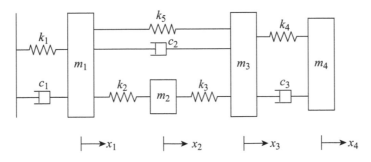

Figure 7.105: *System for Problem 7.59*

7.60 There exist different options to reduce the vibration or acoustic pressure levels in a given vibroacoustical system, and one of them being the factorization among responses by means of what is called the global transfer direct transfer (GTDT) method or, alternatively, by means of path-blocking techniques. Guasch [42] applied this method to a discrete mechanical system consisting of springs, dampers, and masses. In this problem, we are just concerned with the modeling of such a mechanical system, shown in Fig. 7.106. Derive the equations of motion of the system.

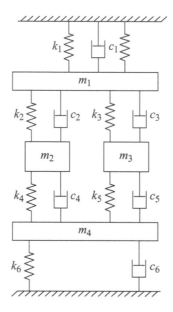

Figure 7.106: *System for Problem 7.60*

7.61 Determine the natural frequencies and mode shapes of the system shown in Fig. 7.107. Assume that the rope passing over the cylinder does not slip.

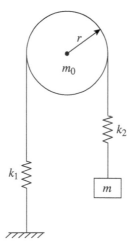

Figure 7.107: *System for Problem 7.61*

7.62 The construction of many high-speed train (HST) lines in European countries and Asia has pushed the railway engineering community to conduct much research related to different aspects of the technology in recent years. Vehicle, track, soil, and structure dynamic behaviors have been studied, leading to the conclusion that dynamic effects are much more important for HST than for the conventional trains. These effects require a deeper analysis in order to maintain security and comfort in the trains and to avoid problems due to vibrations induced in nearby constructions by waves transmitted through the soil. One of the first steps in the study of vibrations induced by HST is accurate modeling. Figure 7.108 shows an MDOF model developed by Galvin et al. [43], where x_c, x_b, x_w represent the car body, bogie, and wheel displacements, respectively. The load transmitted by an axle depends on the car body mass m_c, the bogie mass m_b, the mass of the wheelset m_w, the primary suspension k_1 and c_1, and the secondary suspension k_2 and c_2. Derive the equations of motion of the system.

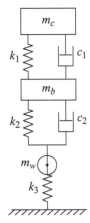

Figure 7.108: *High-speed train model*

7.63 A rigid rod of mass m and length $4a$ is centered on a uniform rigid cylinder of mass $\frac{4}{3}m$, and radius a, as shown in Fig. 7.109. Derive the equation of motion and linearize the equation for small oscillations.

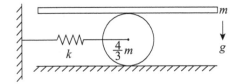

Figure 7.109: *Rod rocking on a rolling cylinder*

7.64 Consider the system shown in Fig. 7.110, which consists of two masses that are connected by rigid massless rods. Determine the equation of motion.

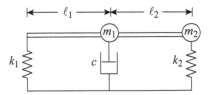

Figure 7.110: *Forced MDOF system*

7.65 A container of mass $m_2 = 5\,\text{kg}$ is suspended by two cables of length $\ell = 1\,\text{m}$, as shown in Fig. 7.111. Inside the container, a mass $m_1 = 1.5\,\text{kg}$ is supported by a spring of stiffness $k = 10\,\text{N/m}$. The cable is made of steel, and is under a uniform tension of 20 N. Determine the natural frequencies and mode shapes.

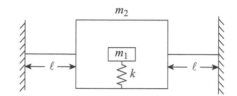

Figure 7.111: *System for Problem 7.65*

7.66 Consider the vibration of a seated human being, an important ergonomic problem we looked at with one-DOF models in earlier chapters, and as two-DOF models in Problem 7.41. Now, let us look at the more complex models (this is really going to be *complex*— hang on for a wild ride!). If this interests you, you should read a nice survey paper [44].

(a) Here is a four-DOF model from the article by Wan and Schimmels [45], shown in Fig. 7.112. The parameter values are as follows: $m_1 = 36\,\text{kg}$, $m_2 = 5.5\,\text{kg}$, $m_3 = 15\,\text{kg}$, $m_4 = 4.17\,\text{kg}$, $k_1 = 49.34\,\text{kN/m}$, $k_2 = 20.00\,\text{kN/m}$, $k_3 = 10.00\,\text{kN/m}$, $k_4 = 134.40\,\text{kN/m}$, $k_5 = 192.00\,\text{kN/m}$, $c_1 = 2475\,\text{Ns/m}$, $c_2 = 330\,\text{Ns/m}$, $c_3 = 2475\,\text{Ns/m}$, $c_4 = 2475\,\text{Ns/m}$, and $c_5 = 909.1\,\text{Ns/m}$. Determine the natural frequencies and mode shapes; interpret the mode shapes with respect to the actual system.

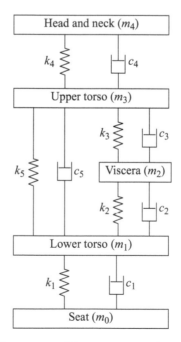

Figure 7.112: *Four-DOF model for a seated human being (see Liang and Feng Chiang [44], Wan and Schimmels [45])*

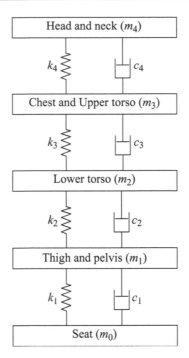

Figure 7.113: *Four-DOF model for a seated human being (see Liang and Feng Chiang [44], Boileau and Rakheja [46])*

(b) Another four-DOF model is from the article by Boileau and Rakheja [46], and is shown in Fig. 7.113. The parameter values are as follows: $m_1 = 12.78$ kg, $m_2 = 8.62$ kg, $m_3 = 28.49$ kg, $m_4 = 5.31$ kg, $k_1 = 90.0$ kN/m, $k_2 = 162.8$ kN/m, $k_3 = 183.0$ kN/m, $k_4 = 310.0$ kN/m, $c_1 = 2064$ Ns/m, $c_2 = 4585$ Ns/m, $c_3 = 4750$ Ns/m, and $c_4 = 400$ Ns/m. Determine the natural frequencies and mode shapes; interpret the mode shapes with respect to the actual system.

7.67 Consider a sophisticated 11-DOF model for a seated human being from the article by Qassem et al. [47], shown in Fig. 7.114. The parametric values were estimated from either models or experiments and are shown in Table 7.1. Determine the natural frequencies and mode shapes; interpret the mode shapes with respect to the actual system.

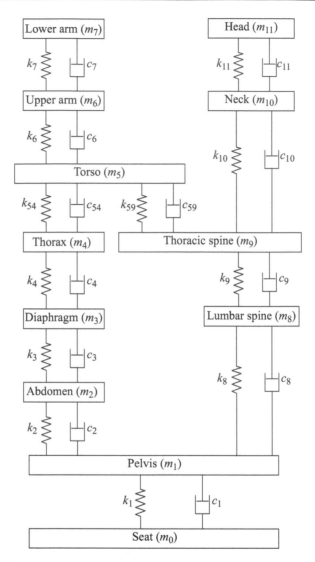

Figure 7.114: *11-DOF model for a seated human being*

7.68 Consider an 11-DOF model for a seated pregnant human female from the article by Qassem and Othmam [48], shown in Fig. 7.114. The parametric values were estimated

Mass (kg)		Stiffness (kN/m)		Damping coefficient (Ns/m)	
m_1	27.23	k_1	25.016	c_1	370.8
m_2	5.906	k_2	877	c_2	292.3
m_3	0.454	k_3	877	c_3	292.3
m_4	1.362	k_4	877	c_4	292.3
m_5	32.697	k_5	877	c_5	292.3
		k_{54}	877	c_{54}	292.3
		k_{59}	52.621	c_{59}	3581.6
m_6	5.470	k_6	67.542	c_6	3581.6
m_7	5.297	k_7	67.542	c_7	3581.6
m_8	2.002	k_8	52.621	c_8	3581.6
m_9	4.806	k_9	52.621	c_9	3581.6
m_{10}	1.084	k_{10}	52.621	c_{10}	3581.6
m_{11}	5.445	k_{11}	52.621	c_{11}	3581.6

Table 7.1: *Parameter values for seated male*

from either models or experiments and are shown in Table 7.2. These studies are important to design automobiles for safety of all kinds of passengers.

Mass (kg)		Stiffness (kN/m)		Damping coefficient (Ns/m)	
m_1	16.304	k_1	25.016	c_1	370.8
m_2	3.544	k_2	877	c_2	292.3
m_3	0.272	k_3	877	c_3	292.3
m_4	0.817	k_4	877	c_4	292.3
m_5	19.628	k_5	877	c_5	292.3
		k_{54}	877	c_{54}	292.3
		k_{59}	52.621	c_{59}	3581.6
m_6	3.282	k_6	67.542	c_6	3581.6
m_7	3.178	k_7	67.542	c_7	3581.6
m_8	1.201	k_8	52.621	c_8	3581.6
m_9	2.884	k_9	52.621	c_9	3581.6
m_{10}	0.650	k_{10}	52.621	c_{10}	3581.6
m_{11}	3.267	k_{11}	52.621	c_{11}	3581.6

Table 7.2: *Parameter values for a seated pregnant female*

(a) Determine the natural frequencies and mode shapes; interpret the mode shapes with respect to the actual system.
(b) Carry out a study by increasing mass due to pregnancy by distributing a mass of 10.35–12.15 kg on the abdomen (m_2) and thorax (m_4) in the ratio of their masses.

8 MDOF Systems: Forced Response

The scientific theory I like best is that the rings of Saturn are composed entirely of lost airline luggage.
—Mark Russell

In this chapter we will consider the forced vibration of multiple-degree-of-freedom (MDOF) systems. Again, we will restrict detailed presentations to two-DOF systems, although the concepts and the methods are equally valid for systems with an indefinite number of degrees of freedom. Just as in the case of single-DOF (SDOF) systems, the forces could be arbitrary functions of time. However, harmonic excitations are the most common and will be considered first. In addition, analysis of harmonic excitation will enable us to get responses to periodic excitation, and ultimately to any kinds of forces using some basic principles of mathematics.

8.1 Harmonic Excitation

We will first consider harmonic forces acting on a two-DOF system. Although the method is quite general, we would like to start with an example of a specific system; we will eventually extend it to a general system defined with matrices.

8.1.1 Undamped Systems

Again, to make things easier to understand, we will start with an undamped system. We will consider the same system we analyzed in the previous chapter on free vibration. Figure 8.1 shows a two-DOF model with two masses interconnected by springs. They are modeled as two particles and hence we have two free body diagrams to contend with, as shown in Fig. 8.2. Each of these particles has an external force acting on it.

The kinematics is quite straightforward:

$$a_1 = \ddot{x}_1 \tag{8.1}$$

$$a_2 = \ddot{x}_2 \tag{8.2}$$

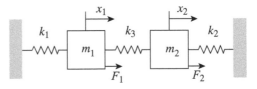

Figure 8.1: *A two-DOF-system model for forced vibration*

Writing Newton's II law for both the particle masses,

$$\vec{F}_i = m_i \vec{a}_i \tag{8.3}$$

we get

$$m_1 \ddot{x}_1 = -k_1 x_1 - k_3 (x_1 - x_2) + F_1(t) \tag{8.4}$$
$$m_2 \ddot{x}_2 = -k_2 x_2 - k_3 (x_2 - x_1) + F_2(t) \tag{8.5}$$

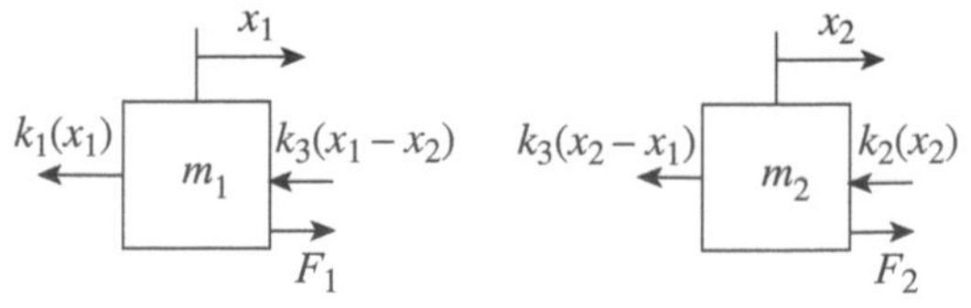

Figure 8.2: *FBDs of the two-DOF-system model*

These two equations can be written in a matrix form,

$$
\begin{bmatrix} m_1 & 0 \\ 0 & m_2 \end{bmatrix}
\begin{bmatrix} \ddot{x}_1 \\ \ddot{x}_2 \end{bmatrix}
+
\begin{bmatrix} k_1 + k_3 & -k_3 \\ -k_3 & k_2 + k_3 \end{bmatrix}
\begin{bmatrix} x_1 \\ x_2 \end{bmatrix}
=
\begin{bmatrix} F_1(t) \\ F_2(t) \end{bmatrix}
\tag{8.6}
$$

These form a set of two linear, coupled ordinary differential equations which we will need to solve for the two unknown quantities, $x_1(t)$ and $x_2(t)$. They are nonhomogeneous because there are external forces acting on the system. They are coupled because clearly the motion of each mass influences the other.

They can be written more compactly in the form

$$\mathbf{M\ddot{x}} + \mathbf{Kx} = \mathbf{F}(t) \tag{8.7}$$

Here, $\mathbf{M}$ is the mass matrix and $\mathbf{K}$ is the stiffness matrix, and they are the same as before [compare with Eq. (7.6)]. In addition, we also have $\mathbf{F}$, the force vector that is presumed known.

Solution

We are interested only in the particular solution (since we have already solved for the homogenous solution in the previous chapter). As always, the particular solution depends upon the nature of the right-hand side; hence, we have to describe the forces before we can attempt a solution. We will specify that the forces are both harmonic with the same frequency, ω.

$$\mathbf{F}(t) = \begin{bmatrix} F_{10} \\ F_{20} \end{bmatrix} \sin \omega t = \mathbf{F}_0 \sin \omega t \tag{8.8}$$

The EOM becomes

$$\mathbf{M\ddot{x}} + \mathbf{Kx} = \mathbf{F}_0 \sin \omega t \tag{8.9}$$

We seek a solution in the form

$$\mathbf{x}(t) = \mathbf{x}_c \cos \omega t + \mathbf{x}_s \sin \omega t \tag{8.10}$$

Then,

$$\ddot{\mathbf{x}}(t) = -\omega^2 \mathbf{x}_c \cos \omega t - \omega^2 \mathbf{x}_s \sin \omega t \tag{8.11}$$

Substituting into the differential equation,

$$(-\omega^2 \mathbf{M} + \mathbf{K}) \mathbf{x}_c \cos \omega t + (-\omega^2 \mathbf{M} + \mathbf{K}) \mathbf{x}_s \sin \omega t = \mathbf{F}_0 \sin \omega t \tag{8.12}$$

This leads to two algebraic equations:

$$(-\omega^2 \mathbf{M} + \mathbf{K})\mathbf{x}_c = \mathbf{0} \tag{8.13}$$
$$(-\omega^2 \mathbf{M} + \mathbf{K})\mathbf{x}_s = \mathbf{F}_0 \tag{8.14}$$

If $\omega = \omega_1$ or ω_2, one of the two natural frequencies, then $|-\omega^2 \mathbf{M} + \mathbf{K}| = 0$; in that case, $(-\omega^2 \mathbf{M} + \mathbf{K})^{-1}$ will not exist, and the above system of equations does not have a solution. If the forcing frequency ω is *not* one of those values, we can solve for the unknown coefficients.

$$\mathbf{x}_c = \mathbf{0} \tag{8.15}$$
$$\mathbf{x}_s = \mathbf{Z}^{-1}\mathbf{F}_0 \tag{8.16}$$

where $\mathbf{Z} = (-\omega^2 \mathbf{M} + \mathbf{K})$ and is often called the *dynamic stiffness matrix*. Hence, the solution is

$$\mathbf{x}(t) = \mathbf{x}_0 \sin \omega t, \quad \mathbf{x}_0 = \mathbf{Z}^{-1}\mathbf{F}_0 \tag{8.17}$$

In order to make better sense of the solution let us choose simple parametric values for the system: $m_1 = m_2 = m$, $k_1 = k_2 = k_3 = k$, $F_{20} = 0$. That is, the masses and springs are identical, and there is a harmonic force acting only on the first mass. Then, the solution is given by

$$\mathbf{x}_0 = \begin{bmatrix} -\omega^2 m + 2k & -k \\ -k & -\omega^2 m + 2k \end{bmatrix}^{-1} \mathbf{F}_0 = \frac{1}{|\mathbf{Z}|} \begin{bmatrix} -\omega^2 m + 2k & k \\ k & -\omega^2 m + 2k \end{bmatrix} \mathbf{F}_0 \tag{8.18}$$

where

$$|\mathbf{Z}| = (-\omega^2 m + 2k)^2 - k^2 \tag{8.19}$$

The frequency response of the two particles is shown in Fig. 8.3; clearly, both of them experience resonance at both the natural frequencies even though the force is only acting on the first mass. It is very important to note this. A single force acting anywhere in a system could cause resonance, which means that typically *all* the degrees of freedom would experience large values.

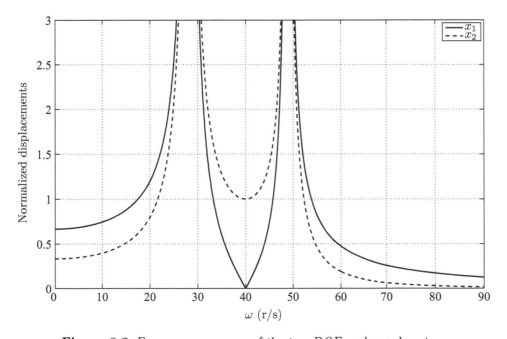

Figure 8.3: *Frequency response of the two-DOF undamped system*

8.1.2 Damped Systems

Suppose we consider the same system, but with linear viscous damping, as shown in Fig. 8.4. The FBD is shown in Fig. 8.5 and now includes the damping forces. Following the same procedure as before, the equations of motion are found to be as follows.

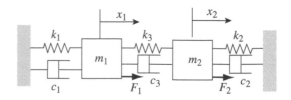

Figure 8.4: *A two-DOF damped model for forced vibration*

$$\begin{bmatrix} m_1 & 0 \\ 0 & m_2 \end{bmatrix} \begin{bmatrix} \ddot{x}_1 \\ \ddot{x}_2 \end{bmatrix} + \begin{bmatrix} c_1 + c_3 & -c_3 \\ -c_3 & c_2 + c_3 \end{bmatrix} \begin{bmatrix} \dot{x}_1 \\ \dot{x}_2 \end{bmatrix} + \begin{bmatrix} k_1 + k_3 & -k_3 \\ -k_3 & k_2 + k_3 \end{bmatrix} \begin{bmatrix} x_1 \\ x_2 \end{bmatrix} = \begin{bmatrix} F_1(t) \\ F_2(t) \end{bmatrix}$$

(8.20)

In matrix form,

$$\mathbf{M}\ddot{\mathbf{x}} + \mathbf{C}\dot{\mathbf{x}} + \mathbf{K}\mathbf{x} = \mathbf{F} = \mathbf{F}_c \cos\omega t + \mathbf{F}_s \sin\omega t \tag{8.21}$$

where we have assumed the more general form for the harmonic excitation. Again, we seek a solution in the form

$$\mathbf{x}(t) = \mathbf{x}_c \cos\omega t + \mathbf{x}_s \sin\omega t \tag{8.22}$$

Then,

$$\dot{\mathbf{x}}(t) = -\omega\mathbf{x}_c \sin\omega t + \omega\mathbf{x}_s \cos\omega t \tag{8.23}$$

$$\ddot{\mathbf{x}}(t) = -\omega^2\mathbf{x}_c \cos\omega t - \omega^2\mathbf{x}_s \sin\omega t \tag{8.24}$$

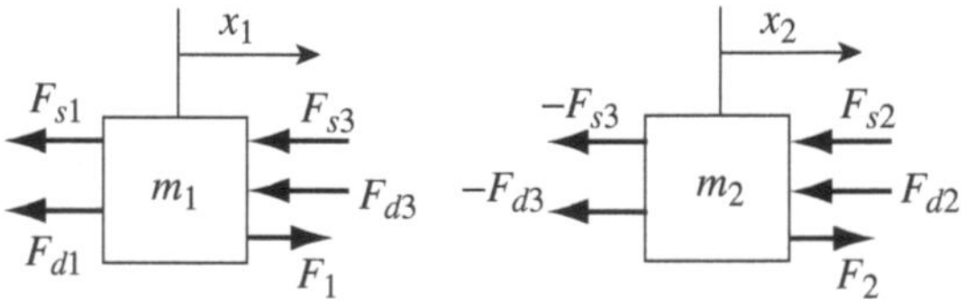

Figure 8.5: *FBDs of the two-DOF damped model*

Substituting into the differential equation, we get the following.

$$-\omega^2\mathbf{M}\left[\mathbf{x}_c \cos\omega t + \mathbf{x}_s \sin\omega t\right] + \omega\mathbf{C}\left[-\mathbf{x}_c \sin\omega t + \mathbf{x}_s \cos\omega t\right]$$
$$+\mathbf{K}\left[\mathbf{x}_c \cos\omega t + \mathbf{x}_s \sin\omega t\right] = \mathbf{F}_c \cos\omega t + \mathbf{F}_s \sin\omega t \tag{8.25}$$

Separating the cosines and sines we get a set of algebraic equations that we can put in the matrix form.

$$\begin{bmatrix} -\omega^2\mathbf{M} + \mathbf{K} & \omega\mathbf{C} \\ -\omega\mathbf{C} & -\omega^2\mathbf{M} + \mathbf{K} \end{bmatrix} \begin{bmatrix} \mathbf{x}_c \\ \mathbf{x}_s \end{bmatrix} = \begin{bmatrix} \mathbf{F}_c \\ \mathbf{F}_s \end{bmatrix} \tag{8.26}$$

This is now a fourth-order $(2 \times \text{DOF})$ linear algebraic equation. Note that the vectors are in the following order.

$$x_{\text{coef}} = \begin{bmatrix} x_{1c} \\ x_{2c} \\ x_{1s} \\ x_{2s} \end{bmatrix}, \quad F_{\text{coef}} = \begin{bmatrix} F_{1c} \\ F_{2c} \\ F_{1s} \\ F_{2s} \end{bmatrix} \tag{8.27}$$

The above system of four algebraic equations can be solved for the unknown coefficients on a computer (also by hand of course!). Once solved, the solution is given by

$$\mathbf{x}(t) = \mathbf{x}_c \cos\omega t + \mathbf{x}_s \sin\omega t \tag{8.28}$$

This will lead to the individual displacements,

$$x_1(t) = x_{10} \sin(\omega t - \phi_1) \tag{8.29}$$
$$x_2(t) = x_{20} \sin(\omega t - \phi_2) \tag{8.30}$$

where the amplitudes are given by

$$x_{10} = (x_{1c}^2 + x_{1s}^2)^{1/2}, \quad x_{20} = (x_{2c}^2 + x_{2s}^2)^{1/2} \tag{8.31}$$

and the phase angles by

$$\phi_1 = \tan^{-1}\left[\frac{-x_{1c}}{x_{1s}}\right], \quad \phi_2 = \tan^{-1}\left[\frac{-x_{2c}}{x_{2s}}\right] \tag{8.32}$$

Example 8.1 A harmonically forced MDOF system

In the above example, let $m = 25$ kg, $k = 20$ kN/m, $c = 150$ Ns/m, and $F_{10} = 100 \sin \omega t$ (N); i.e., there is a sinusoidal force acting only on the first mass. Substituting the numerical values, we get the amplitude of the frequency response, as shown in Fig. 8.6.

Compared to the undamped case, the damped response is more realistic, and shows bounded amplitudes at resonance. In any case, the resonant peaks are clearly seen at the two natural frequencies (28 and 49 rad/s). Again, it should be emphasized that irrespective of which particle was acted upon by the harmonic force, the effect of resonance is seen in the response of *both* the particles. It is also interesting to compare the response of this system with that of the undamped system (Fig. 8.3). Note that they have the same parameters except for damping. In particular, the effect of the second resonance is not felt by m_1. It should be borne in mind that this is by no means obvious, as it is a fairly complex function of all the parameters. The phases of the two coordinates are shown in Fig. 8.7; note that they are quite different from each other. The most relevant code in MATLAB is shown in Listing 8.1.

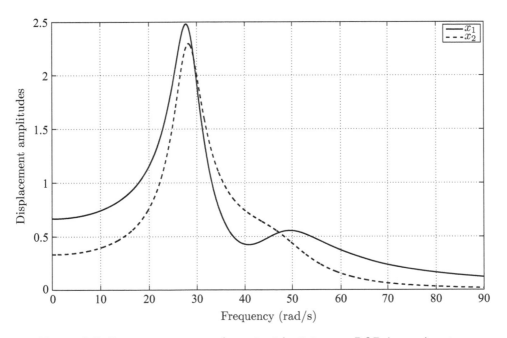

Figure 8.6: *Frequency response (magnitude) of the two-DOF damped system*

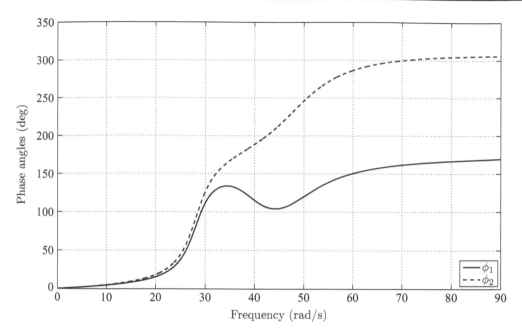

Figure 8.7: *Frequency response (phase) of the two-DOF damped system*

Listing 8.1: *MDOF forced response (part of the code)*

```
% forced vibration of 2 DOF system example

    frequency_range =(0:0.1:90);
    stiff=[k1+k3  −k3;  ...
           −k3      k3+k2];
    damp= [c1+c3  −c3;  ...
           −c3      c3+c2];
    mass=[m1  0;  ...
          0   m2];
    fc =[0;  0];
    fs =[F0;  0];

    faug=[fc;fs];          % = [fc1; fc2; fs1; fs2]
    for ifreq=1:length(frequency_range),
        freq = frequency_range(ifreq);
%       dynamic stiffness matrix
        z = −freq^2*mass+stiff;
        amat=[z              freq*damp;  ...
               −freq*damp  z];
        xaug=inv(amat) * faug;  % = [xc1; xc2; xs1; xs2]
%       displacement amplitudes
        x10(ifreq) = sqrt(xaug(1)^2+xaug(3)^2);
        x20(ifreq) = sqrt(xaug(2)^2+xaug(4)^2);
%       phase lags
        phi1(ifreq) = atan2(−xaug(1),xaug(3));
        phi2(ifreq) = atan2(−xaug(2),xaug(4));
    end;
```

8.2   Dynamic Vibration Absorber

Consider a mass-spring system subjected to a harmonic force, Fig. 8.8. We know that it will have a harmonic displacement whose amplitude goes through resonance when the forcing frequency is equal to the natural frequency. We wish to reduce the vibrational amplitude of the mass; we could accomplish this by adding damping as we know. However, what follows here is another procedure by which we can reduce the vibration theoretically to zero. Stop here for a moment to look at Fig. 8.3; do you see the amplitude, x_{10} going to zero at a certain frequency? You might want to think about how we might exploit that before you read on.

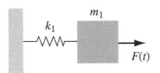

Figure 8.8: *Oscillatory system with harmonic force*

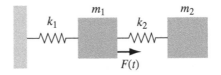

Figure 8.9: *Mass and spring added to the original system*

Suppose we add a mass (m_2) and spring (k_2) to the original system, as shown in Fig. 8.9. Now, we have a two-DOF system subjected to a harmonic force. The free body diagrams are shown in Fig. 8.10. The equations of motion are

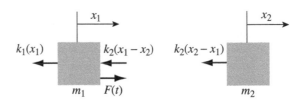

Figure 8.10: *FBD of new system*

$$m_1 \ddot{x}_1 = -k_1 x_1 - k_2(x_1 - x_2) + F_1(t) \tag{8.33}$$
$$m_2 \ddot{x}_2 = -k_2(x_2 - x_1) \tag{8.34}$$

In matrix form, and assuming a harmonic force,

$$F_1(t) = F_0 \sin \omega t \tag{8.35}$$

$$\begin{bmatrix} m_1 & 0 \\ 0 & m_2 \end{bmatrix} \begin{bmatrix} \ddot{x}_1 \\ \ddot{x}_2 \end{bmatrix} + \begin{bmatrix} k_1 + k_2 & -k_2 \\ -k_2 & k_2 \end{bmatrix} \begin{bmatrix} x_1 \\ x_2 \end{bmatrix} = \begin{bmatrix} F_0 \\ 0 \end{bmatrix} \sin \omega t \tag{8.36}$$

They can be written more compactly in the form

$$\mathbf{M}\ddot{\mathbf{x}} + \mathbf{K}\mathbf{x} = \mathbf{F}_0 \sin \omega t \tag{8.37}$$

Solving as before,

$$\mathbf{x}(t) = \mathbf{x}_0 \sin \omega t \tag{8.38}$$

where

$$\mathbf{x}_0 = \left(-\omega^2 \mathbf{M} + \mathbf{K}\right)^{-1} \mathbf{F}_0 \tag{8.39}$$

Substituting for $\mathbf{M}$ and $\mathbf{K}$,

$$\begin{bmatrix} x_{10} \\ x_{20} \end{bmatrix} = \frac{1}{|\mathbf{Z}|} \begin{bmatrix} -\omega^2 m_2 + k_2 & k_2 \\ k_2 & -\omega^2 m_1 + k_1 + k_2 \end{bmatrix} \begin{bmatrix} F_0 \\ 0 \end{bmatrix} \tag{8.40}$$

where

$$|\mathbf{Z}| = [m_1 m_2]\omega^4 - [k_1 m_1 + k_2 m_1 + k_2 m_2]\omega^2 + [k_2 k_1] \tag{8.41}$$

The individual displacement amplitudes are

$$x_{10} = \frac{-\omega^2 m_2 + k_2}{|\mathbf{Z}|} \frac{F_0}{k_1} \tag{8.42}$$

$$x_{20} = \frac{k_2}{|\mathbf{Z}|} \frac{F_0}{k_1} \tag{8.43}$$

A typical response is shown in Fig. 8.11. As expected, the denominator $|\mathbf{Z}|$ goes to zero at the two natural frequencies, and will cause unbounded displacements. Hence, it looks like we took an SDOF system with one resonant frequency and, by adding a DOF, we added another possible frequency at which it can resonate. It looks as though we made the situation worse!

But, wait! Look at the numerator of x_{10}. Keeping in mind that we have not specified the parameters of the system we are adding (i.e., k_2 and m_2), we notice that we can make x_{10} zero by choosing them such that

$$\frac{k_2}{m_2} = \omega^2 \tag{8.44}$$

This then is the principle of a dynamic vibration absorber: if we know the forcing frequency (ω) well enough, we could attach a mass and a spring to the original system in such a fashion that they satisfy the above relationship, and drive the response of the original system at that particular frequency to zero.

There are two practical issues that will impact the selection of these parameters. The first issue is that we usually do not know the exact forcing frequency; i.e., practically speaking, the forcing

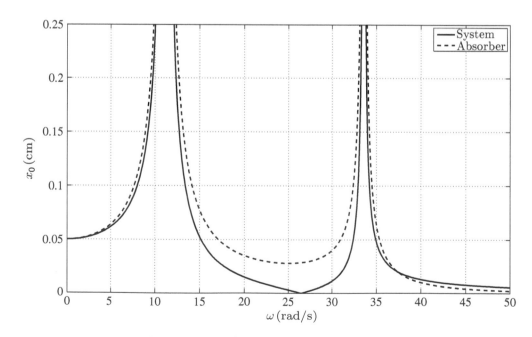

Figure 8.11: *Response of the new system*

frequency will change over a range of values and is not fixed. This means that we cannot choose a single operating value for the absorber, but instead have to deal with a *range* of values, and be prepared to settle for a vibration amplitude that is not zero. For example, in the response shown in Fig. 8.11, we would probably have acceptable performance if the frequency is between 20 and 30 rad/s.

The second practical issue is a constraint with respect to the amount of mass we can add to the original system. Most of the time, real constraints would dictate that we keep this mass to a small fraction of the original system's mass. This factor would decide m_2, and Eq. (8.44) would decide k_2, as you will see in the numerical example to follow.

Dynamic vibration absorbers are used in many places. A popular application is a so-called Stockbridge damper used to suppress vibration in transmission lines. A typical application is shown in Fig. 8.12; a close-up of the actual absorber is shown in Fig. 8.13.

Figure 8.12: *Transmission lines with Stockbridge dampers (see Wikimedia [49])*

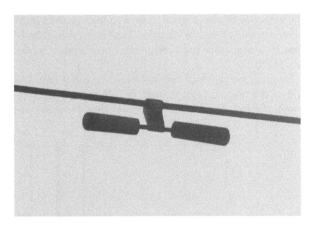

Figure 8.13: *Close-up of the Stockbridge damper (see Wikimedia [50])*

If you think about it, the idea of the dynamic vibration absorber is clever like the dickens (sure, anybody can suggest adding damping, but *adding* a mass-spring system to reduce vibration?!). But why does it really work? Draw a free body diagram of the two masses at the design frequency to show that the *net* force on the first mass is zero, which is why of course the displacement is zero.

Example 8.2 Dynamic vibration absorber for a motor

An electric motor of mass 22 kg is mounted at the midspan of a simply supported steel beam of rectangular cross section. The beam dimensions are: length = 1 m, width = 0.15 m, thickness = 4 cm. The harmonic force from the motor is known to be 50 N in the vertical direction at 30 Hz. We wish to design a dynamic vibration absorber; the added mass should not exceed 10% of the mass of the beam.

Solution

The motor-beam system can be modeled as a mass-spring system subjected to a harmonic force with $m_1 = 22$ kg, $k_1 = 48EI/\ell^3 = 8.06 \times 10^5$ N/m, $F_0 = 50$ N, and $\omega = 60\pi$ rad/s. Its response amplitude (as an SDOF system) is given by

$$x_{10} = \frac{F_0}{k} \frac{1}{-r^2 + 1} \tag{8.45}$$

Next, we design an absorber by selecting the absorber mass to be $1/10$ of the system mass, and calculate the added stiffness based on Eq. (8.44):

$$m_2 = m_1/10 = 2.2 \text{ kg} \tag{8.46}$$

$$k_2 = \omega^2 m_2 = 3.1267 \times 10^5 \text{ N/m} \tag{8.47}$$

The response of the original system as well as that with the absorber is shown in Fig. 8.14. Note how we have succeeded in making the vibration amplitude (theoretically) zero at the running speed; however, it is still somewhat close to a new resonant speed. Hence, our design is effective only if the motor speed is constant and does not vary much.

Suppose we redesign the system using a different absorber mass relaxing the constraint that it be less than 10% of the original system mass; i.e., let $m_2 = 0.3m_1$. Then, $k_2 = 9.3801 \times 10^5$ N/m, and the response is now as shown in Fig. 8.15. Clearly, it is better than the previous design.

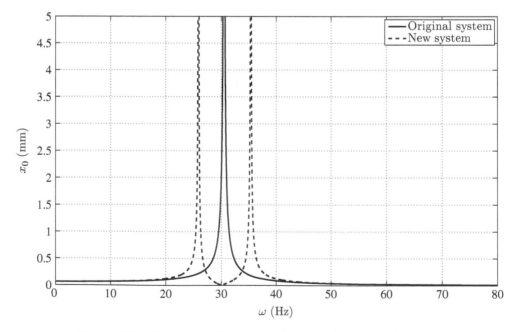

Figure 8.14: *Response of the original and newly designed systems*

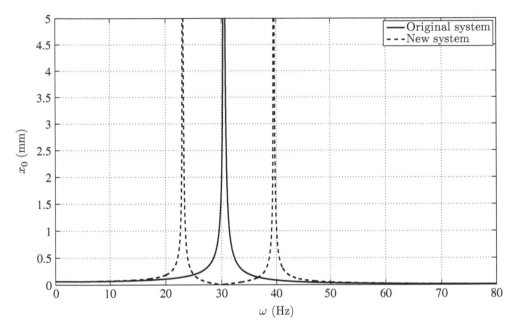

Figure 8.15: *Response of the original and the redesigned systems*

Exercise 8.1

We redesigned using a larger mass in the previous example. Now, experiment to see what the effect of *reducing* the mass would be. Can you get the second natural frequency to move away from 40 Hz? Why not?

Example 8.3 Dynamic vibration absorber for a hydraulic system

A hydraulic rotary pump is mounted on the same bench as a precision positioning system that it is supplying oil to. The pump's vibration is interfering with the positioning system. The mass of the pump is 25 kg and its normal operating speed is 3600 rpm. The resonant speed of the system was found to be 5000 rpm. Design a dynamic absorber with a mass of 1.25 kg to mitigate this problem.

Solution

We have the following given data: $m_1 = 25$ kg, $\omega = 3600$ rpm, $k_1 = \omega_n^2 m = 6.8539 \times 10^6$ N/m, $m_2 = 1.25$ kg. Again, we select k_2 based on Eq. (8.44):

$$k_2 = \omega^2 m_2 = 1.7765 \times 10^5 \text{ N/m} \tag{8.48}$$

The response of the designed system (nondimensionalized) is shown in Fig. 8.16. Comparison of the original system and the new system with the absorber shows that the response, that used to be $2F_0/k_1$, is now zero. However, in practice, the speed of the pump fluctuates around the nominal value of 3,600 rpm. So it is of interest to determine the speed range over which the absorber is effective. From the results it follows that $k_1 x_{10}/F_0$, or the nondimensional amplitude of the system, is less than 1 as long as the running speed is in the range of rpm: $3,571 < \omega < 3,686$.

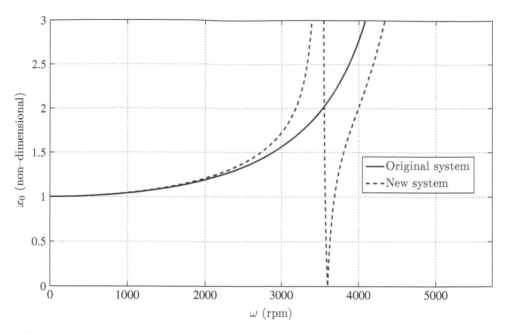

Figure 8.16: *Response of the redesigned system compared to the original response*

8.3   Modal Analysis

Recall that we used modal analysis to develop general response expressions for MDOF systems in Section 7.4.4. It turns out that we can use the same approach to get a response to arbitrary forces. This is especially useful both as a means of analysis as well as to enhance our understanding of MDOF systems. Again, consider the general equation of motion we need to solve. We will look at undamped systems first.

$$\mathbf{M}\ddot{\mathbf{x}} + \mathbf{K}\mathbf{x} = \mathbf{F}(t) \tag{8.49}$$

We restate the expansion theorem here for convenience. Given a set of linearly independent vectors, it follows from linear algebra that they form a basis for the N-dimensional vector space, and hence, any vector $\mathbf{u}$ can be expressed as a combination of the eigenvectors as follows.

$$\mathbf{u} = c_1\mathbf{u}_1 + c_2\mathbf{u}_2 + \cdots + c_n\mathbf{u}_n = \sum_{r=1}^{N} c_r\mathbf{u}_r \tag{8.50}$$

In terms of vibration of a dynamic system, this means that any possible motion of the system can be described by a linear combination of the eigenvectors.

This means that we first need to solve the free response problem. This will lead to the eigenvalue problem

$$\mathbf{K}\mathbf{u} = \omega^2 \mathbf{M}\mathbf{u} \tag{8.51}$$

Solution leads to n eigenvalues, square roots of which are the n natural frequencies. We will also get the eigenvectors, u_r, which are arbitrary to a multiplicative constant. They are orthogonal with respect to the stiffness and mass matrices; this orthogonality relationship can be written in terms of the modal matrices as follows.

$$\mathbf{U}^T \mathbf{M} \mathbf{U} = \mathbf{I} \tag{8.52}$$

$$\mathbf{U}^T \mathbf{K} \mathbf{U} = \Lambda \tag{8.53}$$

where $\Lambda = \mathrm{diag}(\omega_r^2)$ and we have assumed that the eigenvectors have been normalized.

Using the expansion theorem as a basis, we seek a solution to the forced response using

$$\mathbf{x}(t) = \mathbf{U}\eta(t) \tag{8.54}$$

$\eta_r(t)$ is the same old modal (or normal) coordinate. The above equation can also be thought of as simply a linear transformation between two different sets of generalized coordinates. Substituting Eq. (8.54) into Eq. (8.49),

$$\mathbf{M}\mathbf{U}\ddot{\eta} + \mathbf{K}\mathbf{U}\eta = \mathbf{F}(t) \tag{8.55}$$

Next, we premultiply both sides by $\mathbf{U}^T$ and watch the fun.

$$\mathbf{U}^T \mathbf{M} \mathbf{U}\ddot{\eta} + \mathbf{U}^T \mathbf{K} \mathbf{U}\eta = \mathbf{U}^T \mathbf{F}(t) \tag{8.56}$$

It is time now to use that much-talked-about orthogonality principle (Eq. 8.53).

$$\mathbf{I}\ddot{\eta} + \Lambda\eta = \mathbf{U}^T \mathbf{F}(t) \stackrel{\text{def}}{=} \mathbf{N}(t) \tag{8.57}$$

Clearly, the equations are decoupled. The right-hand side, $\mathbf{N}(t)$, is called the *generalized force vector*, and, an element of it, $N_r(t)$ would be called the rth modal generalized force. Hence, we now have N independent equations of the form:

$$\ddot{\eta}_r(t) + \omega_r^2 \eta_r(t) = N_r(t), \quad r = 1, 2, \dots N \tag{8.58}$$

Each of these equations is nonhomogeneous, and has a free and forced response. We already discussed the free response in Section 7.4, so let us just focus on the forced response here. To determine the forced response, we note now that we have already done this for an SDOF system in Section 6.2, and can just use the result in the form of the convolution theorem. That is,

$$\eta_r(t) = \frac{1}{\omega_r} \int_0^t N_r(\tau) \sin\left[\omega_r(t - \tau)\right] d\tau \tag{8.59}$$

The final response of the system could then be obtained from

$$\mathbf{x}(t) = \sum_{r=1}^{n} \mathbf{u}_r \eta_r(t) \tag{8.60}$$

Example 8.4 Forced response of the coupled pendulums

Let us use our familiar example of the coupled pendulums to understand the implications of the modal analysis to get the forced response of an MDOF system. Let us recall the system description. The system matrices are given as follows: [from Eq. (7.41)].

$$\mathbf{M} = \begin{bmatrix} m\ell^2 & 0 \\ 0 & m\ell^2 \end{bmatrix}, \quad K = \begin{bmatrix} ka^2 + mg\ell & -ka^2 \\ -ka^2 & ka^2 + mg\ell \end{bmatrix} \tag{8.61}$$

Recall that the natural frequencies are given by

$$\omega_1 = \sqrt{\frac{g}{\ell}}, \quad \omega_2 = \sqrt{\frac{g}{\ell} + 2\frac{k}{m}\frac{a^2}{\ell^2}} \tag{8.62}$$

and the normalized eigenvectors by

$$\mathbf{u}_1 = \frac{1}{\sqrt{2m\ell^2}} \begin{bmatrix} 1 \\ 1 \end{bmatrix}$$

$$\mathbf{u}_2 = \frac{1}{\sqrt{2m\ell^2}} \begin{bmatrix} 1 \\ -1 \end{bmatrix} \tag{8.63}$$

Now, suppose we subject both the pendulum masses to an identical horizontal step force of magnitude F_0. Or,

$$\mathbf{F}(t) = \begin{bmatrix} F_0 \mathcal{U}(t) \\ F_0 \mathcal{U}(t) \end{bmatrix} = F_0 \ell \begin{bmatrix} 1 \\ 1 \end{bmatrix} \mathcal{U}(t) \tag{8.64}$$

Note that the 'force' vector actually has torques as it should since the equation of motion is a moment equilibrium equation. This is also a good time to realize that the force vector may in general consist of forces, torques, or combinations of both. First, we calculate $N_r(t)$, the generalized forces.

$$N_1(t) = \mathbf{u}_1^T \mathbf{F}(t) = \sqrt{\frac{2}{m}} \frac{F_0}{\ell} \mathcal{U}(t) \tag{8.65}$$

$$N_2(t) = \mathbf{u}_2^T \mathbf{F}(t) = 0 \tag{8.66}$$

Clearly, the second generalized force is zero because the force being applied is essentially the first eigenvector. Hence, the second modal response will be identically zero because of the orthogonality property of the eigenvectors. The system response will, therefore, consist of only the first mode. We would say that the forces acting on the system are only *exciting* the first mode.

We can determine the response by substituting the generalized force expressions into the convolution theorem, Eq. (8.59).

$$\eta_1(t) = F_0 \sqrt{\frac{2\ell}{mg}} \int_0^t \sin\left[\omega_1(t - \tau)\right] d\tau \tag{8.67}$$

$$\eta_2(t) = 0 \tag{8.68}$$

The system response can be calculated from the superposition of the modal responses.

$$\mathbf{x}(t) = \sum_{r=1}^{n} \mathbf{u}_r \eta_r(t) = \eta_1(t)\mathbf{u}_1 \tag{8.69}$$

This leads to identical responses for both the pendulums, as shown in Fig. 8.17.

$$\theta_1(t) = \theta_2(t) = \frac{F_0 \ell}{g} \sqrt{\frac{2}{m}} \left[1 - \cos\left(\sqrt{\frac{g}{\ell}} t\right)\right] \tag{8.70}$$

Next, let us subject each of the masses to step forces which are equal in magnitude and opposite in direction (this may be tricky to do in practice; it is a good thing that this is a mental experiment!). Now,

$$\mathbf{F}(t) = \begin{bmatrix} F_0 \ell \mathcal{U}(t) \\ -F_0 \ell \mathcal{U}(t) \end{bmatrix} = F_0 \ell \begin{bmatrix} 1 \\ -1 \end{bmatrix} \mathcal{U}(t) \tag{8.71}$$

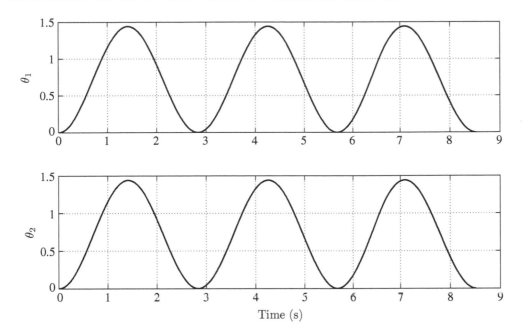

Figure 8.17: *Coupled pendulum response with $F_1 = F_2 = F_0 \mathcal{U}(t)$*

The generalized forces are now:

$$N_1(t) = u_1^T F(t) = 0 \tag{8.72}$$

$$N_2(t) = u_2^T F(t) = \sqrt{\frac{2}{m}} F_0 \mathcal{U}(t) \tag{8.73}$$

This time the first generalized force is zero, because the force being applied is similar to the second eigenvector. Hence, the first modal response will be identically zero because of the orthogonality principle, and the system response will resemble the second mode. We can determine the response by substituting the generalized force expressions into the convolution theorem.

$$\eta_1(t) = 0 \tag{8.74}$$

$$\eta_2(t) = F_0 \sqrt{\frac{2\ell}{mg}} \int_0^t \sin\left[\omega_2(t-\tau)\right] d\tau \tag{8.75}$$

This leads to opposite responses for the pendulums, as shown in Fig. 8.18.

$$\theta_1(t) = -\theta_2(t) = \frac{F_0}{\omega_2^2} \sqrt{\frac{2}{m}} \left[1 - \cos \omega_2 t\right] \tag{8.76}$$

Exercise 8.2

In the previous example, show that the first forcing condition leads to zero response for the second mode because of orthogonality. Then, do the same for the second case.

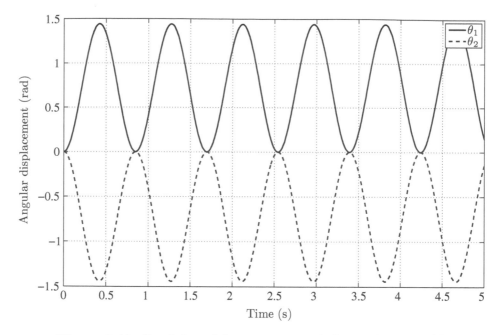

Figure 8.18: *Coupled pendulum response with $F_1 = -F_2 = F_0 \, \mathcal{U}(t)$*

Carry out a dimensional analysis of every equation in this section to convince yourself that the equations are right, at least from the dimensionality point of view. What is the dimension of $\eta(t)$ here? How about $N(t)$?

8.3.1 Modal Analysis with the Damped System

The coupled pendulum response analyzed earlier is clearly not realistic since it ignored damping. Let us look at the same problem but introduce some damping. There are many options here such as: numerical solution of the full damped system, solving for damped modes (as we discussed in Chapter 7), or simply using modal damping. In other words, we can solve the same undamped eigenvalue problem and simply add modal damping – this is fairly accurate for small values of damping.

So, the normal coordinate equation will now look as follows.

$$\ddot{\eta}_r(t) + 2\zeta_r \omega_r \dot{\eta}_r(t) + \omega_r^2 \eta_r(t) = N_r(t), \quad r = 1, 2, \cdots n \tag{8.77}$$

The forced modal response to the step function then follows from the convolution theorem.

$$\eta_r(t) = \frac{1}{\omega_{rd}} \int_0^t N_r(\tau) e^{-\zeta \omega_r (t-\tau)} \sin\left[\omega_{rd}(t-\tau) + \phi_r\right] \, d\tau \tag{8.78}$$

where

$$\phi_r = \tan^{-1}\left[\frac{\sqrt{1-\zeta_r^2}}{\zeta}\right] \tag{8.79}$$

Example 8.5 Forced response of coupled pendulums with damping

Let us consider the coupled pendulums again, and add some modal damping this time: $\zeta_1 = \zeta_2 = 0.1$. The modal responses to the step function is then given by

$$\eta_r(t) = \frac{N_0}{\omega_r^2} \left\{ 1 - \frac{1}{\sqrt{1 - \zeta_r^2}} e^{-\zeta \omega_r t} \sin\left(\omega_{rd} t + \phi_r\right) \right\} \tag{8.80}$$

where

$$\phi_r = \tan^{-1}\left[\frac{\sqrt{1 - \zeta_r^2}}{\zeta_r} \right] \tag{8.81}$$

For the first case ($F_1 = F_2 = F_0 \mathcal{U}(t)$), the real displacements are given by

$$\theta_1(t) = \theta_2(t) = \frac{N_0}{m\ell\omega_1^2} \left\{ 1 - \frac{1}{\sqrt{1 - \zeta_1^2}} e^{-\zeta \omega_1 t} \sin\left(\omega_{2d} t + \phi_1\right) \right\} \tag{8.82}$$

Note that $\omega_1 = 2.2147$ rad/s, and $\omega_2 = 7.4098$ rad/s. The damped natural frequencies are very close to them: $\omega_{d1} = 2.2036$ rad/s and $\omega_{d2} = 7.3726$ rad/s.

For the second case, when $F_1 = F_2 = -F_0 \mathcal{U}(t)$, the displacements are given by

$$\theta_1(t) = -\theta_2(t) = \frac{N_0}{m\ell\omega_2^2} \left\{ 1 - \frac{1}{\sqrt{1 - \zeta_2^2}} e^{-\zeta \omega_2 t} \sin\left(\omega_{2d} t + \phi_2\right) \right\} \tag{8.83}$$

Substituting into the expressions, we get the response, as shown in Figs. 8.19 and 8.20 for the two cases. You may want to refer to Section 6.2.2 for a review on step response.

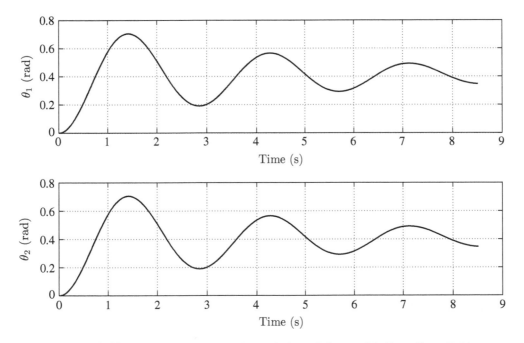

Figure 8.19: *Damped response of coupled pendulums with $F_1 = F_2 = F_0 \mathcal{U}$*

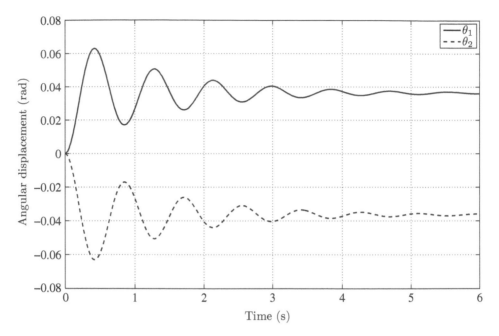

Figure 8.20: *Damped response of coupled pendulums with $F_1 = -F_2 = F_0\,\mathcal{U}(t)$*

Example 8.6 Design of an overhead traveling crane

An overhead traveling crane consists of a trolley that travels on two girders (beams) and is shown in Fig. 8.21. The trolley carries a load by means of a wire rope. Each of the two girders has a square cross section and the wire rope has a circular cross section. They are all made of steel. An electric motor is located in the trolley and has a nominal operating speed of 1500 rpm. The parameter values are as shown in Table 8.1. Design the girders and the wire ropes.

Dynamic model

The load and the trolley can be considered to be rigid masses with masses, m_1 and m_2 respectively. The wire rope can be modeled as a massless spring with stiffness, k_1; each of the girders can also be modeled as a massless spring with stiffness, k_2. Then, a model for dynamic studies would be as shown in Fig. 8.22. Note that the motor gives rise to a harmonic force on the trolley.

From the usual process of kinematics, kinetics (see Fig. 8.23 for the free body diagram) and Newton's laws, we get the matrix equation of motion:

$$\mathbf{M}\ddot{x} + \mathbf{K}x = \mathbf{F} \qquad (8.84)$$

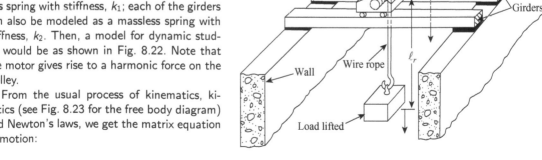

Figure 8.21: *Overhead traveling crane*

Parameter	Symbol	Value	Units
Load	m_1	4,500	kg
Trolley	m_2	18,000	kg
Rope length	ℓ_r	6	m
Rope radius	r_r	varies	m
Span of the girders	ℓ_g	9	m
Girder cross-sectional dimension	b	varies	m
Modulus of elasticity	E	2.06×10^{11}	N/m^2
Ultimate tensile strength	σ_{ut}	4.0×10^8	N/m^2
Force amplitude	F_0	1	kN
Motor operating speed	ω	1,500	r/s
Acceleration due to gravity	g	9.81	m/s^2

Table 8.1: *Parameter values for crane design*

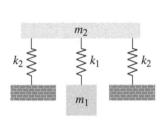

Figure 8.22: *Dynamic model for the overhead crane*

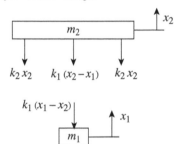

Figure 8.23: *Free body diagram of the crane system*

The matrices are as follows.

$$\mathbf{M} = \begin{bmatrix} m_1 & 0 \\ 0 & m_2 \end{bmatrix} \tag{8.85}$$

$$\mathbf{K} = \begin{bmatrix} k_1 & -k_1 \\ -k_1 & k_1 + 2k_2 \end{bmatrix} \tag{8.86}$$

$$\mathbf{F} = \begin{bmatrix} 0 \\ F_0 \end{bmatrix} \sin \omega t \tag{8.87}$$

Stiffness

The next step is to relate the stiffness in the above models to the physical parameters in the girders and the rope. The primary mode of deformation in the rope is longitudinal (tension) and that in the girders is flexural (bending).

Hence, the stiffness of the rope is

$$k_1 = \frac{A_r E_r}{\ell_r} \tag{8.88}$$

where E_r is the elastic modulus of the material, A_r is the area of cross section of the rope, and ℓ_r is the length of the rope. Also,

$$A_r = \pi r_r^2 \tag{8.89}$$

since the rope has a circular cross section of radius r_r.

The stiffness of the girder depends upon the position of the trolley. Suppose we denote the position by using a nondimensional quantity,

$$\alpha = \frac{x_t}{\ell_g} \tag{8.90}$$

where x_t is the dimensional position of the trolley that can vary from zero to the span of the girders, ℓ_g. Then, the stiffness of the girders is given by

$$k_2 = \frac{3E_g I_g^3}{\ell_g} \frac{1}{\alpha^2(1-\alpha)^2} \tag{8.91}$$

where E_g is the modulus of elasticity, and I_g is the area moment of inertia of the cross section of the girder and is given by

$$I_g = \frac{bh^3}{12} = \frac{b^4}{12} \tag{8.92}$$

since the cross section is a square of side b.

Static considerations

We should first determine the stresses that would arise in the structures from static considerations and determine the minimum dimensions that would be necessary to avoid failure. We will use a small factor of safety and assume that failure would not occur if the peak stresses (due to static loads) are below the ultimate tensile strength.

For the rope,

$$A_r > \frac{m_1 g}{\sigma_{ut}} \tag{8.93}$$

For the girders,

$$b > \left(\frac{6M_{max}}{\sigma_{ut}}\right)^{\frac{1}{3}} \tag{8.94}$$

where the maximum bending moment M_{max} would occur when the trolley is at the center.

Procedure

Since there is no analytical way to predict how different geometrical parameters will affect the natural frequencies, mode shapes or forced response amplitudes, the procedure has to be iterative.

We will use values for the radius of the rope and the cross-sectional dimension of the girder to be a range of values starting with the minimum permissible values as determined by static analysis. Then, we determine the stiffness values as given earlier, and use them in stiffness matrices to solve the eigenvalue problem (to determine the natural frequencies and the mode shapes) as well as the forced response problem.

Numerical results

The minimum value for the radius of the rope turns out to be 8.5 mm, and the minimum thickness of the girders is 0.26 m. Carrying out the eigenvalue solution as a function of the rope radius yields results shown in Figs. 8.24 and 8.25.

Note that as the rope radius increases, it increases k_1, and both the natural frequencies increase. Since the operating speed of the motor is 1500 rpm (or, 157 rad/s), we might try to make the system stiff; or, we would like both the natural frequencies to be above that frequency. In that case, the radius of the rope has to be larger than about 32 mm.

Next, we will use a value of 35 mm for the rope radius and study the natural frequencies as a function of the girder thickness. The results are shown in Figs. 8.26 and 8.27. Again, since the stiffness (k_2) increases, both natural frequencies increase. Note the relative flattening of the first natural frequency beyond a certain value of the girder thickness. Again, if we want the system to have natural frequencies greater than 157 rad/s, the girder thickness has to be at least 0.85 m (that is about 33 inches!).

Figure 8.28 shows the changes in the first mode shape as the girder thickness is changed.

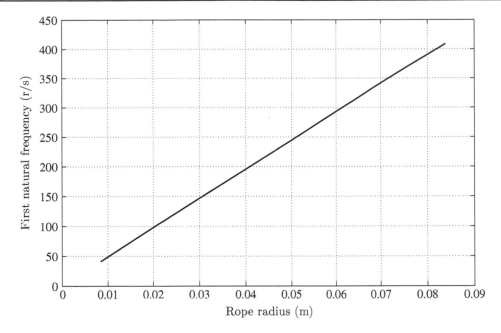

Figure 8.24: *First natural frequency versus rope radius*

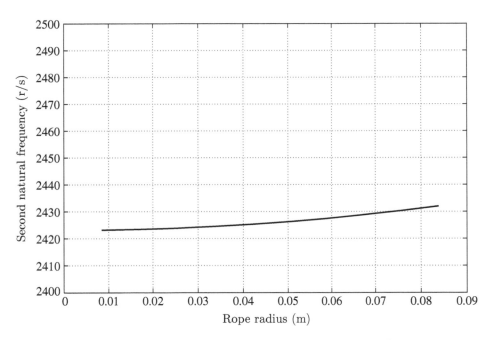

Figure 8.25: *Second natural frequency versus rope radius*

The next step is to check and see how such a design would perform as the trolley moved from one end of the girders to the other. This would change the stiffness, k_2, and results in natural frequencies shown in Figs. 8.29 and 8.30. Clearly, the lowest natural frequencies occur when the trolley is at the center.

Next, we carry out a forced response analysis to determine the response amplitudes of vibration. Suppose we assume that the forcing frequency is exactly the same as the motor operating speed, and analyze the vibrational

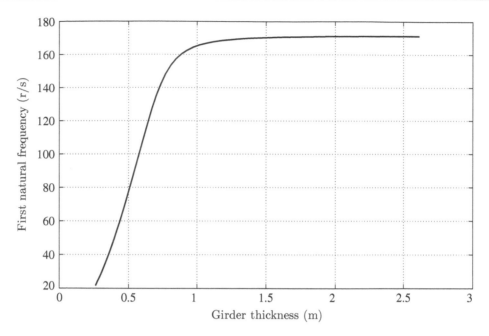

Figure 8.26: *First natural frequency versus girder thickness*

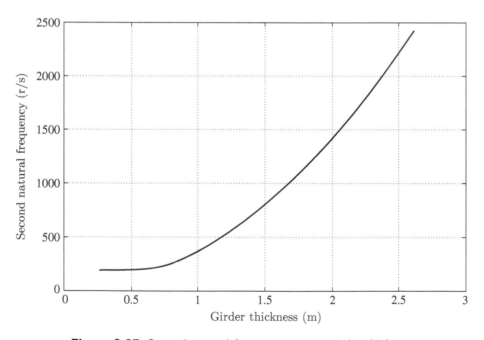

Figure 8.27: *Second natural frequency versus girder thickness*

displacements as a function of the girder thickness, we get the results shown in Fig. 8.31 in which resonance is clearly visible. With a girder thickness of 1.5 m, we would get an acceptable value of forced vibration.

With that value of the girder thickness, Fig. 8.32 shows the forced response amplitude as a function of the position of the trolley.

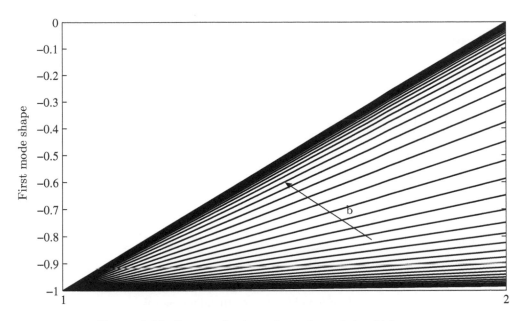

Figure 8.28: *First mode shape for various girder thicknesses*

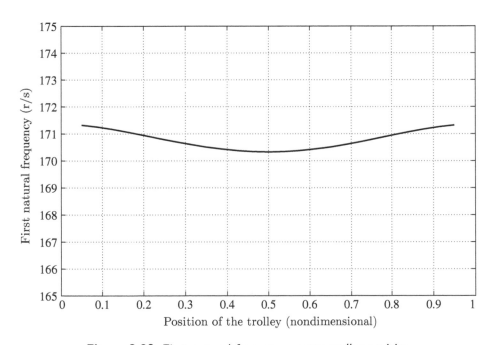

Figure 8.29: *First natural frequency versus trolley position*

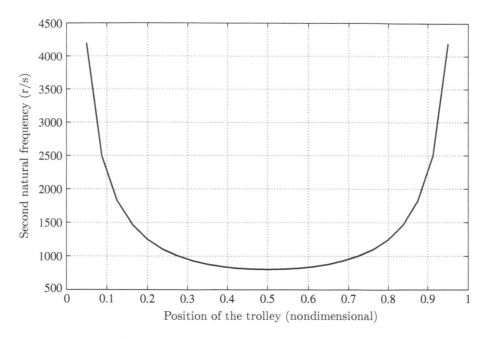

Figure 8.30: *Second natural frequency versus trolley position*

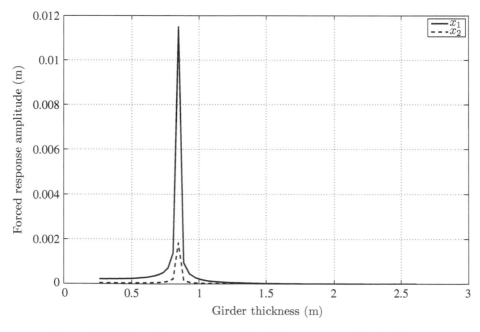

Figure 8.31: *Forced response amplitude for various girder thicknesses*

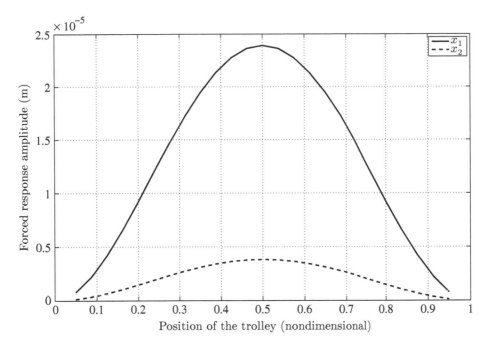

Figure 8.32: *Forced response amplitude for various trolley positions*

8.4 Review Questions

1. In a two-particle mass system considered in the text, the first mass gets excited at the first natural frequency, and the second at the second natural frequency. True or False? Explain.

2. In what ways does damping change the frequency response curve?

3. What is the basic principle of the dynamic vibration absorber?

4. In what fundamental way does the dynamic vibration absorber change the original system?

5. Practically speaking, is it possible to get a zero displacement value? Why or why not?

6. In the design of the dynamic vibration absorber, why do we need to consider a frequency range rather than a single frequency?

7. How does damping alter the performance of the dynamic vibration absorber?

8. Define the modal generalized force.

9. How would you extend the results in this chapter to periodic excitation of MDOF systems?

10. How would you extend the results in this chapter to transient excitation of MDOF systems?

Problems

8.1 Consider Problem 7.1 repeated in Fig. 8.33 for convenience. Determine the forced response of the system if there is a force $F_1 = F_0 \sin \omega t$ on the first mass, and $F_2 = F_0 \cos \omega t$ on the

Figure 8.33: *Forced two-DOF system*

second mass. Assume the following numerical values and plot the response. $m = 2\,\text{kg}$, $k = 2000\,\text{N/m}$, $\omega = 45\,\text{rad/s}$, $F_0 = 2\,\text{N}$.

8.2 Consider Problem 7.2 shown in Fig. 8.34 with a sinusoidal force acting on the first mass ($F_1 = F_0 \cos \omega t$). Determine the forced response for three values of n: 1, 0.1, and 10. Assume $0 < \omega < 5\sqrt{k/m}$.

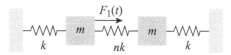

Figure 8.34: *Two-DOF system*

8.3 Consider Problem 7.3 shown in Fig. 7.49. Determine the forced response if $F_1 = F_0 \sin \omega t$ and $F_2 = 2F_0 \sin \omega t$. Use the following numerical values to plot the response: $m = 200\,\text{kg}$, $k = 100\,\text{kN/m}$, $F_0 = 20\,\text{kN}$, $0 < \omega < 100\,\text{rad/s}$.

8.4 Consider Problem 8.1 again. Determine the forced response of the system if there is a force $F_1 = F_0 \sin \omega t$ on the first mass and $F_2 = F_0 \sin 2\omega t$ on the second mass.

8.5 For each of the following shown in general form in Fig. 8.35, determine the forced response over the frequency ranges shown.

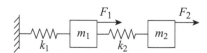

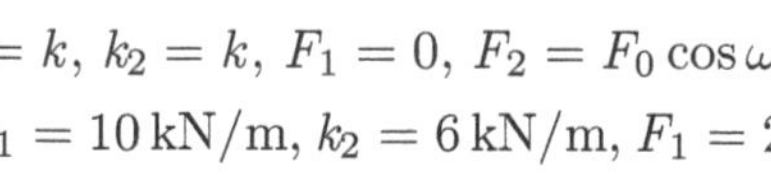

Figure 8.35: *Two-DOF system*

(a) $m_1 = m$, $m_2 = 2m$, $k_1 = k$, $k_2 = 2k$, $F_1 = 0$, $F_2 = F_0 \sin \omega t + 2F_0 \cos \omega t$, $0 < \omega < 5\sqrt{k/m}$

(b) $m_1 = m_2 = m$, $k_1 = 20k$, $k_2 = k$, $F_1 = F_0 \sin (\omega t + \phi)$, $\phi = \pi/4$, $0 < \omega < 12\sqrt{k/m}$

(c) $m_1 = 5m$, $m_2 = m$, $k_1 = k$, $k_2 = k$, $F_1 = 0$, $F_2 = F_0 \cos \omega t$, $0 < \omega < 3\sqrt{k/m}$

(d) $m_1 = 5\,\text{kg}$, $m_2 = 8\,\text{kg}$, $k_1 = 10\,\text{kN/m}$, $k_2 = 6\,\text{kN/m}$, $F_1 = 20 \sin \omega t\,\text{kN}$, $\omega = 60\,\text{rad/s}$, $F_2 = 0$

(e) $m_1 = 200\,\text{kg}$, $m_2 = 50\,\text{kg}$, $k_1 = 220\,\text{kN/m}$, $k_2 = 600\,\text{kN/m}$, $F_1 = F_2 = 2.5 \sin \omega t\,\text{kN}$, $0 < \omega < 150\,\text{rad/s}$

(f) $m_1 = 20\,\text{mg}$, $m_2 = 4\,\text{mg}$, $k_1 = 100\,\text{kN/m}$, $k_2 = 60\,\text{kN/m}$, $F_1 = 350 \cos \omega t\,\text{N}$, $F_2 = 1 \sin \omega t\,\text{kN}$, $0 < \omega < 5\,\text{kHz}$

8.6 For each of the systems shown in Fig. 8.35, determine the forced response for the following cases. Use the numerical values shown in Problem. 8.5. Recall that $\mathcal{U}(t)$ is a unit step function and that $\delta(t)$ is a unit impulse.

(a) $F_1 = F_0 \mathcal{U}(t),\ F_2 = 0$

(b) $F_1 = F_0 \mathcal{U}(t),\ F_2 = \mathcal{U}(t - t_0)$

(c) $F_1 = 0,\ F_2 = F_0 \delta(t)$

(d) $F_1 = 25\delta(t)(\,\mathrm{kN}),\ F_2 = 0$

(e) $F_1 = -F_2 = 100\mathcal{U}(t);\ F_1 = 100\mathcal{U}(t),\ F_2 = 100\mathcal{U}(t - 0.2)\ (\,\mathrm{kN})$

(f) $F_1 = F_2 = 500\mathcal{U}(t)\ (N)$

8.7 For each of the following shown in general form in Fig. 8.36, determine the forced response over the frequency ranges shown.

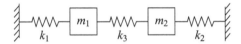

Figure 8.36: *Two-DOF system*

(a) $m_1 = m,\ m_2 = 2m,\ k_1 = k,\ k_2 = k_3 = k,\ F_1 = F_0 \sin\left(\omega t - \pi/8\right),\ F_2 = 0,\ 0 < \omega < 5\sqrt{k/m}$

(b) $m_1 = m_2 = m,\ k_1 = 200k,\ k_2 = 200k,\ k_3 = k,\ F_1 = 0,\ F_2 = F_0 \cos\omega t,\ 0 < \omega < 50\sqrt{k/m}$

(c) $m_1 = 5m,\ m_2 = m,\ k_1 = k,\ k_2 = k,\ k_3 = 100k,\ F_1 = F_0 \sin\omega t,\ F_2 = 2F_0 \cos\omega t,\ 0 < \omega < 12\sqrt{k/m}$

(d) $m_1 = 5\,\mathrm{kg},\ m_2 = 8\,\mathrm{kg},\ k_1 = 10\,\mathrm{kN/m},\ k_2 = 6\,\mathrm{kN/m},\ k_3 = 20\,\mathrm{kN/m},\ F_1 = 250\sin\omega t\,\mathrm{N},\ F_2 = 400\sin\omega t\,\mathrm{N},\ 0 < \omega < 100\,\mathrm{rad/s}$

(e) $m_1 = 200\,\mathrm{kg},\ m_2 = 50\,\mathrm{kg},\ k_2 = 220\,\mathrm{kN/m},\ k_2 = 600\,\mathrm{kN/m},\ k_3 = 440\,\mathrm{kN/m},\ F_1 = 50\sin\omega t\,\mathrm{N},\ F_2 = 0,\ 5 < \omega < 25\,\mathrm{Hz}$

(f) $m_1 = 20\,\mathrm{mg},\ m_2 = 4\,\mathrm{mg},\ k_2 = 100\,\mathrm{kN/m},\ k_2 = 60\,\mathrm{kN/m},\ k_3 = 100\,\mathrm{kN/m},\ F_1 = 300\cos\omega t\,\mathrm{N},\ F_2 = 50\cos\omega t\,\mathrm{N},\ 3 < \omega < 50\,\mathrm{kHz}$

8.8 For each of the systems shown in Fig. 8.36, determine the forced response for the following cases. Use the numerical values shown in Problem. 8.7. Recall that $\mathcal{U}(t)$ is a unit step function and that $\delta(t)$ is a unit impulse.

(a) $F_1 = F_0 \mathcal{U}(t),\ F_2 = 0$

(b) $F_1 = F_0 \mathcal{U}(t),\ F_2 = \mathcal{U}(t - t_0)$

(c) $F_1 = 0,\ F_2 = F_0 \delta(t)$

(d) $F_1 = 25\delta(t)(\,\mathrm{kN}),\ F_2 = 0$

(e) $F_1 = -F_2 = 100\mathcal{U}(t);\ F_1 = 100\mathcal{U}(t),\ F_2 = 100\mathcal{U}(t - 0.2)\ (\,\mathrm{kN})$

(f) $F_1 = F_2 = 500\mathcal{U}(t)\ (N)$

8.9 Consider Problem 7.7 where we modeled a double rod pendulum. Suppose there is a base motion in the horizontal direction given by $u(t) = u_0 \sin\omega t$, as shown in Fig. 8.37. Assuming small amplitudes about the vertical equilibrium position determine the forced response.

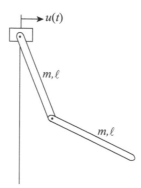

Figure 8.37: *Double rod pendulum with base motion*

8.10 Two masses m_1 and m_2 are attached to a light string with tension T, as shown in Fig. 8.38. Assume that the displacement is small and that the tension remains unchanged when the masses are displaced normal to the string. Assume $m_1 = m$, $m_2 = 2m$. Now, determine the forced response in the following situations.

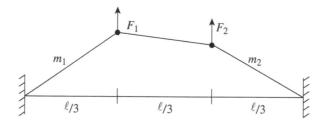

Figure 8.38: *Masses on a string*

(a) $F_1 = F_0 \sin \omega t$, $F_2 = F_0 \sin \omega t$

(b) $F_1 = F_0 \sin \omega t$, $F_2 = -F_0 \sin \omega t$

(c) $F_1 = F_0 \sin \omega t$, $F_2 = F_0 \cos \omega t$

(d) $F_1 = F_0 \sin \omega t$, $F_2 = 2F_0 \sin \omega t$

(e) $F_1 = F_0 \sin \omega t$, $F_2 = 2F_0 \cos \omega t$

8.11 Obtain the forced response to the string problem (Problem 8.10) in the following cases using modal analysis.

(a) $F_1 = F_2 = F_0 \mathcal{U}(t)$

(b) $F_1 = -F_2 = F_0 \mathcal{U}(t)$

(c) $F_1 = F_0 \mathcal{U}(t)$, $F_2 = 0$

(d) $F_1 = 0$, $F_2 = F_0 \mathcal{U}(t)$

(e) $F_1 = F_0 \mathcal{U}(t)$, $F_2 = 0.5 F_0 \mathcal{U}(t)$

8.12 Obtain the forced response to the string problem (Problem 8.10) in the following cases using modal analysis.

(a) $F_1 = F_0\delta(t)$, $F_2 = 0$

(b) $F_1 = 0$, $F_2 = F_0\delta(t)$

(c) $F_1 = \delta(t)$, $F_2 = F_0\delta(t - t_0)$

8.13 Obtain the forced response to the string problem (Problem 8.10) to a pulse force delivered to the first mass; it has magnitude F_0 and acts over a length of time t_0.

8.14 Consider Problem 7.9 where we analyzed coupled pendulums. Suppose the base of the pendulums is subjected to a horizontal motion, $u(t) = u_0 \sin \omega t$. Derive the equations of motion and determine the forced response.

8.15 Consider Problem 7.9 where we analyzed coupled pendulums. Suppose the base of the pendulums is subjected to a horizontal acceleration, $u(t) = a_0 \mathcal{U}(t)$. Derive the equations of motion and determine the forced response.

8.16 Consider Problem 7.26 where we looked at the lower body modeled as a double pendulum. Now we are interested in the effect of vibration on the human body. Derive the equation of motion assuming a base motion, $u(t)$ in the vertical direction. Determine the forced response if $u(t) = u_0 \sin \omega t$. Assume $u_0 = 2.5$ cm, and $0 < \omega < 30$ Hz.

8.17 Consider the problem of balancing a stick on the palm of your hand. This is often called the "broom-balancing" problem. Do not dismiss this as a silly problem; in fact, it is a very important problem in the theory of dynamic systems as it relates to stabilizing what is inherently an unstable system. Without delving into the subject of control theory, let us look at the dynamics. Figure 8.39 shows a model for the hand (or a mechanism) along with the inverted pendulum. Derive the equations of motion assuming that the pendulum stays in the vicinity of the vertical position. Also, determine the forced response if the base force is harmonic. Then, go back to the nonlinear system and use a numerical method to simulate the response to the base input of varying magnitudes and frequencies. Use any parameter values that you fancy.

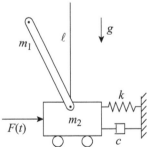

Figure 8.39: *Inverted pendulum with base motion*

8.18 The model shown in Fig. 8.40 simulates the effect of earthquakes on buildings. The building is assumed to be rigid and its base is connected to the ground through two springs: k_h, a translational spring, and k_r, a torsional spring (not shown). Note that the ground motion during an earthquake is a function of time. The mass of the building is m, and its center of mass C is at a distance ℓ from the bottom. Assume that the motion at the base is purely horizontal and sinusoidal. Derive the equation of motion.

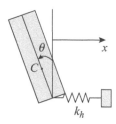

Figure 8.40: *A two-DOF model of a building*

8.19 Consider the model of the seated human being we looked at in the previous chapter (Problem 7.41). Assume a base acceleration of $\ddot{u} = a_0 \sin \omega t$, with $a_0 = 5 \, \text{m/s}^2$, and $0 < \omega < 10(\text{Hz})$. Determine the forced response. Hint: Assuming zero initial conditions, the base displacement is also sinusoidal with an amplitude of a_0/ω^2.

8.20 Consider the four-DOF model of the automobile we considered in Problem 7.32. Analyze the forced vibration over a range of speeds due to support motion. Assume that the road roughness is a sinusoid with a wavelength of 1.8 m and an amplitude of 4 cm.

8.21 Consider the problem of the car with the driver considered in Problem 7.35. Analyze the forced vibration over a range of speeds due to support motion. Assume that the road roughness is a sinusoid with a wavelength of 1.2 m and an amplitude of 6 cm. In particular, what is the vibrational amplitude suffered by the driver?

8.22 A force F is applied to a uniform rod AB of length ℓ which is supported by two springs with spring constants k_1 and k_2, as shown in Fig. 8.41. Derive the equation of motion of the system, where C is the center of mass of the rod. Solve for the forced response; given that $F = F_0 \sin \omega t$, $k_1 = 10 \, \text{kN/m}$, $k_2 = 14 \, \text{kN/m}$, $m = 20 \, \text{kg}$, $\ell = 3.5 \, \text{m}$, $F_0 = 450 \, \text{N}$, $0 < \omega < 100 \, \text{rad/s}$.

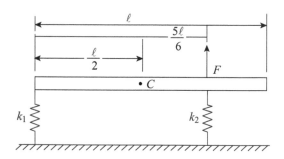

Figure 8.41: *An MDOF system*

8.23 Consider the system shown in Fig. 8.42.

(a) Derive the equation of motion for the system.

(b) Assume the following numerical values: $\theta = 30^0$, $k_1 = 5 \, \text{N/m}$, $k_2 = 8 \, \text{N/m}$, and $k_3 = 12 \, \text{N/m}$, $m_1 = 59 \, \text{kg}$, $m_2 = 68 \, \text{kg}$, and $m_3 = 86 \, \text{kg}$. Determine the natural frequencies of the system.

(c) Determine the forced response if $F = 5 \sin \omega t$ N. Plot the amplitude response for a range of values of ω.

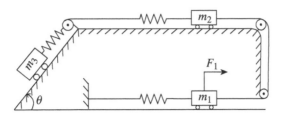

Figure 8.42: *Forced MDOF system*

8.24 Consider the system shown in Fig. 8.43.

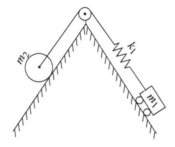

Figure 8.43: *Forced MDOF system*

(a) Derive the equation of motion and determine the natural frequencies of oscillation.

(b) Determine the forced response of the system if there is a harmonic force on mass m_1.

8.25 For the translational mechanical system shown in Fig. 8.44, derive the equations of motion. F_{v1}, F_{v2}, and F_{v3} are frictional forces that are acting between the masses and can be modeled by equivalent linear viscous damping coefficients. Assume $m_1 = 1\,\text{kg}$, $m_2 = 1\,\text{kg}$, $m_3 = 1\,\text{kg}$, $k_1 = k_2 = k_3 = 20\,\text{kN/m}$, $c_{v1} = c_{v2} = c_{v3} = 500\,\text{Ns/m}$ and $c_1 = 200\,\text{Ns/m}$. Solve for the natural frequencies and obtain the forced response if $F(t) = 500\sin\omega t\,\text{N}$, with $0 < \omega < 100\,\text{rad/s}$.

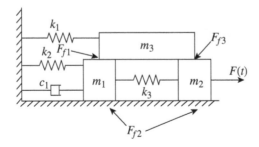

Figure 8.44: *Forced MDOF system for Problem 8.25*

8.26 For the system shown in Fig. 8.45, the inertia, J, of radius, r, is constrained to move only about the axis perpendicular to the paper. The cylinder rolls without slipping. If an external force, $F(t)$, is applied to the mass, derive the equations of motion. Determine the natural frequencies and the forced response if $F(t)$ is harmonic.

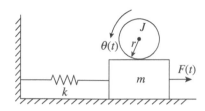

Figure 8.45: *Forced MDOF system for Problem 8.26*

8.27 Derive the equations of motion for the system shown in Fig. 8.46 if $F(t) = F_0 \sin \omega t$. Determine the response if $m_1 = 200 \, \text{kg}$, $m_2 = 50 \, \text{kg}$, $k_1 = 22 \, \text{kN/m}$, $k_2 = 15 \, \text{kN/m}$, $c_2 = 3 \, \text{kNs/m}$, $F_0 = 500 \, \text{N}$, $0 < \omega < 1000 \, \text{rad/s}$.

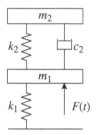

Figure 8.46: *Forced MDOF system for Problem 8.27*

8.28 Derive the equations of motion for the system shown in Fig. 8.47 if $F(t) = F_0 \sin \omega t$. Determine the response if $m_1 = 10 \, \text{mg}$, $m_2 = 8 \, \text{mg}$, $k_1 = 150 \, \text{N/m}$, $k_2 = 80 \, \text{N/m}$, $k_3 = 120 \, \text{N/m}$, $c_1 = c_2 = 10 \, \text{Ns/m}$, $F_0 = 350 \, \text{mN}$, $0 < \omega < 150 \, \text{rad/s}$.

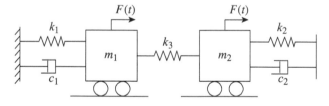

Figure 8.47: *Forced MDOF system for Problem 8.28*

8.29 Derive the equations of motion for the system shown in Fig. 8.48, and determine the forced response. $m_1 = m_2 = m_3 = m$, $k_1 = k_2 = k_3 = k_4 = k_5 = k$, $c_1 = c_2 - c_3 = c$, $F_1 = F_3 = F_0 \sin \omega t$, $F_2 = 0$, and $0 < \omega < 25\sqrt{k/m}$.

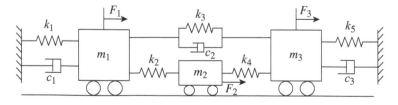

Figure 8.48: *Forced MDOF system for Problem 8.29*

8.30 In the system shown in Fig. 8.49, a pendulum is attached at the center of mass of m_1 and a force $F(t)$ is applied as shown. Derive the nonlinear equation of motion. Linearize for small motions, and solve to determine the natural frequencies. Also obtain the forced response for harmonic excitation. Assume $m_1 = 5m_2 = 5m$, $k = 1.5mg/\ell$, and $0 < \omega < 5\sqrt{k/m}$.

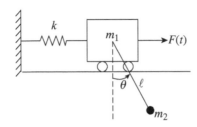

Figure 8.49: *Forced MDOF system for Problem 8.30*

8.31 A crankshaft has an equivalent mass moment of inertia of 0.15 kgm^2, radius of 3 cm, length of 0.4 m, and is made of steel. It is subject to torsional vibration in the range of frequencies, $0 < \omega < 5000\,\text{rpm}$. Design a dynamic vibration absorber for it; given that its mass moment of inertia cannot exceed 0.1 kgm^2.

8.32 An air-conditioning unit is attached to the end of a cantilever beam and is causing a lot of vibration that needs to be mitigated. The compressor runs at 250 rpm. The beam is made of aluminum, is 0.5 m long, and has a square cross section of size 10×2 (cm). The mass of the beam support can be neglected; the mass of the a/c unit is 70 kg. Design a dynamic vibration absorber; the mass of the added unit cannot exceed 1/10 the mass of the a/c unit. Determine the subsequent maximum static displacement in the support.

8.33 A handheld machine tool can be modeled as a mass-spring-damper system with the following properties: $m = 8\,\text{kg}$, $k = 15\,\text{kN/m}$, $c = 3000\,\text{Ns/m}$. The speed range of operation is $0 < \omega < 400\,\text{rpm}$. Design a dynamic vibration absorber with the stipulation that the added mass may not exceed 200 g.

8.34 An electric razor can cause vibration on the hands of the user resulting in unpleasant shaving experiences including unintended blood donations! It can be modeled as a damped oscillator with $m = 50\,\text{g}$, $k = 800\,\text{N/m}$. It has negligible damping. The running speed of the razor is 60 Hz. Design a dynamic vibration absorber; the added mass cannot exceed 1/10 the mass of the razor.

8.35 Figure 8.50 represents a schematic of actuator-driven reciprocating diaphragm air-compressors. They are extensively used in areas such as health care and aerospace. The reciprocating motion of the shuttle results in vibrational forces being transmitted to the supporting structure and may result in a high level of acoustic noise. This cannot be tolerated in many applications; therefore, an effective means of controlling the transmitted vibration energy between the compressor and its surroundings must be employed. One way to minimize the problem is to support the compressor on elastomeric antivibration (AV) mounts. Assume a shuttle mass of 37×10^{-3} kg, compressor case mass of 1.19 kg, compressor bellow stiffness of 2700 N/m, compressor bellow damping of 0.4 Ns/m, antivibration mount stiffness of 1500 N/m and antivibration mount damping of 6.3 Ns/m. You might want to read the article by Rens et al. [51] for additional information.

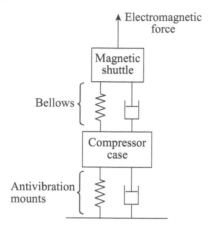

Figure 8.50: *Lumped parameter model of compressor and AV mounts*

(a) Figure 8.51 represents a lumped parameter model, in which the actuator shuttle (m_1) and compressor case (m_2) are represented as lumped masses and the bellows (k_1, c_1) and anti vibration mounts (k_2, c_2) are modeled as damped springs. Derive the equation of motion for the system. Solve for the response given that the harmonic electromagnetic force $F_0 = 50 \sin \omega t (\mathrm{N})$ and the frequency varies: $0 < \omega < 30 \, \mathrm{rad/s}$. Determine the harmonic response.

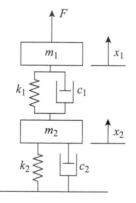

Figure 8.51: *Lumped parameter model of compressor and AV mounts*

(b) In order to improve on the vibration isolation performance of AV mounts, a dynamic vibration absorber can be employed. Figure 8.52 shows the block diagram of the extended model of the reciprocating compressor which has been extended to include an additional mass-compliance system, and Fig. 8.53 represents the lumped parameter model. Determine the parameters of the vibration absorber if the primary operating frequency is 20 rad/s, and the mass of the vibration absorber should be less than 20 g. Determine the forced response of the resulting system.

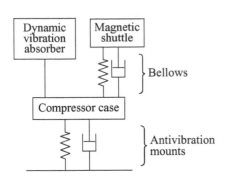

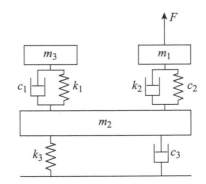

Figure 8.52: *Lumped parameter model of compressor and AV mounts*

Figure 8.53: *Lumped parameter model of compressor with dynamic vibration absorber*

(c) In another technique to mitigate the vibration transmission from the reciprocating compressor, an additional actuator F_2 is employed in parallel with the passive antivibration mounts, as shown in Fig. 8.54. This additional actuator counteracts the vibrational forces which are transmitted through the antivibration mounts so as to dynamically isolate the vibrating compressor from the support. Determine the response of the system assuming $F_2 = 60 \sin \omega t \, \text{N}$.

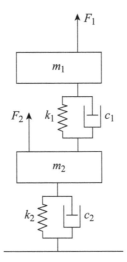

Figure 8.54: *Lumped parameter model of actively mounted compressor*

8.36 Vibro-elastography is a new medical imaging method that identifies the mechanical properties of tissue by measuring tissue motion in response to a multifrequency external vibration source. In vibro-elastography multiple ultrasound images are captured while the tissue is vibrating, and each image is compared to the previous image to provide a grid of local displacement measurements. These displacements fields are then used to determine the visco-elastic properties of the tissue at each grid location by measuring the tissue response at different frequencies and different time instances. This method is then used to extract viscosity and density from the dynamic response of the tissue [52]. Clinical acceptance of vibro-elastography would be aided by the development of a handheld probe that includes both the imaging device (ultrasound probe) and an excitation source that vibrates the tissue. Figure 8.55 shows an early proposal for such a handheld device. Two actuators jointly vibrate the tissue and the attached ultra-sound probe images the tissue motion. For further reading you should consult the article by Rivaz and Rohling [52].

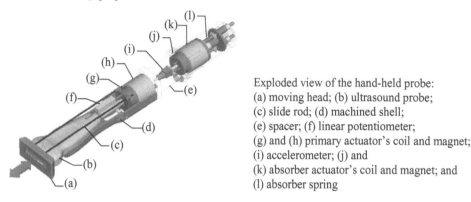

Exploded view of the hand-held probe:
(a) moving head; (b) ultrasound probe;
(c) slide rod; (d) machined shell;
(e) spacer; (f) linear potentiometer;
(g) and (h) primary actuator's coil and magnet;
(i) accelerometer; (j) and
(k) absorber actuator's coil and magnet; and
(l) absorber spring

Figure 8.55: *A handheld probe (see Rivaz and Rohling [52])*

(a) Held by hand, the probe can move in all possible degrees of freedom. The operator's hand is also a distributed mass system that vibrates around the nonlinear visco-elastic joints at the wrist, elbow, and shoulder. Since the reaction forces exerted on the device and the amplitude of tissue vibration is small, the greatest motion is axial. Therefore, the handheld device can be modeled as a one degree-of-freedom linear mass-spring-damper system as shown in Fig. 8.56. The term primary system refers to the combined effective mass, stiffness, and damping of the handheld assembly, i.e., including the operator. The external forces from the hand are represented by a spring and a damper. F_1 is the actuator force. Derive the differential equations of motion. Note that the tissue modeled with m_2 is subject to a base motion, $u(t)$.

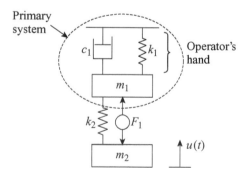

Figure 8.56: *Handheld probe model*

(b) For a handheld probe, the reaction forces between the tissue and the vibrator act as an external disturbance on the probe assembly and will cause unwanted vibration. These vibrations should be minimized to maintain the comfort of the operator and the accuracy of the vibro-elastography calculations. Therefore, the probe design should use some form of vibration absorption. Figure 8.57 shows a dynamic vibration absorber attached to the primary system. Derive the mathematical model.

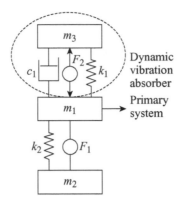

Figure 8.57: *Model of the dynamic vibration absorber for the handheld probe*

(c) An actively controlled dynamic vibration absorber with acceleration feedback is developed and is shown in Fig. 8.58. The control force F_2 of a PI controller with acceleration feedback is given as $F_2 = K_p \ddot{x}_a$, where K_p is called the proportional gain coefficient and x_a is the position of the absorber mass. Derive the mathematical model.

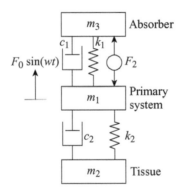

Figure 8.58: *Actively controlled dynamic vibration absorber*

8.37 Consider the system shown in Fig. 8.59. Determine the forced response if $F = F_0 \sin \omega t$, with $F_0 = 2\,\text{N}$, $m_1 = 4\,\text{kg}$, $k = 15\,\text{N/m}$, and $m_2 = 1$ kg.

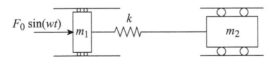

Figure 8.59: *MDOF system*

8.38 This is a problem in active vibration isolation which is an extension of the concept of dynamic vibration absorbers, which, on the other hand, are passive [53]. The model of an SDOF primary system with the active absorber attached to it is shown in Fig. 8.60. The primary structure is an SDOF undamped oscillator having mass m_1 attached to a spring of stiffness k_1. The absorber consists of a secondary mass m_2 attached to the primary structure by a spring of stiffness k_2. An actuator placed in between the primary and secondary mass exerts the control force $F_c = K_p x_2 - K_d \dot{x}_2$. x_1 and x_2 are the absolute displacements of the primary and secondary mass, respectively, and a harmonic force $F(t) = F_0 \sin \omega t$ is acting on mass m_1. Determine the equations of motion. Note how a damping matrix is introduced due to the controller. You should read an interesting paper by Chatterjee [53] discussing optimal selection of controller parameters to suppress vibration.

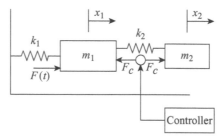

Figure 8.60: *Optimal vibration absorber*

8.39 Vibration transmitted by a machinery system is reduced by using rubber and hydraulic mounts. A model for this system consisting of springs and masses is shown in Fig. 8.61, where m is the effective mass of inertia of the track fluid column, c_2 is the effective viscous damping of the track fluid, k_2 is the effective stiffness of the upper hydraulic chamber, and k_3 is the equivalent stiffness of the lower chamber, k_1 and c_1 are the effective stiffness and damping coefficients of the rubber mount respectively and F_T is the transmitted force. Assume $m_1 = 200$ kg, $m_2 = 81$ kg, $c_2 = 4.1 \times 10^3$ N s/m, $k_2 = 8.1 \times 10^5$ N/m, $k_3 = 8.4 \times 10^3$ N/m, $k_1 = 2 \times 10^5$ N/m and $c_1 = 496$ Ns/m. Also assume $F(t) = F_0 \sin \omega t$, where, $F_0 = 1$ kN, and ω varies from 100 r/s. Derive the equation of motion and solve to obtain the response. Also solve for the transmitted forces. Interested in reading more about it? Look up the article by Yoon and Singh [54].

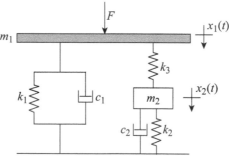

Figure 8.61: *Problem 8.39 machinery mount system*

8.40 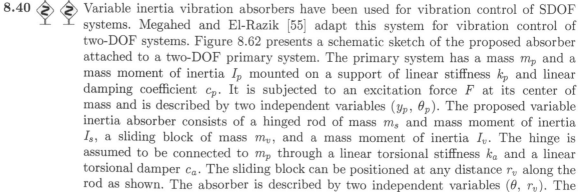 Variable inertia vibration absorbers have been used for vibration control of SDOF systems. Megahed and El-Razik [55] adapt this system for vibration control of two-DOF systems. Figure 8.62 presents a schematic sketch of the proposed absorber attached to a two-DOF primary system. The primary system has a mass m_p and a mass moment of inertia I_p mounted on a support of linear stiffness k_p and linear damping coefficient c_p. It is subjected to an excitation force F at its center of mass and is described by two independent variables (y_p, θ_p). The proposed variable inertia absorber consists of a hinged rod of mass m_s and mass moment of inertia I_s, a sliding block of mass m_v, and a mass moment of inertia I_v. The hinge is assumed to be connected to m_p through a linear torsional stiffness k_a and a linear torsional damper c_a. The sliding block can be positioned at any distance r_v along the rod as shown. The absorber is described by two independent variables (θ, r_v). The complete system has four degrees of freedom (y_p, θ_p, θ, and r_v). Derive the equation of motion of the system.

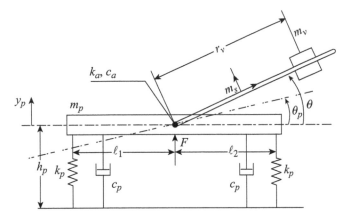

Figure 8.62: *Variable inertia vibration absorber*

8.41 Noise and vibration are of concern with many mechanical systems including industrial machines, home appliances, transportation vehicles, and building structures. Many such structures are composed of beam- and plate-like elements. The vibration of beam and plate systems can be reduced by the use of passive damping once the system parameters have been identified. One such passive damping approach utilizes momentum transfer between a primary system and auxiliary masses during collisions in order to reduce severe vibrations. For dynamic analysis the impact damper is modeled as a two-DOF model as shown in Fig. 8.63. Note that the mass m_1 rolls without slipping. Derive the equations of motion. For further reading look up the article by Park et al. [56].

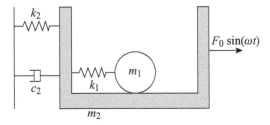

Figure 8.63: *Impact damper*

8.42 Grinding wheels used in the ball screw grinding invariably possess a certain degree of unbalance and deviation in roundness due to material inhomogeneity and manufacturing errors. This unbalance causes severe ball screw vibration and then results in undesirable surface roughness on the ball screw. In order to suppress the excessive vibrations caused by the grinding wheel especially on a long ball screw, steady rest supports are utilized, as shown in Fig. 8.64. They are fixed on the machine bed and distributed periodically along the ball screw. The installation of these supports is time consuming, especially when one needs to change the ball screw with a different length.

In this problem an alternative ball screw support made of a vibration absorber is introduced. The dynamic vibration absorber is an effective device to reduce the undesired vibration of a machine. This absorber can be regarded as a mass-spring-damper system which has its best performance in vibration reduction when the excitation frequency coincides with its natural frequency, i.e., when the absorber creates a vibration node on the location at which it is bonded on to, and the machine vibration can be reduced due to the node acting as a fixed support for the machine [57].

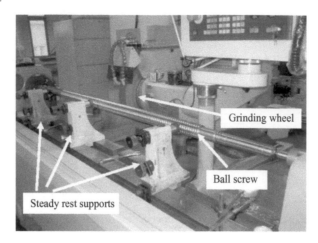

Figure 8.64: *Grinding wheel setup (see Cheng et al.[57])*

(a) Consider the ball screw as a damped oscillator of mass m, damping constant c, and a spring constant k subjected to a harmonic excitation $F(t)$, as shown in Fig. 8.65. Derive the equations of motion.

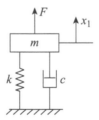

Figure 8.65: *Ball screw model*

(b) A physical model for a ball screw with a vibration absorber is shown in Fig. 8.66. The ball screw is modeled as a nonrotating beam due to its slenderness and small rotational speed. Here, we can model the beam as a rigid mass m_2 and an equivalent stiffness, k_2 which depends on the distance, ℓ_1 as show and the absorber parameters are modeled by m_1, k_1 and c_1. Derive the equations of the motion assuming the harmonic force on m_2.

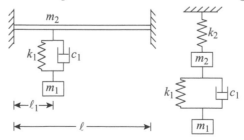

Figure 8.66: *Ball screw with dynamic vibration absorber*

(c) Figure 8.67 illustrates a more sophisticated model of the interaction between the three subsystems, i.e., ball screw, grinding wheel, and vibration absorber. Derive the equations of motion assuming the harmonic force on m_b.

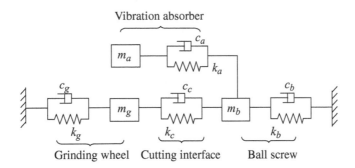

Figure 8.67: *Grinding wheel, ball screw, and vibration absorber*

8.43 Bridges are dynamically loaded when vehicles travel on them. In practice, usually a moving static load with a dynamic amplification factor or dynamic load coefficient is used to model the vehicle force on the bridges. However, in reality, the dynamic load due to the interaction between the vehicle and the bridge is a complex problem and is affected by many factors such as vehicle speed, road surface roughness and dynamic properties of both the vehicle and bridge. Ding et al. [59] present an evolutionary spectral method to evaluate the dynamic loads on bridges due to the passage of a vehicle along a rough bridge surface at a constant speed. The response of the bridge and vehicle is described by two separate sets of equations, which are coupled by the interaction force at the location of their contact point. The equations are then combined to form a fully coupled system. The results showed that the road surface roughness has a significant influence on the dynamic vehicle bridge interaction. Derive the equation of the vehicle model shown in Fig. 8.68. For a heavier workout for your brain, model the bridge with *moving* forces from the vehicle and derive the equation of motion for the bridge as well. (It is best to do this part after studying Chapter 9.)

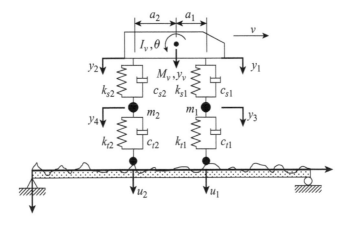

Figure 8.68: *Model for vibrations of a bridge*

8.44 The control of the stationary random response of a two-DOF quarter car vehicle with nonlinear passive elements traversing a homogenous rough road with sky-hook damper control strategy is considered in the article by Rao and Narayanan [60]. Figure 8.69 shows the quarter car model. In the figure, $m_1 = 1000$ kg and $m_2 = 100$ kg are sprung and unsprung masses, respectively, $c = 1000$ Ns/m is the damping coefficient of the linear viscous damper connecting the sprung and unsprung masses, k_1 is the stiffness of the suspension spring, F is the control force, u is the road input, and $k_1 = k_2 = 36000$ N/m are the tire stiffnesses. The control force is of the form $F = K_{g1}x_1 + K_{g2}x_2$, where K_1 and K_2 are the control gains. The control force results in a force that acts upwards on mass m_1, and downwards on mass m_2. Assume $u = u_0 sin\omega t$. Also, assume $u_0 = 0.01$ m, $0 < \omega < 60$ r/s, and $K_{g1} = 30000$ N/m, and $K_{g2} = 20000$ N/m. Derive the equation of motion and solve for the uncontrolled and controlled responses. What happens to the natural frequencies after control is applied?.

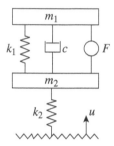

Figure 8.69: *Control of car vibration*

8.45 Tuned mass dampers consisting of a mass, a spring, and a damper are quite effective for vibration control of large structures in many practical situations due to their simplicity and high level of reliability. However, the single-tuned mass-damper system has a main drawback – its performance may worsen due to a mistuned frequency or non-optimal damping. A possible remedy to this is using more than one-tuned mass damper with different dynamic characteristics, as proposed in the article by Ok et al. [61]. Figure 8.70 shows a model for a structure with bi-tuned mass dampers. Derive the equation of motion.

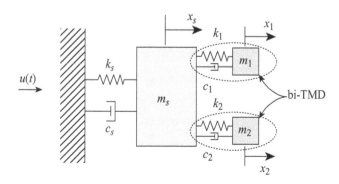

Figure 8.70: *Tuned mass damper*

8.46 In Fig. 8.71, the primary system consisting of mass m_2 is subjected to base excitation u. The absorber mass m_1 has two planar degrees of freedom: translation x_1 and rotation θ. The rotational inertia about its center of mass C is $I = m_1 r_g^2$, where r_g is the radius of gyration. Mass m_1 is connected to mass m_2 at distances d_1 and d_2 from its center of mass via springs k_1 and k_2 and dampers c_1 and c_2. Derive the equation of motion. Ignore gravity.

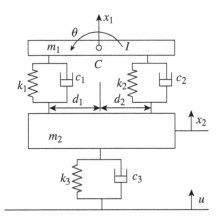

Figure 8.71: *Forced MDOF system*

8.47 The system shown in Fig. 8.72 has mass m_1 coupled to a spring and a damper and is supported on another mass m_2 which is subjected to an excitation $F(t) = F_0 \sin \omega t$. The sliding frictional force F_f due to sliding has a coefficient of friction μ. Derive the equations of motion of the system. Also compute the forced response for the following parameter values: $m_1 = 5\,\text{kg}$, $m_2 = 15\,\text{kg}$, $k_1 = 2\,\text{kN/m}$, $k_2 = 2.5\,\text{kN/m}$, $c_1 = 630\,\text{Ns/m}$, $c_2 = 725\,\text{Ns/m}$, $F_0 = 200\,\text{N}$, $0 < \omega < 100\,\text{rad/s}$, and $\mu = 0.1$.

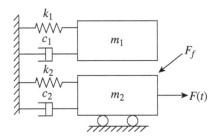

Figure 8.72: *Forced MDOF system*

8.48 Consider the elaborate 11-DOF model we considered in the previous chapter for the vibrational response of a seated male (Problem 7.67). Assume a base acceleration of $\ddot{u} = u_{a0} \sin \omega t$, with $u_{a0} = 5\,\text{m/s}^2$, and $0 < \omega < 10$ (Hz). Determine the forced response.

8.49 Consider the elaborate 11-DOF model we considered in the previous chapter for the vibrational response of a seated pregnant female (Problem 7.68). Assume a base acceleration of $\ddot{u} = u_{a0} \sin \omega t$, with $u_{a0} = 5$ m/s^2, and $0 < \omega < 10$ (Hz). Determine the forced response. Also estimate the forces transmitted to the abdomen.

Design Problems

8.1 Design of an Automobile Suspension System

In this problem you will continue to design a suspension system for a typical[1] automobile shown in Fig. 8.73, from a consideration of its ride quality. This is a continuing design problem that we looked at with a simple SDOF model ("Model 1") in Chapter 5. This time, since now that we know better, we will use more sophisticated models. Please review the problem statement for parametric data and an overview.

Figure 8.73: *Race car on suspensions*

Model 2   Here, we stick to the quarter model, but separate the sprung and unsprung masses, as shown in Fig. 8.74.

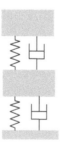

Figure 8.74: *Model 2*

Model 3   This is an extended model of one half of the car and is used to predict the bouncing as well as the pitching motion of the car (Fig. 8.75). Note that the front and rear suspension can have different parameter values. In this case, the excitation from the road to a moving vehicle will affect the front wheels first and the rear wheels later. Consequently, there is a (speed-dependent) time lag between excitation at the front and that at the rear. This results in a pitching motion which can be very annoying (more so than bouncing).

---

[1]To be honest, this picture is hardly representative of a 'typical' automobile!

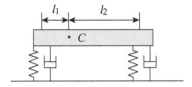

Figure 8.75: *Model 3*

Model 4   This is a combination of Models 2 and 3 and results in four degrees of freedom (Fig. 8.76).

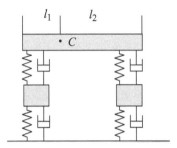

Figure 8.76: *Model 4*

Analysis and design procedure

Typically, you would start with an assumed value for the spring stiffness. Then, analyze the problem using all the four models, and then check to see if the performance satisfies the design specifications listed earlier. Finally, reanalyze the problem with a different value of stiffness, and check to see if the vibration levels are lower or higher. An important parameter in your analysis is the speed of the car.

The kinds of analyses you will need to carry out are:

- Determine the natural frequencies and mode shapes;
- Determine the forced response due to the road roughness.

The following points would be of interest (in addition to many more you can probably think of).

(a) Is there a particular value of stiffness for each suspension that would be very satisfactory from all considerations?

(b) How would the performance degrade with age? (What happens when the damping coefficient of the shock absorber goes down?)

(c) How do the three models compare in their predictions of vibrational levels and natural frequencies?

(d) What are the mode shapes? How are they affected by different stiffnesses? How do they affect the performance?

(e) How do your models compare on the two different road surfaces?

(f) Can you select different suspension stiffnesses for the front and the rear to minimize the pitching motion?

(g) Are there certain specific speeds where the vibration is very high? Is one likely to travel on the roads at those specific speeds?

(h) Which inflation pressure is better for the tires from the (somewhat narrow) point of view of this project? Should the front and rear tires be inflated to different pressures?

(i) Do one or more of the models predict vibrational levels that violate the recommended standards?

8.2 Design of a Test Rotor

In this problem you will be concerned with the design of a turbine test rotor. The problem is based on a recent design requirement of a turbomachinery company. The intended application of the rotor is in a test setup for studying blade vibrations in turbine wheels. (This is a problem we looked at using simpler SDOF models in Chapter 5.)

Design configuration

The rotor will consist of a rigid disk on a flexible shaft. Some of the parameters of the system are known; the essential design parameters that you will need to select are the shaft geometric properties such as diameter and length, and the bearing properties. Note that the shaft need not be uniform. The rotor will be supported on two identical bearings that can be assumed to be simple supports. Figure 8.77 shows the overall configuration.

Figure 8.77: *Test rotor configuration*

Numerical values

The following numerical values are known.

m_d	mass of the disk	600 kg
ω	running speed	4000–5200 rpm
E	Youngs modulus	2.06×10^{11} Pa
G	Shear modulus	8.5×10^{10} Pa
ρ	material density	7850 kg/m^3
$m_0 e$	disk unbalance	15 kg mm

Design requirements

The following requirements are to be met.

- The running speeds should be between the first and second natural frequencies.
- The unbalance response at the location of the disk at the running speed should not exceed 0.0254 mm (1 mil).
- The forces transmitted at the running speed should not exceed 150 N.
- None of the torsional natural frequencies should be in the range of running speeds.

The design parameters should be within certain limits.

- Shaft diameter: 0.13 m $< d_s < 0.19$ m.
- Shaft length: 1.8 m $< \ell < 2.2$ m.

- Bearing stiffness: $0.2 \text{ MN/m} < k_b < 0.3 \text{ MN/m}$.
- Bearing damping coefficient: $1.5 \text{ kNs/m} < c_b < 2.5 \text{ kNs/m}$.

Analyses

The following analyses are required for each of the models.

- Natural frequencies and mode shapes.
- Unbalance response of the system over a speed range from zero to twice the operating speed.
- Forces transmitted to the foundation through the bearings over a speed range from zero to twice the operating speed.
- Torsional natural frequencies and mode shapes.

Procedure

Described below are a series of models for the *same* system. For each of them, select the design parameters to meet the design requirements and carry out the required analyses. Each model should serve as a starting point for the next – more complicated – model.

Model 1   Ignore the mass of the shaft and model it as a flexible beam. Ignore the bearing stiffness (assume that they are rigid).

Model 2   Lump half the mass of the shaft with that of the disk. Note that this is still a one-DOF model.

Model 3   Lump the mass of the shaft into three equal parts and space them equally along the axis of the shaft, Fig. 8.78. This makes it a three-DOF system.

Figure 8.78: *Rotor model with three DOFs*

Model 4   Lump the mass of the shaft into five equal parts and space them equally along the axis of the shaft. This makes it a five-DOF system.

9 Continuous Systems

From Infinity, Infinity has come into existence. From Infinity, when Infinity is taken away, Infinity remains.
—The Upanishads

Real systems are continuous, a fact that has been alluded to many times in the previous chapters. In particular, a vibrating system's mass and elasticity are not discretized elements but are an integral part of the whole system.

Hence, the displacement field of the vibrating system depends on space as well as time. This means that space introduces independent variables additional to time. That will lead to mathematical models that have partial differential equations as opposed to ordinary differential equations that we have dealt with so far. In this chapter, we will look at some quintessential models of continuous systems dealing with transverse, longitudinal and torsional vibrations. However, this is just an introduction to this complex and interesting topic, and you are advised to consult an advanced text for an in-depth treatment.

9.1 Transverse Vibration of Strings

In this section, we analyze the transverse vibration of thin strings. Obvious examples that spring to mind are systems that consist of real *strings* such as those in musical instruments (Fig. 9.1).

A very interesting analysis with application to musical instruments can be seen in an old article by the Nobel laureate C. V. Raman [62]. However, there are additional reasons why we want to study string vibration. Strings are the simplest continuous systems from the mathematical point of view and we need to understand these before we can look at more complicated systems; in addition, strings are useful models for various practical engineering devices such as belts and elevator cables.

Consider a string tightly stretched between two fixed points as shown in Fig. 9.2. It is under tension, T, that could change with the axial position along the string. It has a length,

©Photos.com

Figure 9.1: *A guitar string in euphonious vibration*

ℓ, and a linear mass density (mass per unit length), $\rho(x)$. Its transverse displacement is $v(x, t)$, and it has a distributed force, $f(x, t)$, acting on it.

We assume that the string moves only in the y-direction or that there is no longitudinal displacement. In addition, we assume that the tension does not change due to the motion. All of these are consistent with the small-displacement assumption we will make in order to obtain a linearized model.

Figure 9.3 shows the free body diagram (FBD) of an infinitesimal section of the string. Note that tension could be a function of x (but not of t). Writing Newton's second law in the y-direction,

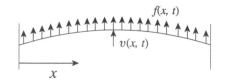

Figure 9.2: *Transverse vibration of a string*

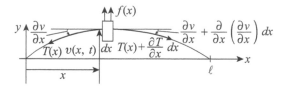

Figure 9.3: *FBD of an infinitesimal section of the string*

$$\left[T(x) + \frac{\partial T(x)}{\partial x}\, dx\right]\left[\frac{\partial v(x, t)}{\partial x} + \frac{\partial^2 v(x, t)}{\partial x^2}\, dx\right] - T(x)\frac{\partial v(x, t)}{\partial x} + f(x, t)\, dx = \rho(x)\, dx\, \frac{\partial^2 v(x, t)}{\partial t^2}$$

$$(9.1)$$

Ignoring the second-order terms,

$$\frac{\partial T}{\partial x}\frac{\partial v}{\partial x}\, dx + T\frac{\partial^2 v}{\partial x^2}\, dx + f(x)\, dx = \rho\, dx\, \frac{\partial^2 v}{\partial t^2} \qquad (9.2)$$

This can be written in the form

$$\frac{\partial}{\partial x}\left[T(x)\frac{\partial v(x, t)}{\partial x}\right] + f(x, t) = \rho(x)\frac{\partial^2 v(x, t)}{\partial x^2} \qquad 0 < x < \ell \qquad (9.3)$$

The above equation is the equation of motion (EOM) for the motion of the string valid over the open interval $0 < x < \ell$. It is a variable coefficient linear PDE, second order in x, and second order in t. It is a very important equation that governs all kinds of phenomena, and is called the *wave equation*. Examples are sound waves along a single dimension, electromagnetic waves in a channel, and a couple of more that we will see later in this chapter.

At the end points, we have the imposed boundary conditions that the string has zero displacement:

$$v(0, t) = 0 \qquad (9.4)$$
$$v(\ell, t) = 0 \qquad (9.5)$$

To complete the problem definition, we also need two initial conditions:

$$v(x, 0) = g(x) \qquad (9.6)$$
$$\left.\frac{\partial v(x, t)}{\partial t}\right|_{t=0} = h(x) \qquad (9.7)$$

9.1.1   Free Vibration

We set the distributed force to zero: $f(x, t) = 0$. Then, the governing equation becomes

$$\frac{\partial}{\partial x}\left[T(x)\frac{\partial v(x, t)}{\partial x}\right] = \rho(x)\frac{\partial^2 v(x, t)}{\partial x^2} \quad 0 < x < \ell \tag{9.8}$$

We seek a solution by separation of variables. Let

$$v(x, t) = F(x)G(t) \tag{9.9}$$

Substituting into the PDE and separating the x and t functions, we get

$$\frac{\ddot{G}(t)}{G(t)} = \frac{[T(x)F'(x)]'}{\rho(x)F(x)} = -\omega^2 \tag{9.10}$$

The first of these is a function of t only, and the second is a function of x only. The only way they can be equal to each other is if they are both equal to a constant – called the separation constant – which is a function of neither x nor t. We have chosen this separation constant to be a negative real quantity knowing that it will give us an oscillatory solution. Note that primes denote derivatives with respect to x, and dots represent derivatives with respect to t.

This leads to two ODEs in x and t:

$$\ddot{G} + \omega^2 G = 0 \tag{9.11}$$

$$[T(x)F'(x)]' = -\omega^2\rho(x)F(x) \tag{9.12}$$

The first of these is an initial value problem (IVP) and the second is a boundary value problem (BVP). The boundary conditions (BCs) become

$$F(0) = 0, \quad F(\ell) = 0 \tag{9.13}$$

Note that the equation we have to solve, Eq. (9.12), is a variable coefficient equation, and is not always amenable to a closed form equation. So, before we can solve this, we will simplify the system description, and solve the simpler problem first.

Uniform string

Let us assume that the string is uniform and the tension is also uniform. Then, $\rho(x) = \rho$ and $T(x) = T$ and the BVP becomes

$$F'' + \beta^2 F = 0 \tag{9.14}$$

where

$$\beta \stackrel{\text{def}}{=} \omega\sqrt{\frac{\rho}{T}} \tag{9.15}$$

Equation (9.14) is solved easily:

$$F(x) = C_1 \sin \beta x + C_2 \cos \beta x \tag{9.16}$$

Using the first of the BCs, Eq. (9.13), we get $C_2 = 0$. Applying the BC at the other end,

$$C_1 \sin \beta\ell = 0 \tag{9.17}$$

Now, if $C_1 = 0$, we end up with a trivial solution; so, we have to have

$$\sin \beta\ell = 0 \tag{9.18}$$

This is called the characteristic equation for the system, and has infinite solutions called *characteristic roots* (β_n):

$$\beta_n = \frac{n\pi}{\ell} \quad n = 1, 2, \ldots \tag{9.19}$$

This in turn corresponds to infinite values for the separation constant

$$\omega_n = \frac{n\pi}{\ell} \sqrt{\frac{T}{\rho}} \tag{9.20}$$

Getting back to the temporal problem we begin to see the significance of the characteristic roots

$$\ddot{G}_n + \omega_n^2 G_n = 0$$

Solving,

$$G_n(t) = D_{1n} \sin \omega_n t + D_{2n} \cos \omega_n t \tag{9.21}$$

Hence, it is clear that the separation constant, ω_n, actually corresponds to the natural frequency of oscillation of free vibration. The characteristic roots are therefore indirectly related to the natural frequencies of oscillation. Note that the continuous system model has infinite natural frequencies compared to an N DOF model, which has N natural frequencies.

Of course, the story does not end here. We need to substitute the eigenvalues back in the solution to get the *eigenfunctions*. We get that from Eq. (9.16), after substituting for C_2 and β_n. Clearly, since there are infinite values for β_n, there will be infinite eigenfunctions:

$$F_n(x) = C_n \sin \beta_n x = C_n \sin \left(\frac{n\pi x}{\ell} \right) \quad n = 1, 2, \ldots \tag{9.22}$$

Corresponding to each of these spatial solutions, we have a temporal solution, $G_n(t)$, given earlier; or we have the nth modal solution given by

$$v_n(x, t) = F_n(x) G_n(t) = \sin \left(\frac{n\pi x}{\ell} \right) (D_{1n} \sin \omega_n t + D_{2n} \cos \omega_n t) \tag{9.23}$$

where we have absorbed the constant C_n into the other constants without loss of generality.

Solving our governing differential equation has hence resulted in infinite solutions. Since the equation is linear, the total solution can be written simply as a superposition of all the solutions, or

$$v(x, t) = \sum_{n=1}^{\infty} v_n(x, t) = \sum_{n=1}^{\infty} \sin \left(\frac{n\pi x}{\ell} \right) (D_{1n} \sin \omega_n t + D_{2n} \cos \omega_n t) \tag{9.24}$$

The unknown constants would of course be determined from the initial conditions just as we did in the case of multiple degree of freedom (MDOF) systems.

Again, as in the case of the MDOF systems, without looking at the composite solution, it makes a lot more sense to look at the individual modal solutions. We can plot each of the mode shapes and imagine that what we are looking at is the snapshot of the harmonic vibration that occurs at the respective natural frequency. Figures 9.4–9.6 show the first three mode shapes, which are of course simple sines that you could have plotted in your sleep (MATLAB is surely an overkill for this). Note the locations of the *nodes* (1 for Mode 2 and 2 for Mode 3), which are points that experience no vibration in their respective modes.

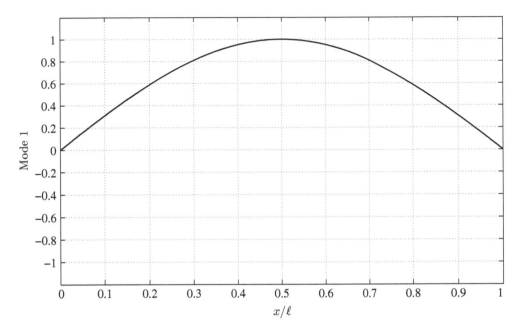

Figure 9.4: *Mode 1 for transverse vibration of the string*

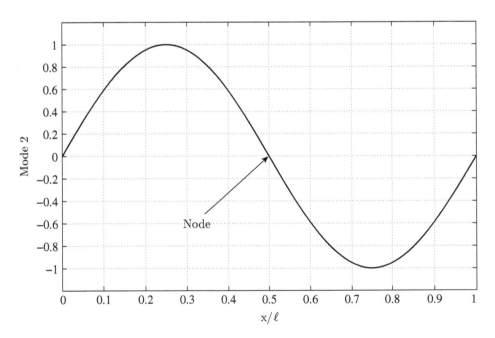

Figure 9.5: *Mode 2 for transverse vibration of the string*

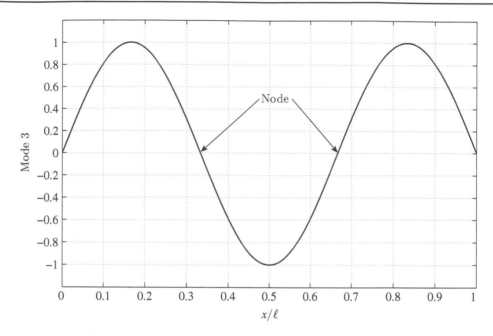

Figure 9.6: *Mode 3 for transverse vibration of the string*

RUNNING PROBLEM: STRING VIBRATION

Comparison with discrete models

Now that we have an accurate model for the string vibration, it is time to look back at all those simplistic models we had come up with and to compare the results. This will give us a valuable insight into the concept of modeling. Recall that we started with a lumped one-DOF model, and the natural frequency turned out to be

$$\omega_n = \frac{2}{\ell}\sqrt{\frac{T}{\rho}} \tag{9.25}$$

The exact solution for the first natural frequency is

$$\omega_n = \frac{\pi}{\ell}\sqrt{\frac{T}{\rho}} \tag{9.26}$$

Comparing the two, we have an error of approximately 36%.

Next, for the lumped two-DOF model, the answers we predicted were (remember that we now get *two* natural frequencies):

$$\omega_1 = \sqrt{6}\sqrt{\frac{T}{\rho\ell^2}}, \quad \omega_2 = 3\sqrt{2}\sqrt{\frac{T}{\rho\ell^2}} \tag{9.27}$$

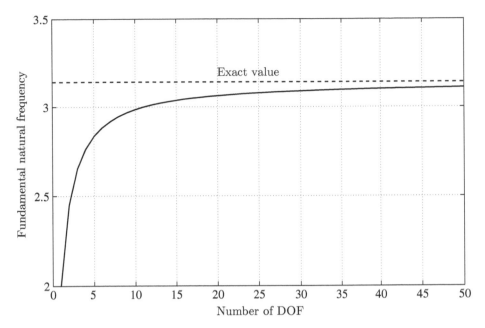

Figure 9.7: *Prediction of fundamental natural frequency with discrete models*

Now, the error in the first natural frequency is 22%, and that in the second natural frequency is 32%. Note that how the estimate of the first natural frequency got better.

For the lumped three-DOF model, the answers we predicted were as follows:

$$\omega_1 = 2.6513\sqrt{\frac{T}{\rho\ell^2}}, \quad \omega_2 = 4.8890\sqrt{\frac{T}{\rho\ell^2}}, \quad \omega_3 = 6.4008\sqrt{\frac{T}{\rho\ell^2}} \tag{9.28}$$

Now, the error in the first natural frequency is 16%, that in the second natural frequency is 22%, and that in the third natural frequency is 32%. Note, how the estimate of the first natural frequency got even better; so did the estimate of the second natural frequency.

Numerical experiments show that it takes a 10-DOF system to get an answer that has less than 5% error in the first natural frequency. Figure 9.7 shows the fundamental natural frequency computed by using an increasingly higher number of discrete degrees of freedom (DOF).

It is also interesting to compare the mode shape predictions from the discrete models with the "exact" one we just obtained. Figures 9.8 and 9.9 show the first two modes obtained using two lumped masses compared with the continuous mode we obtained in this chapter.

Even though we obtained a satisfying convergence from an intuitive process of lumping the masses, it is really not the best way to go about it. By the way, a better word for "lumping" is *discretization*, as we are converting a continuous system into a discrete one. A rigorous technique for discretization is the Rayleigh Ritz method, based on some excellent robust properties of what is called the Rayleigh quotient. In fact, the popular finite element method (FEM) is firmly based on the Rayleigh Ritz method, and hence shares the same nice properties. For an excellent discussion of various discretization techniques including FEM, read *Computational Methods in Structural Dynamics* by Meirovitch [63].

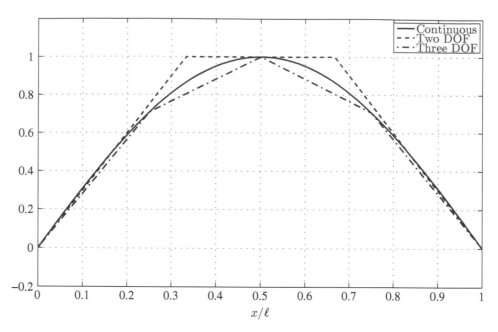

Figure 9.8: *First mode with two- and three-DOF discrete models compared with the continuous model*

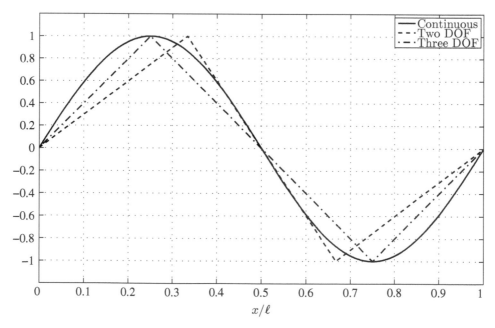

Figure 9.9: *Second mode with two- and three-DOF discrete models compared with the continuous model*

9.2   Longitudinal Vibration of Rods

Consider a thin rod subject to axial (or longitudinal) vibration, as shown in Fig. 9.10. This would be the dynamic equivalent of the tension (or compression) problem you would have seen in a mechanics of materials course. The rod has a linear mass density $m(x)$, length ℓ, area of cross section $A(x)$, and Young's modulus, $E(x)$. Note that we have intentionally let area, mass density, and Young's modulus be functions of the axial coordinate x. We will continue to do so to preserve generality in our derivations as much as possible. We designate the vibrational displacement in the x-direction by $u(x,\ t)$. Note that we are not using x as x has already been used to represent the axial coordinate or the distance from the left end. It is best not to confuse the *displacement* with the *coordinate*. Assume also that the rod is subject to some distributed force, $f(x,\ t)$. Note that u is a function of two independent variables, x and t, which is going to lead us to a partial differential equation.

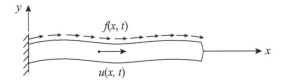

Figure 9.10: *Thin rod in longitudinal vibration*

Clearly, there will be internal stresses in the rod. As an approximation we only consider the normal stresses in the x-direction; we denote it by $\sigma(x,\ t)$. If we assume small displacements, we can invoke linearity and use Hooke's law:

$$\sigma(x,\ t) = E(x)\epsilon(x,\ t) \tag{9.29}$$

where ϵ is the normal strain in the rod. Note again that stress and strain are functions of x and t. We have assumed that they are uniform at a cross section; i.e., they are not functions of the other coordinates, y and z. These are decent assumptions for thin rods and small displacements. The strain is related to the displacement by

$$\epsilon(x,\ t) = \frac{\partial}{\partial x}[u(x,\ t)] \tag{9.30}$$

Of course, we have to use the partial derivative because the displacement is a function of two variables.

Consider an infinitesimal section of the rod for which we will write the FBD in Fig. 9.11, where the stress resultant, P is shown. The internal force, P, is related to stress by

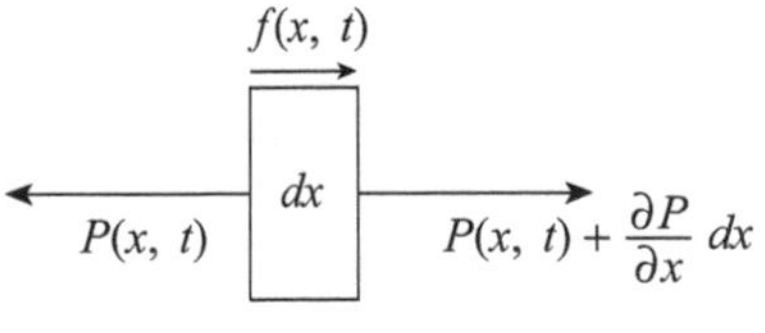

$$P(x,\ t) = \int_A \sigma(x,\ t)\,dA$$
$$= \sigma(x,\ t)A(x) \tag{9.31}$$

Figure 9.11: *FBD of rod in longitudinal vibration*

because of the assumed uniformity of stress across the cross section. From the FBD, the net force acting on the infinitesimal section is

$$F = P(x,\ t) + \frac{\partial P(x,\ t)}{\partial x}dx - P(x,\ t) + [f(x,\ t)\,dx]$$
$$= \frac{\partial P(x,\ t)}{\partial x}dx + [f(x,\ t)\,dx]$$

$$= \frac{\partial}{\partial x} [\sigma(x,\, t) A(x)]\, dx + [f(x,\, t)\, dx]$$

$$= \frac{\partial}{\partial x} \left[E(x) A(x) \frac{\partial u(x,\, t)}{\partial x} \right] dx + [f(x,\, t)\, dx] \qquad (9.32)$$

Note that $f(x,\, t)$ is a distributed force; its dimensions are force/length, which should explain the multiplication by dx.

Kinematics is very easy as the acceleration is only in the x-direction.

$$a = \frac{\partial^2}{\partial t^2} [u(x,\, t)] \qquad (9.33)$$

From Newton's second law, and putting the above expressions together, the EOM can be written as follows:

$$[m(x)\, dx] \frac{\partial^2}{\partial t^2} [u(x,\, t)] = \frac{\partial}{\partial x} \left[E(x) A(x) \frac{\partial u(x,\, t)}{\partial x} \right] dx + [f(x,\, t)\, dx] \qquad (9.34)$$

Simplifying,

$$\frac{\partial}{\partial x} \left[E(x) A(x) \frac{\partial u}{\partial x} \right] + f(x,\, t) = m(x) \frac{\partial^2 u}{\partial t^2} \qquad (9.35)$$

This is a linear partial differential equation (PDE), second order in x, and second order in t. For free vibration, we drop the forcing term

$$\frac{\partial}{\partial x} \left[E(x) A(x) \frac{\partial u}{\partial x} \right] = m(x) \frac{\partial^2 u}{\partial t^2} \qquad (9.36)$$

Note that this equation is exactly the same as the string equation, with EA in place of T, and a longitudinal displacement here instead of the transverse displacement in the case of the string. Hence, the solution follows the same process as before (we will skip a few steps this time).

For completeness, the equation needs boundary conditions (on x) and initial conditions (on t). For example, the BCs could be

$$u(0,\, t) = A(t), \quad u(\ell,\, t) = B(t) \qquad (9.37)$$

and the initial conditions would be

$$u(x,\, 0) = g(x), \quad \left. \frac{\partial u}{\partial t} \right|_{t=0} = h(x) \qquad (9.38)$$

Note that the boundary conditions could possibly be functions of time, and the initial conditions could be functions of the axial coordinate in general.

The general equation given here does not always have an analytical solution even though it is *linear* (contrast with the lumped-parameter model given by ODEs). For instance, if the area of cross section varied with x or if the boundary conditions were complex, there is no guarantee that we would get an analytical solution. Here, we will solve a simple problem to get an idea of what the solution might look like.

Uniform rod fixed at the left end

Consider the case when the rod is uniform: $A(x) = A$, $E(x) = E$, $m(x) = m$. Also, assume that the rod is fixed at the left end and is free at the right end (Fig. 9.12). Then, the EOM simplifies to the following:

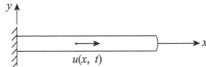

$$EA\frac{\partial^2 u}{\partial x^2} = m\frac{\partial^2 u}{\partial t^2} \qquad (9.39)$$

The boundary condition at the left end is simple and obvious: zero displacement for all time.

Figure 9.12: *Uniform rod fixed at the left end*

$$u(0, t) = 0 \qquad (9.40)$$

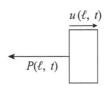

How about the right end? Now we need to use some elementary mechanics. Let us draw an FBD of an infinitesimal section at the very end of the rod (at $x = \ell$); this is shown in Fig. 9.13. Writing Newton's second law for this element,

Figure 9.13: *Element FBD at the free end*

$$[m\,dx]\frac{\partial^2}{\partial t^2}[u(\ell, t)] = -P(x, t)|_{x=\ell} \qquad (9.41)$$

In the limit, as we shrink the element, we are left with

$$P(\ell, t) = 0 \quad \Rightarrow \quad EA(\ell)\frac{\partial}{\partial x}[u(x, t)]\,|_{x=\ell} = 0 \qquad (9.42)$$

where we have used the expression for the internal force that we derived earlier. You might remember this from your mechanics of materials course as a statement that the internal forces are zero on a free surface. For the uniform rod, this reduces to

$$\frac{\partial u}{\partial x}\bigg|_{x=\ell} = 0 \qquad (9.43)$$

Hence, at the free end, the *derivative* of the displacement is zero.

Now we are ready to solve this PDE. As before, we seek to solve this constant coefficient ODE using the *method of separation of variables*. Essentially, we seek a solution for the displacement that is a product of individual functions of x and t:

$$u(x, t) = U(x)G(t) \qquad (9.44)$$

Substituting into the EOM,

$$EA\frac{d^2 U}{dx^2}G = mU\ddot{G} \qquad (9.45)$$

Reorganizing and separating the x and t functions,

$$\frac{\ddot{G}(t)}{G(t)} = \frac{EA U''(x)'}{mU(x)} = -\omega^2 \qquad (9.46)$$

where we have chosen the separation constant to be a negative real quantity knowing that it will give us an oscillatory solution.

This leads to two ODEs in x and t:

$$\ddot{G} + \omega^2 G = 0 \tag{9.47}$$

$$[EAU'(x)]' = -\omega^2 m U(x) \tag{9.48}$$

The first of these is an IVP and the second one is a BVP. The boundary conditions become

$$U(0) = 0$$
$$U'(\ell) = 0 \tag{9.49}$$

We can rewrite the BVP in a more convenient form

$$U'' + \beta^2 U = 0 \tag{9.50}$$

where

$$\beta \stackrel{\text{def}}{=} \omega \sqrt{\frac{m}{EA}} \tag{9.51}$$

This is a good time to go back and check the similarity with—and differences from—the string problem.

Equation (9.50) is solved easily:

$$U(x) = C_1 \sin \beta x + C_2 \cos \beta x \tag{9.52}$$

Using the first of the BC's, Eq. (9.49), we get $C_2 = 0$. Applying the BC at the other end,

$$C_1 \beta \cos \beta \ell = 0 \tag{9.53}$$

If $C_1 = 0$, we end up with a trivial solution; so, we have to have

$$\cos \beta \ell = 0 \tag{9.54}$$

which is the characteristic equation.

Zero eigenvalue?

One possible solution of Eq. (9.53) is of course,

$$\beta = 0 \tag{9.55}$$

If we did let this happen though, the solution we just got (Eq. 9.52) is not valid because that solution process assumed $\beta \neq 0$. In fact, to check the possibility of $\beta = 0$, we will need to go back and let $\beta = 0$ in Eq. (9.50) and re-solve the equation. Let us do that quickly to make sure that we are not making a terrible mistake.

$$U'' = 0 \tag{9.56}$$

The solution is then

$$U(x) = C_1 + C_2 x \tag{9.57}$$

Using the first BC, $C_1 = 0$ and using the second BC, $C_2 = 0$. So, the eigenvalue $\beta = 0$ leads us to a trivial solution this time, and we can safely drop this eigenvalue from future consideration. We should be careful, however, and check the possibility of a zero eigenvalue every time, because there are instances when we do get a nontrivial solution. See Problem 9.7 for a case when that is so.

Back to the characteristic equation, Eq. (9.54). As before, it has infinite eigenvalues or character-istic roots (β_n):

$$\beta_n = \frac{(2n-1)\pi}{2\ell} \quad n = 1, 2, \ldots \tag{9.58}$$

This also corresponds to infinite values for the separation constant

$$\omega_n = \frac{(2n-1)\pi}{2\ell} \sqrt{\frac{EA}{m}} \tag{9.59}$$

Getting back to the temporal problem we again see the significance of the characteristic roots

$$\ddot{G}_n + \omega_n^2 G_n = 0$$

Solving,

$$G_n(t) = D_{1n} \sin \omega_n t + D_{2n} \cos \omega_n t \tag{9.60}$$

Hence, it is clear that ω_n corresponds to the frequency of oscillation of free vibration.

The corresponding eigenfunctions (or mode shapes) are given by

$$U_n(x) = C_n \sin\left(\frac{\beta x}{\ell}\right) = C_n \sin\left[(2n-1)\frac{\pi x}{2\ell}\right] \quad n = 1, 2, \ldots \tag{9.61}$$

where C_n is an arbitrary multiplicative constant. Figure 9.14 shows the first three modes plotted versus the scaled length, x/ℓ. Note that the first mode has zero nodes, the second has one at $2\ell/3$, and the third has two nodes at 0.4ℓ and 0.8ℓ. All of them have a slope of zero at the free end. Should it be so? Why or why not?

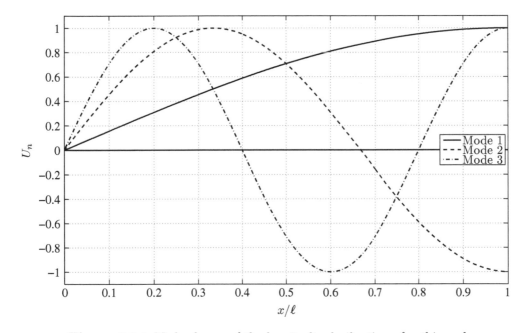

Figure 9.14: *Mode shapes of the longitudinal vibration of a thin rod*

The complete general solution can be written summing all the (infinite) modes as follows:

$$u(x,\,t) = \sum_{n=1}^{\infty} U_n(x)\,G_n(t)$$

$$= \sum_{n=1}^{\infty} \sin\left[(2n-1)\frac{\pi x}{2\ell}\right] [C_n \cos\omega_n t + D_n \cos\omega_n t] \qquad (9.62)$$

where ω_n is given by Eq. (9.59). Note that we have absorbed the arbitrary constant associated with the eigenfunction with the other constants, C_n and D_n. These constants would have to be determined from the initial conditions.

9.3   Torsional Vibration

We consider the torsional vibration of slender shafts in this section. Torsional vibration is present wherever you have a twisting or driving torque; examples abound—driveshafts, crankshafts, motors, turbine shafts, aircraft engines, drills, etc. The treatment here is highly simplified, but complex in a different way in the sense that we will not lump the masses at a single point like we did in the earlier chapters. The mass and stiffness are both continuous properties and will be treated as such.

We assume that the shafts are thin and circular with no warping, and that all displacements are small and linear. Consider a shaft shown in Fig. 9.15 of length ℓ subject to torsional vibration. Its angle of twist is denoted by $\theta(x,\,t)$, and an external distributed twisting moment acting on it is denoted by $T(x,\,t)$. $G(x)$ is the shear modulus of the material, possibly varying along the axis of the shaft, $I(x)$ is the polar mass moment of inertia of the bar per unit length about its cross-sectional center of mass, and $J(x)$ is the polar area moment of inertia.

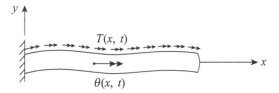

Figure 9.15: *Thin circular shaft in torsional vibration*

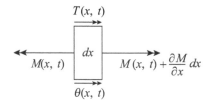

Figure 9.16: *FBD of infinitesimal section of the shaft in torsion*

Taking an infinitesimal section of the shaft, we draw an FBD to show all the moments acting on it (Fig. 9.16). $M(x,\,t)$, shown in the FBD, is the internal stress resultant which we will have to relate to stress and eventually, to the displacement, using principles from mechanics of materials. For now, we can write the net moment on the infinitesimal section as follows:

$$M_{\text{net}} = \left[M(x,\,t) + \frac{\partial M(x,\,t)}{\partial x}\,dx\right] - [M(x,\,t)] + T(x,\,t)\,dx = \frac{\partial M(x,\,t)}{\partial x}\,dx + T(x,\,t)\,dx \quad (9.63)$$

The kinematics is quite simple and obvious:

$$\alpha = \frac{\partial^2 \theta(x,\,t)}{\partial t^2} \tag{9.64}$$

Putting them together using Newton's second law ($M = I\alpha$ about the center of mass),

$$\frac{\partial M(x,\,t)}{\partial x} dx + T(x,\,t) dx = [I(x)\ dx]\frac{\partial^2 \theta(x,\,t)}{\partial t^2} \tag{9.65}$$

Note that the mass moment of inertia, $I(x)$, is specified per unit length. Simplifying, in the limit,

$$\frac{\partial M(x,\,t)}{\partial x} + T(x,\,t) = [I(x)]\frac{\partial^2 \theta(x,\,t)}{\partial t^2} \tag{9.66}$$

Next, if we refer back to our treasure trove of mechanics of materials and use standard assumptions of linear elastic properties, simple kinematic assumption, and strain-displacement relationships, we get

$$M(x,\,t) = G(x)J(x)\frac{\partial \theta(x,\,t)}{\partial x} \tag{9.67}$$

Substituting this into Eq. (9.66), we get the EOM for torsional vibration:

$$\frac{\partial}{\partial x}\left[G(x)J(x)\frac{\partial \theta(x,\,t)}{\partial x}\right] + T(x,\,t) = I(x)\frac{\partial^2 \theta(x,\,t)}{\partial t^2}, \quad 0 < x < \ell \tag{9.68}$$

Surprise! This is again a linear PDE second order in x and second order in t, and has the same shape and form as the one we encountered in the transverse vibration of the string and the longitudinal vibration of the rod. So, the solution is going to be the same, and we will not bother to go through all the steps again.

As in the previous cases, we would need boundary conditions:

$$\theta(0,\,t) = A(t), \quad \theta(\ell,\,t) = B(t) \tag{9.69}$$

and initial conditions:

$$\theta(x,\,0) = g(x), \quad \left.\frac{\partial \theta(x,\,t)}{\partial t}\right|_{t=0} = h(x) \tag{9.70}$$

Free Vibration of the Uniform Shaft

For a uniform shaft, with all properties being constant through the length of the shaft, the EOM simplifies to the following:

$$GJ\frac{\partial^2 \theta(x,\,t)}{\partial x^2} = I\frac{\partial^2 \theta(x,\,t)}{\partial t^2}, \quad 0 < x < \ell \tag{9.71}$$

After the usual process of separation of variables with $\theta(x,\,t) = \Theta(x)G(t)$, we come up with a BVP and an IVP. The BVP is

$$GJ\Theta''(x) + \omega^2 I\Theta(x) \tag{9.72}$$

and the IVP is

$$\ddot{G} + \omega^2 G = 0 \tag{9.73}$$

Again, defining a convenient quantity

$$\beta \overset{\text{def}}{=} \omega \sqrt{\frac{I}{GJ}} \tag{9.74}$$

the equation becomes

$$\Theta'' + \beta^2 \Theta = 0 \tag{9.75}$$

The solution is of course (for the nth time!)

$$\Theta(x) = C_1 \sin \beta x + C_2 \cos \beta x \tag{9.76}$$

The constants need to be evaluated from the boundary conditions.

Fixed-Fixed Shaft

Consider a shaft that is fixed at both ends (Fig. 9.17). Then, the BCs simplify to

$$\theta(0, t) = 0 \Rightarrow \Theta(0) = 0 \tag{9.77}$$
$$\theta(\ell, t) = 0 \Rightarrow \Theta(\ell) = 0 \tag{9.78}$$

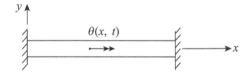

Figure 9.17: *Torsion of shaft fixed at both ends*

Solving, we get $C_2 = 0$, and the characteristic equation

$$\sin \beta \ell = 0 \tag{9.79}$$

It has infinite solutions, resulting in the characteristic roots (β_n):

$$\beta_n = \frac{n\pi}{\ell} \quad n = 1, 2, \ldots \tag{9.80}$$

Again, this corresponds to infinite values for the natural frequencies:

$$\omega_n = \frac{n\pi}{\ell} \sqrt{\frac{GJ}{I}} \tag{9.81}$$

The eigenfunctions, or mode shapes, can be written from above as

$$\Theta_n(x) = C_n \sin\left(\frac{n\pi x}{\ell}\right) \quad n = 1, 2, \ldots \tag{9.82}$$

The first three mode shapes are shown symbolically in Fig. 9.18. Note that the real displacements are angular twists.

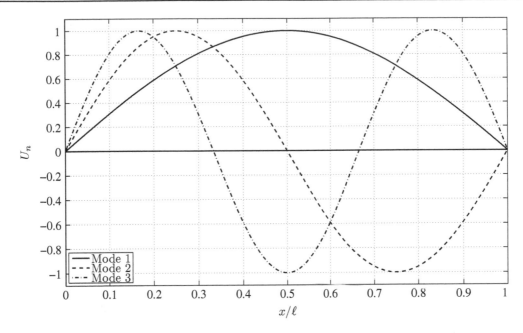

Figure 9.18: *Torsional mode shapes of the shaft fixed at both ends*

Rigid Disk at the End

Now, let us look at a situation which is quite often seen in engineering practice, when we have what we might model as a rigid disk at one end of a flexible shaft. (Can you think of a few practical cases? *Hint*: look at Fig. 9.19.) A schematic is shown in Fig. 9.20.

It does look as though the boundary condition at the end, $x = \ell$, is very complicated. OK, let us not panic! Let us just go back to basics and draw an FBD[1] of the disk connected to the last infinitesimal section of the shaft (if in doubt, follow the rigorous dynamic analysis procedure outlined in Chapter 2—it never fails!). This is shown in Fig. 9.21.

Writing Newton's second law for this system,

$$-M(\ell, t) = [I_D + I(\ell)\, dx] \left. \frac{\partial^2 \theta(x, t)}{\partial t^2} \right|_{x=\ell} \tag{9.83}$$

In the limit, and after substituting for the internal moment M,

Figure 9.19: *A ceiling fan*

$$-\left[G(x)J(x) \frac{\partial \theta(x, t)}{\partial x} \right]\Bigg|_{x=\ell} = I_D \left. \frac{\partial^2 \theta(x, t)}{\partial t^2} \right|_{x=\ell} \tag{9.84}$$

---

[1] *If all else fails, start with a free body diagram*—C. Nataraj (when he achieved *nirvana* at the end of a frustrating day working on a complicated dynamics problem in the 1980s!)

Figure 9.20: *Shaft with a rigid disk at the end*   **Figure 9.21:** *FBD of the end element with the rigid disk*

This does look like a hopeless mess, but let us first simplify it for the case of the uniform shaft.

$$-GJ \left.\frac{\partial\theta(x,\,t)}{\partial x}\right|_{x=\ell} = I_D \left.\frac{\partial^2\theta(x,\,t)}{\partial t^2}\right|_{x=\ell} \tag{9.85}$$

Next, we carry out separation of variables on the *boundary condition*

$$-GJ\Theta'(\ell)\,G(t) = I_D\Theta(\ell)\,\ddot{G}(t) \tag{9.86}$$

Now comes a neat trick, so pay attention. Going back to the initial value problem, Eq. (9.73), we can substitute for $\ddot{G}$ with

$$\ddot{G} = -\omega^2 G \tag{9.87}$$

and simplify the boundary condition

$$-GJ\Theta'(\ell) = -\omega^2 I_D\Theta(\ell) \tag{9.88}$$

Bingo! We got rid of the pesky time function. Mathematically speaking, we were able to do this because this model's boundary conditions are also separable. If they were not, separation of variables would not have worked even if the differential equation itself were separable.

The equation can be further simplified using the relationship between β and ω from Eq. (9.74):

$$\Theta'(\ell) = (\beta^2\ell)\,\frac{I_D}{I\ell}\,\Theta(\ell) \tag{9.89}$$

Note the key nondimensional parameter, $I_D/(I\ell)$, which is the ratio of the "total" mass moment of inertia of the shaft to the mass moment of inertia of the rigid disk. Clearly, this parameter is critical in determining how the system behaves. Let

$$\alpha = \frac{I_D}{I\ell} \tag{9.90}$$

Now for the solution. First, we use the left-end boundary condition, $\Theta(0) = 0$, in Eq. (9.76). This will lead to $C_2 = 0$. Hence, the boundary condition at the right end now leads to the following equation:

$$C_1\beta\cos\beta\ell = \beta^2\ell\alpha C_1\sin\beta\ell \tag{9.91}$$

This leads to the characteristic equation

$$\cot\gamma = \alpha\gamma \tag{9.92}$$

where for convenience, we have defined another nondimensional quantity, $\gamma = \beta\ell$.

Unfortunately, there is no simple straightforward way to solve this equation for a general value of α. So, let us go ahead and solve this for a case of

$$\alpha = \frac{I_D}{I\ell} = 2 \qquad (9.93)$$

So, we need to solve

$$\cot\gamma = 2\gamma \qquad (9.94)$$

This is a transcendental equation and can be solved by several approximate methods. One procedure would be to look up a table where such equations are listed; an excellent one is in *Handbook of Mathematical Functions* by Abramowitz and Stegun [64]. The second method would be to plot the equation on a calculator or a computer and try to find each of the solutions by zooming in on the area where it crosses zero (try it!). Another way is to employ a nonlinear algebraic equation solver; here, we will use MATLAB (although there are many standard routines available in FORTRAN, C, etc.). Listing 9.2 shows the main program, and Listing 9.1 shows the routine that supplies the characteristic equation to be solved. Caution should be used whenever large numbers of roots are being solved for as it is possible for routines to miss some solutions. The guess provided is quite critical in determining which solution such routines converge to. Detailed discussion of such routines—which form a subset of nonlinear optimization programs—is outside the scope of this book, but can be found in any good book on Numerical Methods.

Listing 9.1: *The characteristic equation for the torsion with rigid disk*

```
function y=torsion_disk_chareq(gamma)
%      characteristic equation for torsion with disk at the end

global alpha

y=cos(gamma) - alpha*gamma*sin(gamma);
```

Listing 9.2: *Solution of the characteristic equation for the torsion with rigid disk*

```
% torson_disk_cesoln.m
% torsion of a shaft with a rigid disk
% find the eigenvalues and modeshapes

  global alpha

% nondimensional ratio of moments of inertia
  alpha=2;
% optional argument to equation solver
  tolerance = 1e-10;
% how many modes we need
  order = 5;

% call nonlinear algebraic equation solver for each mode
  for i=1:order ,
%    need to provide a guess for each mode - this is critical
     guess=(i-1)*pi+0.8;
%    'torsion_disk_chareq' is the function that contains the equation
     ce_root(i) = fzero('torsion_disk_chareq',guess,tolerance);
%    examine this to make sure that the error is within bounds of comfort
     error(i) = torsion_disk_chareq(ce_root(i));
  end;
```

```
% compute modeshapes
  x =(0:.02:1);
  for mode=(1:order),
    gamma=ce_root(mode);
    mode_shape(mode,:) = sin(gamma*x);
  end;
% you may want to plot them to make sure they look right
% with appropriate nodes
```

From the numerical solution, the first five eigenvalues turn out to be

$$\gamma = 0.6533, 3.2923, 6.3616, 9.4775, \text{ and } 12.6060.$$ (9.95)

So, for example, the first natural frequency would be

$$\omega_1 = 0.6533\sqrt{\frac{GJ}{I\ell^2}}$$ (9.96)

The first three mode shapes are shown in Fig. 9.22.

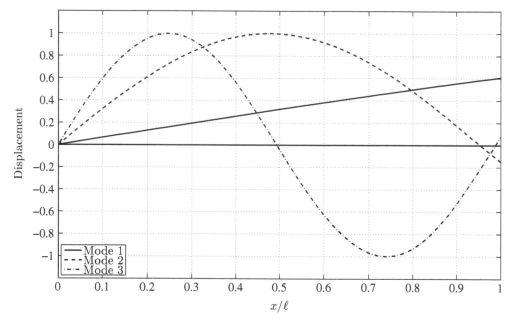

Figure 9.22: *Mode shapes of the shaft with a disk at its end*

Look at the eigenvalues of the torsion of the shaft with a rigid disk at the end. Do you see a convergence pattern? If not, program your computer and run it to get 10 or 15 eigenvalues. Now, do you see it? No? OK, here is a hint: divide the eigenvalues by π. Got it? OK, that was the easy part. Now, tell me why that is so!

9.4 Transverse Vibration of Beams

In this section, we will derive the governing equation for an Euler beam; this is the simplest model for the flexural vibration of a beam where shear deformation and rotatory inertia effects are ignored. It will also result in linear equations since we will ignore higher order effects in curvature as well as assume small displacements.

Consider the beam shown in Fig. 9.23; it has a length, ℓ, mass per unit length, $m(x)$, and flexural rigidity, $EI(x)$. The x-axis is along the length of the beam; we will confine ourselves to the two-dimensional situation (xy-plane); $v(x, t)$ is the transverse displacement of the beam in the y-direction. The beam could be subject to a distributed force, $f(x, t)$, which could also conceivably include concentrated forces.

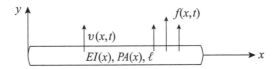

Figure 9.23: *Flexural vibration of an Euler beam*

Figure 9.24 shows the FBD of an incremental section of the beam with the shear force, $Q(x, t)$, bending moment, $M(x, t)$, and the distributed force, $f(x, t)$. Applying Newton's equations of motion (for two-dimensional motion),

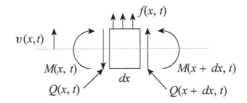

Figure 9.24: *FBD of a beam element*

$$\vec{F} = m\vec{a}, \quad M = I\alpha \tag{9.97}$$

The acceleration is due to motion in the y-direction, and the angular acceleration is ignored:

$$a = \frac{\partial^2}{\partial t^2}[v(x, t)], \quad \alpha = 0 \tag{9.98}$$

Substituting the acceleration values and using the FBD,

$$\frac{\partial}{\partial x}[Q(x, t)]\,dx + f(x, t)\,dx = [m(x)\,dx]\frac{\partial^2}{\partial t^2}[v(x, t)] \tag{9.99}$$

$$\frac{\partial}{\partial x}[M(x, t)]\,dx + \left[Q(x, t) + \frac{\partial}{\partial x}[Q(x, t)]\,dx\right]\frac{dx}{2} + Q(x, t)\frac{dx}{2} + f(x, t)\,dx\frac{dx}{2} = 0 \tag{9.100}$$

Ignoring the higher order terms, the moment equation becomes

$$\frac{\partial}{\partial x}[M(x, t)] + Q(x, t) = 0 \tag{9.101}$$

This equation should be familiar from your background in undergraduate mechanics of materials (remember shear and bending moment diagrams?). Substituting it into the force equation,

$$m(x)\frac{\partial^2}{\partial t^2}[v(x, t)] = -\frac{\partial^2}{\partial x^2}[M(x, t)] + f(x, t) \tag{9.102}$$

From mechanics of materials, using the Euler moment equation (and under a lot of simplifying assumptions about loading conditions),

$$M(x, t) = EI(x)\frac{\partial^2}{\partial x^2}[v(x, t)] \tag{9.103}$$

Substituting this into the force equation, we get the governing EOM.

$$m(x)\frac{\partial^2}{\partial t^2}\left[v(x,\,t)\right] = -\frac{\partial^2}{\partial x^2}\left[EI(x)\frac{\partial^2}{\partial x^2}\left[v(x,\,t)\right]\right] + f(x,\,t) \quad 0 < x < \ell \tag{9.104}$$

It is a PDE, second order in t and *fourth* order in x and is valid at every point in the *interior* of the beam. The formulation is of course not complete without the specification of sufficient boundary conditions and initial conditions. It turns out that we need *two* boundary conditions at every boundary point; recall that we needed *one* boundary condition at every boundary point in the case of the longitudinal vibration and other similar problems. The difference is of course because the equation is now fourth order in x, and not second order.

Boundary Conditions

Here we list the boundary conditions for a few of the common situations. For illustration we assume that the boundary conditions are specified at $x = 0$.

Simply supported

$$\text{Deflection} = 0: \quad v(x,\,t)|_{x=0} = 0 \tag{9.105}$$

$$\text{Moment} = 0: \quad EI(x)\frac{\partial^2}{\partial x^2}\left[v(x,\,t)\right]\Big|_{x=0} = 0 \tag{9.106}$$

Fixed (clamped)

$$\text{Deflection} = 0: \quad v(x,\,t)|_{x=0} = 0 \tag{9.107}$$

$$\text{Slope} = 0: \quad \frac{\partial}{\partial x}\left[v(x,\,t)\right]\Big|_{x=0} = 0 \tag{9.108}$$

Free

$$\text{Shear force} = 0: \quad \frac{\partial}{\partial x}\left[EI(x)\frac{\partial^2}{\partial x^2}\left[v(x,\,t)\right]\right]\Big|_{x=0} = 0 \tag{9.109}$$

$$\text{Moment} = 0: \quad EI(x)\frac{\partial^2}{\partial x^2}\left[v(x,\,t)\right]\Big|_{x=0} = 0 \tag{9.110}$$

Free Vibration

Let us first analyze the free system by dropping the forcing term, $f(x,\,t)$.

$$m(x)\frac{\partial^2}{\partial t^2}\left[v(x,\,t)\right] = -\frac{\partial^2}{\partial x^2}\left[EI(x)\frac{\partial^2}{\partial x^2}\left[v(x,\,t)\right]\right] \quad 0 < x < \ell \tag{9.111}$$

We can attempt to obtain a solution to this equation by separation of variables:

$$v(x,\,t) = F(x)G(t) \tag{9.112}$$

Substitution and separation leads to the following:

$$\frac{\ddot{G}(t)}{G(t)} = \frac{-\left[EI(x)F''(x)\right]''}{m(x)F(x)} = -\omega^2 \tag{9.113}$$

where we have set the separation constant to a negative real value since we expect the temporal solution to be oscillatory. This leads to two ODEs in x and t:

$$\ddot{G} + \omega^2 G = 0 \tag{9.114}$$

$$-[EI(x)F''(x)]'' = \omega^2 m(x) F(x) \tag{9.115}$$

Hence, it is clear that the equation is separable; whether the complete solution is separable or not depends upon the boundary conditions and the initial conditions.

Simply Supported Uniform Beam

Let us consider a uniform beam such that

$$EI(x) = EI, \quad m(x) = m \tag{9.116}$$

where EI and m are constant quantities, not varying with x. Let it also be simply supported (or hinged) at both ends. Then, the EOM becomes

$$m \frac{\partial^2}{\partial t^2} [v(x, \, t)] = -EI \left[\frac{\partial^4}{\partial x^4} [v(x, \, t)] \right] \quad 0 < x < \ell \tag{9.117}$$

The corresponding boundary value problem is

$$-\left[EIF^{IV}(x) \right] = m\omega^2 F(x) \quad 0 < x < \ell \tag{9.118}$$

with the boundary conditions

$$\begin{array}{cc} F(0) = 0 & F(\ell) = 0 \\ F''(0) = 0 & F''(\ell) = 0 \end{array} \tag{9.119}$$

In addition, we have the temporal problem

$$\ddot{G} + \omega^2 G = 0 \tag{9.120}$$

with appropriate initial conditions.

The governing equation can be rewritten as

$$F^{IV}(x) - \beta^4 F(x) = 0 \tag{9.121}$$

where we define β using

$$\beta^4 = \frac{m\omega^2}{EI} \tag{9.122}$$

Solving,

$$F(x) = C_1 \sin \beta x + C_2 \cos \beta x + C_3 \sinh \beta x + C_4 \cosh \beta x \tag{9.123}$$

Applying the boundary equations,

$$C_2 = C_3 = C_4 = 0 \tag{9.124}$$

and

$$\sin \beta \ell = 0 \tag{9.125}$$

This is the characteristic equation for the problem and has an infinite number of discrete solutions:

$$\beta_n \ell = n\pi \quad n = 1, \, 2, \ldots \tag{9.126}$$

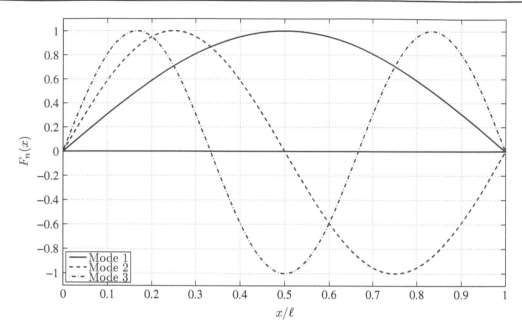

Figure 9.25: *Mode shapes of a simply supported beam*

For each of the values of β_n, we have a natural frequency, ω_n,

$$\omega_n = (n\pi)^2 \sqrt{\frac{EI}{m\ell^4}} \tag{9.127}$$

Solution of the temporal problem for every n leads to

$$G_n(t) = A_n \sin(\omega_n t + \phi_n) \tag{9.128}$$

It is hence clear that each value of β_n corresponds to a modal oscillation occurring at the frequency given by ω_n; in other words, the system has infinite natural frequencies, and therefore could be said to have infinite DOF. The mode shape at each of these natural frequencies is specified by $F_n(x)$:

$$F_n(x) = \sin \beta_n x = \sin\left(\frac{n\pi x}{\ell}\right) \quad n = 1, 2, \ldots \tag{9.129}$$

The first three mode shapes are shown in Fig. 9.25. Note that the first mode does not have a node (point with zero displacement), and that the number of nodes increases for the higher modes.

9.5 Review Questions

1. If a system is mathematically modeled to be continuous, what kind of equations result? How are they different from the models based on discrete systems?

2. What is the wave equation?

3. What are the assumptions and approximations made in the derivation of the string vibration model?

4. What is the principal assumption in the free vibration solution of the string EOM?

5. Which parameters do the natural frequencies of the string depend upon?

6. What is the ratio of successive natural frequencies of the string?

7. How many lumped masses would you feel is a reasonable approximation for the string vibration problem? (Warning: This is a loaded question and does not have a simple answer!).

8. What are the assumptions and approximations made in the derivation of the longitudinal vibration model?

9. What are the similarities and differences between the mathematical models for transverse vibration of the string and the longitudinal vibration of a thin rod?

10. Which parameters do the natural frequencies for longitudinal vibration depend upon?

11. How do the natural frequencies compare for the various boundary conditions for the longitudinal vibration problem?

12. What are the similarities and differences between the transverse vibration of the string and the torsional vibration of a shaft?

13. Which parameters do the natural frequencies for torsional vibration depend upon?

14. What are the similarities and differences between the mathematical models for torsional vibration and longitudinal vibration?

15. How do the natural frequencies compare for the various boundary conditions for the torsional vibration problem?

16. How does a lumped mass at the end of the shaft change the mathematical nature of the boundary value problem?

17. How does a lumped mass at the end of the shaft affect the natural frequencies?

18. What are the assumptions and approximations made in the derivation of the Euler beam vibration?

19. What are the similarities and differences between the mathematical models for flexural vibration and longitudinal vibration?

20. How many boundary conditions do you need at each boundary point for solving the beam vibration problem? How does this compare with the longitudinal and torsional vibration?

21. How do the natural frequencies compare for the various boundary conditions for the flexural vibration problem?

22. What is the ratio of successive natural frequencies of transverse vibration of the beam? How does this compare with the longitudinal vibration?

Problems

9.1 Consider a string fixed between two points located 3 m apart. The tensile force on the string is given as 250 N. Determine the fundamental frequency and the next two natural frequencies. Assume that the string is made of steel and has a diameter of 2 mm.

9.2 A steel string is fixed at both ends and has a length of 3 mm and a diameter of 1 mm. Find the tension required in order to have a fundamental frequency of 2 Hz.

9.3 Find the second and third natural frequencies of a stretched steel cable of length 2 m if the tension is increased by 20%. The fundamental frequency of the original cable is 2000 Hz.

9.4 Consider the string vibration problem with approximate discretization that we carried out as a running problem. Write a computer program to carry out the discretization with an arbitrary number of particle masses, and produce Fig. 9.7, which plots the approximate first natural frequency with the discretization order. Also, plot the second and third natural frequencies. What conclusions can you derive from this analysis?

9.5 Consider a string hanging vertically down; it is fixed at the upper end, and is free at the lower end as shown in Fig. 9.26. Go back to basics and derive the EOM with gravity included, or include it as a distributed force in the equation that was already derived; in either case, you will have to consider an infinitesimal section of the string and write Newton's second law for it. Be careful; this is a very tricky problem as the boundary value problem you end up with will be a variable coefficient one (if you did not get that, it is time to start again!). *Hint*: The solution will need you to dig up your mathematics book on Bessel functions (and you thought Bessel functions were of no possible practical significance!).

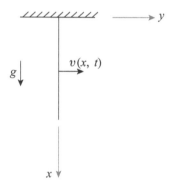

Figure 9.26: *A vertically hanging string*

9.6 Derive the characteristic equation for the longitudinal vibration of a rod fixed at both ends as shown in Fig. 9.27. Solve the equation to determine the first three natural frequencies, and plot the mode shapes.

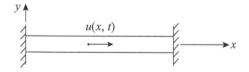

Figure 9.27: *Rod fixed at both ends*

9.7 Derive the characteristic equation for the longitudinal vibration of a rod free at both ends (!) as shown in Fig. 9.28. Solve the equation to determine the first three natural frequencies, and

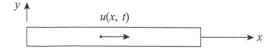

Figure 9.28: *Rod free at both ends*

plot the mode shapes. Can you think of a practical example for this situation? Also, compare your answer with the discussion in Sections 9.2 and 7.3.

9.8 Consider the longitudinal vibration of a rod fixed at the left end, and with a spring of stiffness, k, at the right end as shown in Fig. 9.29. Derive the characteristic equation and solve the equation to determine the first three natural frequencies for each of the following cases. You will need MATLAB or some numerical tool to do this. Also, plot the mode shapes.

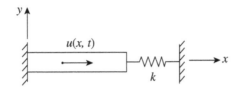

Figure 9.29: *Rod with a spring at the end*

(a) $EA/(k\ell) = 4$
(b) $EA/(k\ell) = 1/4$
(c) $EA/(k\ell) = 1$
(d) $EA/(k\ell) \approx 0$
(e) $EA/(k\ell) \approx \infty$

9.9 Consider the longitudinal vibration of a rod fixed at the left end, and with a rigid mass, M, at the right end, as shown in Fig. 9.30. Derive the characteristic equation and solve the equation to determine the first three natural frequencies for each of the following cases. You will need MATLAB or some numerical tool to do this. Also, plot the mode shapes.

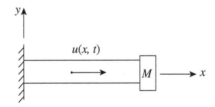

Figure 9.30: *Rod with a mass at the end*

(a) $m\ell/M = 5$
(b) $m\ell/M = 1/5$
(c) $m\ell/M = 1$
(d) $m\ell/M \approx 0$
(e) $m\ell/M \approx \infty$

9.10 Derive the characteristic equation for the longitudinal vibration of the system shown in Fig. 9.31.

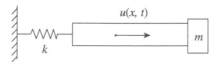

Figure 9.31: *Rod with a mass and a spring*

9.11 Derive the characteristic equation for the longitudinal vibration of the system shown in Fig. 9.32.

Figure 9.32: *Rod with springs*

9.12 A free-free rod is in longitudinal vibration. Suppose it represents a rocket (or missile) in flight, in which case it experiences an acceleration, a, in the x-direction along its longitudinal axis. Assume a uniform rod and derive the EOM from fundamental principles. What is the effect of acceleration on its free vibration properties? An interesting practical case is NASA's Ares launch vehicle shown in Fig. 9.33, which had many longitudinal vibration problems. You might want to read the online article by Warwick [65].

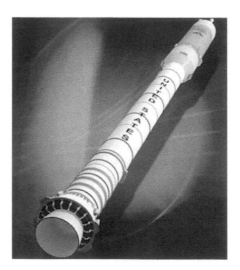

Figure 9.33: *The Ares launch vehicle (see Warwick [65])*

9.13 Derive the characteristic equation for the torsional vibration of a thin circular shaft for the following boundary conditions. Solve the equation to determine the first three natural frequencies, and plot the mode shapes.

(a) Fixed at the left end and free at the right end (Fig. 9.34).

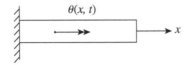

Figure 9.34: *Torsional member fixed at one end, and free at the other*

(b) Free at both ends (see Fig. 9.35). This condition approximates many rotating shafts. Also, compare your answer with the discussion in Section 7.3.

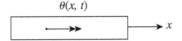

Figure 9.35: *Torsional member free at both ends*

9.14 Consider the torsional vibration of a shaft fixed at the left end and with a torsional spring of stiffness, k, at the right end as shown in Fig. 9.36. Derive the characteristic equation and solve the equation to determine the first three natural frequencies for each of the following cases. You will need MATLAB or some numerical tool to do this. Also, plot the mode shapes.

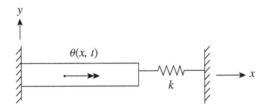

Figure 9.36: *Torsional member fixed at one end and with a spring at the other end*

(a) $GJ/(k\ell) = 2$
(b) $GJ/(k\ell) = 1/2$
(c) $GJ/(k\ell) = 1$
(d) $GJ/(k\ell) \approx 0$
(e) $GJ/(k\ell) \approx \infty$

9.15 Consider again the torsional vibration of a shaft fixed at the right end and with a rigid mass, I_D, at the other end, as shown in Fig. 9.37. Note that, unlike in the example solved in the text on page 379, the rigid mass here is at the left end. Derive the characteristic equation and solve the equation to determine the first three natural frequencies. For this part, you can assume $I\ell/I_D = 5$. Compare with the solution in the solved example. Also, plot the mode shapes. What would happen in the limit as I_D reduces to zero? (You will really need to do this with the *nondimensional* ratio, $I\ell/I_D$, and not with a dimensional quantity, I_D.) Does this agree with the result for the case without the rigid mass?

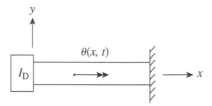

Figure 9.37: *Torsional member with a rigid inertia at the left end*

That was just a warm up! Now, to give a *real* workout for the little gray cells[2], determine the response as $I_D \to \infty$. Physically, what kind of a system would it reduce to?

9.16 Repeat Problem 9.15 for $I\ell/I_D = 1/5$.

9.17 Repeat Problem 9.15 for $I\ell/I_D = 2$.

9.18 Derive the characteristic equation for the transverse vibration of a uniform beam for the following boundary conditions. Solve the equation to determine the first three natural frequencies, and plot the mode shapes.

(a) Fixed at the left end, and free at the right end (Fig. 9.38); what is often referred to as a cantilever beam.

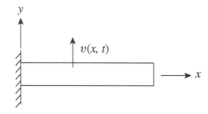

Figure 9.38: *Beam fixed at one end and free at the other*

(b) Fixed at the left end, and hinged at the right (Fig. 9.39).

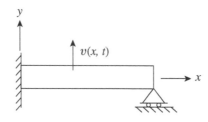

Figure 9.39: *Beam fixed at one end and hinged at the other*

(c) Fixed at both ends (Fig. 9.40).

---

[2]Dr. Watson, this is a three-pipe problem—Sherlock Holmes.

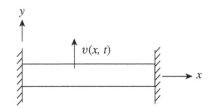

Figure 9.40: *Beam fixed at both ends*

(d) ⚡ Supported on identical springs of stiffness, k, at both ends (Fig. 9.41). Assume $EI/(k\ell^3) = 1$. This would represent rotating shafts on bearings or just beams with some flexibility in the supports.

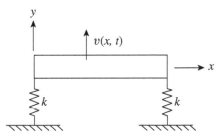

Figure 9.41: *Beam supported on springs at its ends*

(e) ⚡ Fixed at the left end and a rigid mass, M, at the right end (Fig. 9.42). Assume $m\ell/M = 3$.

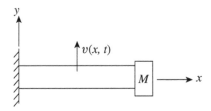

Figure 9.42: *Beam fixed at one end and with a rigid mass at the other end*

(f) ⚡ Free at both ends; see Fig. 9.43. Can you think of a practical example for this situation? Also, compare your answer with the discussion in Sections 7.3 and 9.2.

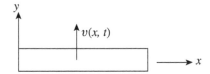

Figure 9.43: *Beam free at both ends*

9.19 Vibrations are always present in manufacturing processes and cause all kinds of problems from minor errors to major failures. Boring operations, shown in Fig. 9.44, are especially associated

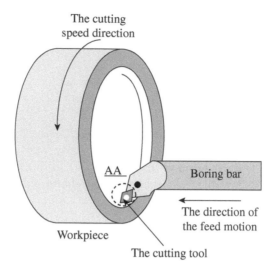

Figure 9.44: *Boring operation (see Andren et al. [66])*

with serious vibration-related problems [66]. In order to avoid the problem, one should know the dynamic properties of the boring bar. A boring bar clamped to one end can be modeled as a simple cantilever beam. Derive the characteristic equation and find the natural frequencies. Use the following numerical values: $E = 206 \times 10^9$ N/m^2, $\ell = 0.2$ m, $I = 1.095 \times 10^{-7}$ m^4, $\rho = 7800$ kg/m^3, $A = 1.236 \times 10^{-3}$ m^2.

10 Experimental Techniques

A theory must be tempered with reality.

—Jawaharlal Nehru

This chapter will give you a brief overview of some experimental techniques and introduce you to some basic underlying principles of vibration instrumentation. A detailed technical description or a technological survey of the vast array of instruments is beyond the scope of this book however, and you will need to consult a handbook such as the one by Piersol and Paez [67] or appropriate manufacturer provided material for that.

In this book you have mostly learnt the *theory* of vibrating systems; but that is only half the story! Real systems demand real measurements after all. How do we know that what we are predicting with sophisticated physics and mathematics is any good if we cannot verify it for the real system?[1] Hence, it is wise to remember that the ultimate truth is always with real measurements. In addition to situations when we can model and predict vibrational behavior, there are many cases that elude simple modeling by virtue of being very complicated, or having components or processes that we do not understand yet very well. For example, many nonlinear phenomena, certain kinds of friction, and newer materials whose constitutive equations have not been established completely are some examples. In general, in almost every case, experimental measurement will give us a satisfying confirmation of our theoretical predictions (or not!), and should be attempted.

10.1   Vibration Transducers

By definition, a transducer is something that converts signals from one type to another. For us, that would be converting standard vibration relevant variables such as displacement, velocity, acceleration and force to an electrically measurable signal such as current or voltage. In general, there are two kinds of motion transducers.

- Seismic transducers. Here, the base of the transducer sits directly and firmly on the vibrating system itself. Examples are accelerometers, and displacement transducers.

---

[1]The colloquial expression for this is "where the rubber hits the road."

- Fixed-reference transducers. In this case, one of the terminals of the transducer is attached to a fixed point, and the other is attached to the point whose motion we need to measure. Examples are laser vibrometers, proximity probes, etc.

First, we explore the principle of the seismic transducers.

10.1.1   Seismic Transducers

Seismic transducers are very versatile as they can just be screwed or glued onto a vibrating structure without worrying about finding a "fixed" point. In principle, they consist of just a damped oscillator, as shown schematically in Fig. 10.1. A mass m is separated from the moving structure by means of a spring (k) and damper (c). The case of the transducer is assumed to have the same displacement as the structure (in other words, it should be fixed firmly to the structure so that there is no relative motion between them). Suppose the structure's motion is $u(t)$, and the displacement of the mass is $x(t)$. We will see how this kind of transducer can be configured to measure any of the three quantities we are interested in: displacement, velocity, or acceleration. The measured quantity will be the relative displacement, $z(t) = x(t) - u(t)$, as shown in the figure.

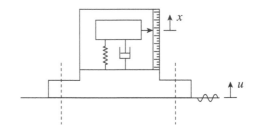

Figure 10.1: *Schematic of the seismic transducer*

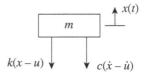

Figure 10.2: *FBD of the seismic transducer*

We now have the same problem as base excitation that we looked at in Chapter 5. We will repeat the analysis here however, since there are some subtle differences in what we are looking for from the analysis. Figure 10.2 shows the free body diagram (FBD) of the mass. Remembering that absolute acceleration should be used in the kinematics ($a = \ddot{x}$), we can write the equation of motion.

$$m\ddot{x} = -c(\dot{x} - \dot{u}) - k(x - u) \tag{10.1}$$

Since the relative displacement, z, is the quantity of interest here (unlike when we looked at the base excitation problem), we will rewrite this equation in terms of z.

$$m\ddot{z} + c\dot{z} + kz = -m\ddot{u} \tag{10.2}$$

Now, let us suppose the support motion to be harmonic. Then,

$$u(t) = u_0 \sin \omega t \tag{10.3}$$

Then,

$$m\ddot{z} + c\dot{z} + kz = -m\omega^2 u_0 \sin \omega t \tag{10.4}$$

This can be solved easily for z

$$z(t) = z_0 \sin(\omega t - \phi) \tag{10.5}$$

where

$$\frac{z_0}{u_0} = \frac{m\omega^2}{\sqrt{(k - m\omega^2)^2 + (c\omega)^2}} = \frac{r^2}{\sqrt{(1 - r^2)^2 + (2\zeta r)^2}} \tag{10.6}$$

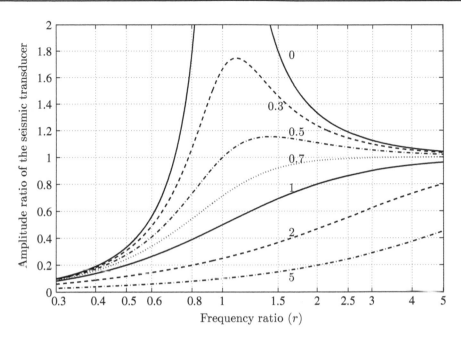

Figure 10.3: *Amplitude ratio for the seismic transducer as a function of the frequency ratio and the damping ratio*

where

$$r = \frac{\omega}{\omega_n}$$

is the frequency ratio. The phase angle is given by

$$\phi = \tan^{-1}\left[\frac{2\zeta r}{1 - r^2}\right] \tag{10.7}$$

Figure 10.3 plots the amplitude as a function of the frequency ratio and the damping ratio, and Fig. 10.4 shows the phase lag. Let us examine the utility of this device to measure the base vibration using these figures as a guide.

Acceleration measurement

Suppose we focus on the low frequency range of the instrument, or when $\omega/\omega_n \ll 1$. Then, the denominator of z_0/u_0 is nearly 1 [scrutinize Eq. (10.6)], and hence,

$$z_0 \approx r^2 u_0 = \left(\frac{1}{\omega_n^2}\right)(\omega^2 u_0) \tag{10.8}$$

In other words, z_0, which is the display of the instrument, is proportional to the *acceleration amplitude* of the base. Hence, we can rewrite this equation as follows:

$$\frac{z_0}{a_{u0}} = \frac{1}{\omega_n^2}\frac{1}{\sqrt{(1 - r^2)^2 + (2\zeta r)^2}} \tag{10.9}$$

where a_{u0} is the amplitude of the acceleration of the base.

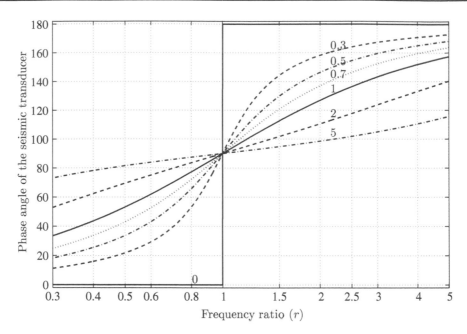

Figure 10.4: *Phase angle for the seismic transducer as a function of the frequency ratio and the damping ratio*

That means that we are now able to measure the base acceleration, or you and I have just come up with the basic principle of the accelerometer. (Unfortunately, it is too late to patent this idea, because the very same idea occurred to some smart people many decades ago!).

To analyze the accuracy of the instrument when used as an accelerometer, let us replot the response as a ratio of what we display (z) to what we intend to measure ($\omega^2 u_0$), but also nondimensionalize it using ω_n^2, which now can be treated just as a scale factor. Or, we want to scrutinize the accelerometer response characteristic defined as

$$\frac{\omega_n^2 z_0}{\omega^2 u_0} = \frac{1}{\sqrt{\left(1 - r^2\right)^2 + (2\zeta r)^2}} \tag{10.10}$$

Figure 10.5 shows this response characteristic as a function of the frequency ratio and the damping ratio. Clearly, the most desirable characteristic of an instrument is a flat response with respect to any varying parameter. In this instance, the most common variation in measurement is the frequency ω, and the amplitude of vibration. So, we want the instrument to be accurate over a range of frequencies, and over a range of amplitudes. The amplitude discussion is quite involved; suffice it to say that a limited amplitude that preserves the linearity of the instrument would give us a reasonably correct measurement. The frequency discussion is more interesting (given an accelerometer with a specific damping ratio, and natural frequency). From the figure it follows that either small or large amounts of damping would lead to large errors. Also, larger frequency ratios would similarly lead to large errors.

Figure 10.6 shows the same result from a different perspective: a percentage error in the measurement plotted versus frequency, and for different damping ratios. We have somewhat arbitrarily truncated the graph at a 5% error assuming that it would be the maximum error we are willing to live with. Then, it is clear that about a frequency ratio of 0.2 would be a reasonable limit for the usefulness of the instrument. One way to extend the accuracy range would be to use a moderate value of the damping ratio. This is more clearly seen from Fig. 10.7. Clearly, an accelerometer designed

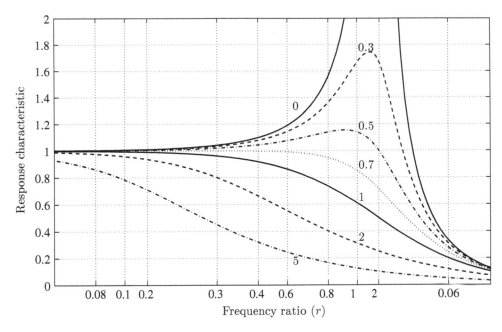

Figure 10.5: *Accelerometer response characteristic*

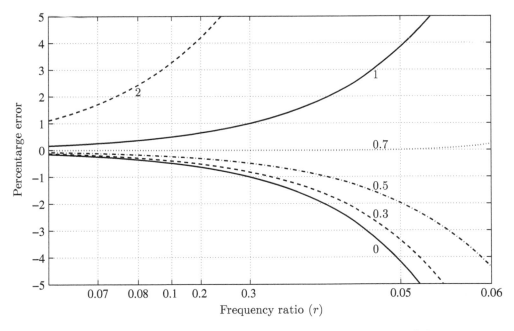

Figure 10.6: *Accelerometer error as a function of frequency ratio and damping ratio*

to have a damping ratio of 0.65 or so could be used up to a frequency equal to 60% of the natural frequency of the instrument with an error of less than 1%.

Now, remembering that we are striving to get an instrument capable of measuring the largest range of vibration frequencies possible, it is clear that the useful frequency range of an accelerometer

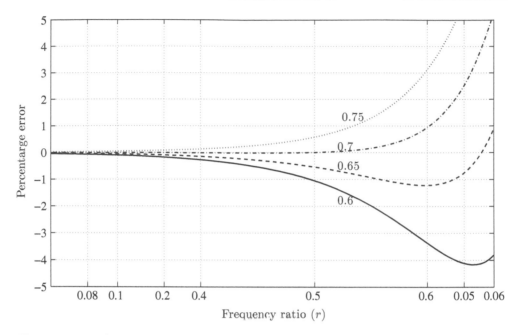

Figure 10.7: *Accelerometer error magnified for a restricted range of damping ratios*

is higher for an instrument with a larger natural frequency. From Eq. (10.9), it follows that the relative vibrational amplitude would be smaller for a system with a higher ω_n, all other parameters (including r) being the same. This essentially reduces the sensitivity of the device requiring larger amplification which may not always be feasible. Hence, to summarize, the design principle of an accelerometer is to try to achieve as large a natural frequency as possible under the constraints of adequate sensitivity. (Read this again to make sure that you understand all the implications!)

10.1.2   Practical Sensors

Practically speaking, the most preferred accelerometers are piezoelectric in nature. The piezoelectric elements produce an electric charge proportional to the strain they are subjected to. Noting that the strain is proportional to the applied load (for small deflections, the linearity assumption is reasonably valid), what we have here is an electrical output that measures the applied load, be it tension, compression, or shear. When subject to vibration, the system is configured to apply a force proportional to the acceleration to the piezoelectric elements. Thus, we have a way of measuring acceleration, subject, of course, to the theoretical limitations we just discussed. Two common configurations are in use: compression and shear types. Commercial examples are shown in Fig. 10.8. They are typically glued or screwed on to the structure whose vibration we wish to measure. Of course, there are many engineering specifications for such a device; Fig. 10.9 lists a typical specification sheet—note the limitation of the frequency that we just discussed. Also, note the other operational constraints such as temperature, peak amplitude, and other practical issues. Web sites of commercial manufacturers give extended information about their practical usage and selection.

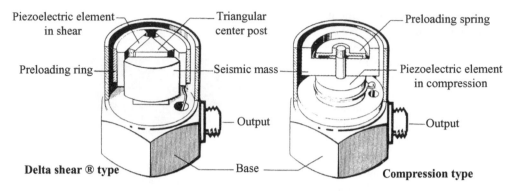

Figure 10.8: *Typical accelerometers (courtesy of Brüel & Kjær)*

Bruel & Kjær 4366 — Accelerometer

Piezoelectric Charge Accelerometer

Charge sensitivity : 5 pC/ms^{-2} or 50 pC/g $\pm$ 2 % **
Voltage sensitivity : 4 mV/ms^{-2} or 40 mV/g $\pm$ 2 % **

Mounted resonance : 16 kHz
Frequency range - 5 % : 0.2 Hz to 5000 Hz
Frequency range - 10 % : 0.1 Hz to 8000 Hz

Capacitance - typical : 1100 pF excluding cable

Max transverse sensitivity : < 2 %
Piezoelectric material : PZ23
Construction : Delta shear

Typical temperature transient sensitivity : 0.02 ms^{-2}/°
Typical magnetic sensitivity (50 Hz to 0.03T) : 1 ms^{-2}/T
Typical acoustic sensitivity (154 dB SPT) : 0.001 ms^{-2}
Minimum leakage resistance at 20 °C : 20 GΩ.
Ambient temperature range : –74 to 250 °C

Maximum operational shock (Peak) : $\pm$ 20 kms^{-2};
Maximum continuous sinusoidal acceleration (Peak) : 20 kms^{-2}
Maximum acceleration (Peak) with mounting magnet : 50 kms^{-2}

Electrical Connector : 10-32 UNF, side entry
Recommended cable : AO 0038

Mechanical
Mounting thread : 10-32 UNF, 3.2 mm deep
Mounting torque : 1 Nm, max = 2 Nm, min = 0.5 Nm
Dimensions - body : 16 mm diameter
Dimensions - overall height: 19 mm
Dimension - across base flats : 16 mm - spanner size
Case Material : Stainless steel
Weight: 28 g

Figure 10.9: *Typical accelerometer specification sheet (courtesy of Brüel & Kjær)*

Noncontact vibration measurement, especially of the displacement, comes in handy in many situations. Again, there are many kinds of these in practice. We will mention just two of them here. The first is the so-called proximity probe, that uses eddy current disturbances to measure fine vibration down to tenths of a mil (1 mil = 0.001 inch = 0.025 mm, is a common unit of vibration measurement in the United States). They are quite popular for measuring whirling displacements in rotating machinery, and are often needed for vibration monitoring to keep tabs on the "health" of machinery. Figure 10.10 shows a commercial proximity probe.

Also, quite popular are the laser vibrometers that use laser techniques for noncontact measurement of the vibrational displacement. They can be very precise; spatial resolution falls off with distance from the object. Figure 10.11 shows a commercial laser vibrometer.

Figure 10.10: *A commercial proximity probe (courtesy of Bently & Nevada)*

Figure 10.11: *A commercial laser vibrometer (courtesy of Keyence)*

10.2   Force Transducers

Often it is of interest to determine the forces producing (or resulting from) the vibrations. Most practical force transducers also use a piezoelectric element, which produces a proportionate electric signal when subjected to a force. The force transducer is mounted in the path of the force transmission. Usually, it is designed with a high stiffness so that it would have a high resonant frequency. Figure 10.12 shows the picture of a commercial force transducer.

Related to the measurement of a force is the concept of *impedance*. Mechanical impedance is defined as the force divided by the velocity in the frequency domain (Laplace or Fourier), and

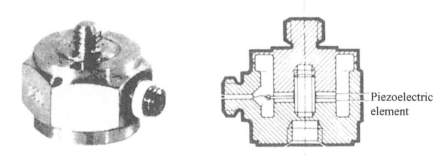

Figure 10.12: *A typical force transducer (courtesy of Brüel & Kjær)*

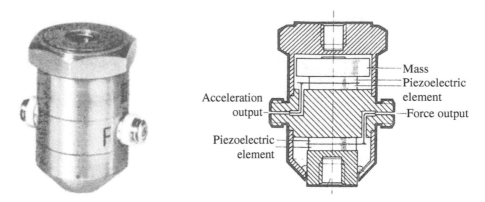

Figure 10.13: *An impedance head, measuring both force and acceleration or velocity (courtesy of Brüel & Kjær)*

essentially gives us information about both cause and effect; from this we would be able to derive the frequency response of the system and can characterize all kinds of important issues such as resonant frequencies, damping ratios, etc. Figure 10.13 shows a commercial impedance head. Impedance heads are very convenient for measuring vibration in light structures and in soft samples.

10.3   Exciters

The easiest way to characterize an unknown system is by applying known forces and by measuring the vibrational response to those forces. In order to apply forces, we use exciters. There are many kinds and sizes of exciters. Bigger ones are often electrodynamic; the smallers ones tend to be ultrasonic or piezoelectric (note that the piezoelectric sensor, described above, also works as an actuator, which is the reverse action of a transducer). They are often driven by a computer or a signal generator in order to apply the forces of a desired range of amplitudes and frequencies. Figures 10.14 and 10.15

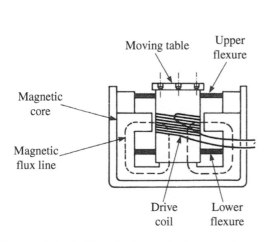

Figure 10.14: *Sketch of a typical electrodynamic exciter (see Broch [68])*

Figure 10.15: *Photo of a commercial electrodynamic exciter (courtesy of Brüel & Kjær)*

Figure 10.16: *A typical impact hammer (courtesy of Brüel & Kjær)*

show some exciters. Also in use is what is called an "impact hammer" with an impedance head to deliver a known impulsive force (Fig. 10.16).[2]

10.4   Data Acquisition and Other Equipment

Accelerometers, proximity probes, and other transducers often produce a very weak signal and need to be amplified. Also, most of the time, they can benefit from signal conditioning to remove noise and other unwanted signals. These, as well as power supplies, are typically adapted and tailored for the specific sensors, and are sold by the same manufacturers who sell the transducers.

Ultimately, all the data generated need to be collected, compared, and analyzed. As with all things, most of this process is nowadays done on a personal computer. However, the frequency analyzer is still in use, especially in many portable applications. Frequency analyzers can essentially display an instantaneous frequency spectrum for the signal being measured produced using Fourier transform theory. This can be quite useful especially in the field conditions, as the Fourier spectrum often carries much more important and useful information than the time data. Figure 10.17 shows a typical frequency analyzer.

If the data are going to be collected and analyzed on a personal computer, they will need paraphernalia to connect to the PC. This may take the form of data acquisition cards with various performance specifications including number of inputs, sampling frequencies, signal strength, etc. Increasingly, this part is just taking the shape of "plug and play" with just cables connecting the equipment to the PC (USB, serial, parallel, ...).

An example of a test arrangement is shown in Fig. 10.18.

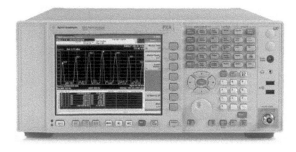

Figure 10.17: *A frequency analyzer (courtesy of Agilent)*

---

[2]Not to be employed on your professors or teaching assistants.

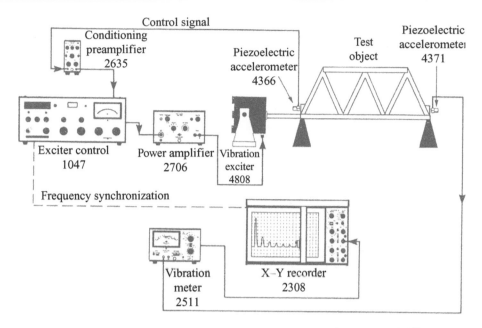

Figure 10.18: *Typical test arrangement from (see Broch [68])*

10.5 Review Questions

1. What is a transducer?
2. What are the two kinds of motion transducers? How do they differ in principle?
3. For what frequency range is an accelerometer accurate? Why?
4. How does damping affect the operation and accuracy of the accelerometer?
5. Summarize the design principles of an accelerometer.
6. How do piezoelectric sensors work?
7. How do proximity probes work, and what is their principal application?
8. How do laser vibrometers work, and where are they used?
9. What is mechanical impedance?
10. How do force transducers work?
11. Why would one need an exciter?
12. What are the different kinds of exciters?
13. What is an impact hammer, and what are its applications (other than tapping your labmate on the head)?
14. What is a frequency analyzer? What kind of information can be gleaned from a frequency analyzer?

11 Beyond This Text

All matter originates and exists only by virtue of a force which brings the particle of an atom to vibration and holds this most minute solar system of the atom together.

—Max Planck

I hope this book has sparked an interest in you to study vibrations further. I have tried to give you a glimpse into the fascinating area, but clearly, there is only so much you can get from a 500-page book! There are many special topics unexplored in this book; we discuss them briefly below. Each of them has sufficient material to merit an individual book; so, we are only going to discuss the overall scope of these topics to help guide you into choosing them for future studies.

Continuous Systems

We did discuss continuous systems briefly in this book (Chapter 9), but did not delve too deeply. All real systems are continuous, in the sense that you do not see lumped masses and massless springs in practice; i.e., they have distributed parameters. The mathematics involved gets so complex that continuous system study to a large extent is restricted to research, and does not get into industrial practice. However, the insight gained from the continuous system modeling can be substantial and can help design new and complex systems. Most of the times, continuous system modeling leads to partial differential equations that cannot be solved analytically, and approximate methods, such as the Rayleigh Ritz or Galerkin procedure, are pressed for the task. And, of course, there is always the popular finite element method (which is really an extension of the Rayleigh-Ritz method), that is used to convert the continuous models into the lumped models.

Nonlinear Vibrations

As has been made clear at various points in this text, all real systems are nonlinear, and linear systems only exist in the engineer's imagination! Clearly, the nonlinear system response would be quantitatively different from that of the linearized system, and the need for precision in modern engineering methods and designs would demand better models. But the story does not end there. It turns out that nonlinear system behavior is *qualitatively* different from the linearized system behavior. Interesting and important phenomena including instabilities, limit cycle oscillations, sub- and

Figure 11.1: *The Tacoma Narrows Bridge torsional oscillations and failure (see Smith [69])*

superharmonic responses, bifurcations, strange attractors and chaotic phenomena are predicted by nonlinear models, which can never be predicted by linearized models. Why is that prediction important? Because these phenomena do occur (in fact, rarely have I *not* see them in practice), and, if we, the engineers, want to design effective, efficient, and safe engineering systems, we will need to consider nonlinear models. The solution process goes beyond simple numerical simulations (which we used heavily in this book), and uses techniques from differential algebra, calculus, and even topology. If you still need convincing, take a look at a dramatic example of a nonlinear phenomenon that destroyed the bridge across Tacoma Narrows in Washington (USA), shown in Fig. 11.1. It was not a *resonant* phenomena (as written wrongly in some text books), but was actually a fluid-induced vibration due to a nonlinear phenomenon called self-excited vibration. If you search on the web, you will also be able to find a video of this spectacular disaster.

Random Vibrations

Most excitation that occurs in vibrating systems has at least an element of stochasticity to it. It is somewhat more prevalent in aerospace situations, but is true in other instances as well, such as in automobiles and in manufacturing. Stochastic excitation simply means that we do not have sufficient understanding or measurement to describe the forces explicitly and deterministically; we only have a rough estimate of its statistical properties. Especially with advances in aeronautical and space technologies, random vibrations has become an important application area. The study of random vibrations involves models and techniques very similar to what we discussed in this book, but adds Fourier transforms to the mix.

Rotor Dynamics

Vibration occurs in all rotating machinery some of which we discussed in this book. Rotating machinery, such as steam and gas turbines, compressors, aircraft and rocket engines, are among the most complex that the engineering world has built, and the problems tend to be equally complicated. Since so many of them are supported on fluid film bearings, an element of fluid mechanics also gets into the analysis; rolling element bearings add their own complexity with Hertzian contact stress theory, and the somewhat novel magnetic bearings add electromagnetic and control concepts. Plus, almost all of these are strongly nonlinear. You get the idea! This is a very complex and specialized field even though what you see in these systems is just vibrations, fundamentally not different from what we have studied in this text. If you like to use advanced theory to bring to bear on a very practical field, this would be an ideal subject to pursue.

The Newfangled Stuff

As of the writing of this book, the newfangled stuff includes nanosystems and biological systems. Nanosystems are systems at very small scales (10^{-9} m); systems are being invented and built using nanostructures and materials at a breakneck pace, although, in the humble opinion of this author, the fundamental understanding (especially with regard to constitutive modeling) is missing largely. The problems can appear strange because of the interaction with forces and systems somewhat alien to the traditional vibration theory; however, most of the vibration aspects of the nanosystems are being handled by using the same concepts we studied in this text. The other new trend is biological systems, which are of course not really new—they started with the first single-cellular life on the planet! What is new is that there is a growing trend to try to understand dynamical phenomena (and other aspects) in biological systems. This again happens to be in an interdisciplinary setting, but the methods we have learnt in this book are very much applicable.

The Bottom Line

The bottom line is that all real systems are nonlinear, stochastic, and continuous, but this book—to a large extent—dealt with linear, deterministic, and discrete models. Before you throw away this book in disgust, consider two facts. One, an overwhelming number of industrial systems designed since the industrial revolution have been designed with linear, deterministic, and discrete models. And, clearly, they work reasonably well: we are able to send a manned spacecraft to the moon and get it back! But we always want more—we want to design ever more efficient and safer systems, and for that, we would need to work with better and ever more accurate models. Two, to understand nonlinear, stochastic, and continuous systems, you will *have* to study and understand the simpler stuff first. And, that is where we started; and if you did justice to this book, you have indeed made a good start.

A Review of Mathematical Fundamentals

Most subjects at universities are taught for no other purpose than that they may be retaught when the students become teachers[1].

—G. C. Lichtenberg

A.1 Elementary Mathematics

A.1.1 Complex Algebra

A complex number z can be expressed in the form

$$z = a + ib \tag{A.1}$$

where a and b are real numbers, and i is the imaginary number, $\sqrt{-1}$. The following simple rules of algebra apply to complex numbers:

$$z_1 = a_1 + ib_1 \quad z_2 = a_2 + ib_2 \tag{A.2}$$
$$z_1 \pm z_2 = (a_1 \pm a_2) + i(b_1 \pm b_2) \tag{A.3}$$
$$cz = (ca) + i(cb) \tag{A.4}$$

Of the many possible representations of complex numbers, a vector analog is particularly useful. For this, we consider a coordinate plane, with the x-axis corresponding to the real numbers, and the y-axis to the imaginary numbers. Such a coordinate system is called the complex plane and any complex number z can be easily represented as shown in Fig. A.1 as a vector drawn from the origin to the point (a, b).

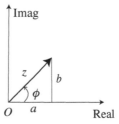

Figure A.1: *Graphical representation of a complex number*

---

[1]Although this is a cynical statement, there is indeed some truth to it, especially with respect to mathematics.

Then, the *magnitude* of the complex number is defined to be the length of the vector, and is hence equal to

$$|z| = \sqrt{(a^2 + b^2)} \tag{A.5}$$

The *phase* of the complex number z is defined to be the angle the vector makes with the x-axis, and is hence equal to

$$\angle z = \tan^{-1}\left(\frac{b}{a}\right) \tag{A.6}$$

Euler formula

$$e^{i\theta} = \cos\theta + i\sin\theta \tag{A.7}$$

where $i = \sqrt{-1}$.

Proof   From Taylor series,

$$e^x = 1 + x + \frac{x^2}{2!} + \frac{x^3}{3!} + \frac{x^4}{4!} + \cdots \tag{A.8}$$

Let $x = i\theta$. Then,

$$e^{i\theta} = 1 + i\theta - \frac{\theta^2}{2!} - i\frac{\theta^3}{3!} + \frac{\theta^4}{4!} + \cdots \tag{A.9}$$

Differentiating,

$$\frac{d}{d\theta}\left(e^{i\theta}\right) = i - \theta - i\frac{\theta^2}{2!} - \frac{\theta^3}{3!} + \cdots$$
$$= i\left[1 + i\theta - \frac{\theta^2}{2!} - i\frac{\theta^3}{3!} + \cdots\right] \tag{A.10}$$

In other words,

$$\frac{d}{d\theta}\left(e^{i\theta}\right) = ie^{i\theta} \tag{A.11}$$

This means $e^{i\theta}$ is a function that satisfies the following first-order linear differential equation:

$$\frac{du}{d\theta} = iu \tag{A.12}$$

Next, consider

$$\frac{d}{d\theta}\left(\cos\theta + i\sin\theta\right) = -\sin\theta + i\cos\theta$$
$$= i\left(\cos\theta + i\sin\theta\right) \tag{A.13}$$

Note that this has exactly the same form as Eq. (A.11).

In summary, the *linear* differential equation, Eq. (A.12), has the two solutions $u_1 = e^{i\theta}$ and $u_2 = \sin\theta + i\cos\theta$. Moreover, $u_1(0) = u_2(0) = 1$. Hence, the two solutions have to be identical. Therefore,

$$e^{i\theta} = \cos\theta + i\sin\theta$$

Binomial series

A special case of Taylor series is for the function of the form $(1 + x)^n$, and can be useful in many instances:

$$f(x) = (1 + x)^n = 1 + nx + \frac{n(n-1)}{2} x^2 + \cdots + \frac{(n(n-1)\cdots(n-k+1)}{k!} x^n + \cdots \qquad \text{(A.14)}$$

Refer to a mathematics text book for more details about convergence properties and such things.

A.1.2   Trigonometric Functions

Elementary trigonometric formulae

$$\sin(\theta_1 \pm \theta_2) = \sin\theta_1 \cos\theta_2 \pm \cos\theta_1 \sin\theta_2 \qquad \text{(A.15)}$$
$$\cos(\theta_1 \pm \theta_2) = \cos\theta_1 \cos\theta_2 \mp \sin\theta_1 \sin\theta_2 \qquad \text{(A.16)}$$
$$\sin(-\alpha) = -\sin(\alpha) \qquad \text{(A.17)}$$
$$\cos(-\alpha) = +\cos(\alpha) \qquad \text{(A.18)}$$

Functional relations

1. $\sin x = \frac{1}{2i}\left(e^{ix} - e^{-ix}\right) = -i\sinh(ix)$

2. $\sinh x = \frac{1}{2}\left(e^x - e^{-x}\right) = -i\sin(ix)$

3. $\cos x = \frac{1}{2}\left(e^{ix} + e^{-ix}\right) = \cosh(ix)$

4. $\cosh x = \frac{1}{2}\left(e^x + e^{-x}\right) = \cos(ix)$

5. $\cos^2 x + \sin^2 x = 1$

6. $\cosh^2 x - \sinh^2 x = 1$

7. $\sin x \pm \sin y = 2\sin\frac{1}{2}(x \pm y)\cos\frac{1}{2}(x \mp y)$

8. $\cos x + \cos y = 2\cos\frac{1}{2}(x + y)\cos\frac{1}{2}(x - y)$

9. $\cos x - \cos y = 2\sin\frac{1}{2}(x + y)\sin\frac{1}{2}(y - x)$

10. $\sinh x \pm \sinh y = 2\sinh\frac{1}{2}(x \pm y)\cosh\frac{1}{2}(x \mp y)$

11. $\cosh x + \cosh y = 2\cosh\frac{1}{2}(x + y)\cosh\frac{1}{2}(x - y)$

12. $\cosh x - \cosh y = 2\sinh\frac{1}{2}(x + y)\sinh\frac{1}{2}(x - y)$

Integrations

$$\int \sin(ax + b)\sin(cx + d)\,dx = \begin{cases} \dfrac{\sin[(a-c)x + b - d]}{2(a-c)} - \dfrac{\sin[(a+c)x + b + d]}{2(a+c)} & [a^2 \neq c^2] \\[2ex] \dfrac{x}{2}\cos(b - d) - \dfrac{\sin(2ax + b + d)}{4a} & [a = c] \end{cases}$$

$$\text{(A.19)}$$

$$\int \sin\left(ax+b\right)\cos\left(cx+d\right)dx = \begin{cases} -\dfrac{\cos\left[(a-c)x+b-d\right]}{2(a-c)} - \dfrac{\cos\left[(a+c)x+b+d\right]}{2(a+c)} & \left[a^2 \neq c^2\right] \\[3mm] \dfrac{x}{2}\sin\left(b-d\right) - \dfrac{\cos\left(2ax+b+d\right)}{4a} & \left[a = c\right] \end{cases}$$

$$(A.20)$$

$$\int \cos\left(ax+b\right)\cos\left(cx+d\right)dx = \begin{cases} \dfrac{\sin\left[(a-c)x+b-d\right]}{2(a-c)} + \dfrac{\sin\left[(a+c)x+b+d\right]}{2(a+c)} & \left[a^2 \neq c^2\right] \\[3mm] \dfrac{x}{2}\cos\left(b-d\right) + \dfrac{\sin\left(2ax+b+d\right)}{4a} & \left[a = c\right] \end{cases}$$

$$(A.21)$$

$$\int \sinh\left(ax+b\right)\sinh\left(cx+d\right)dx = \begin{cases} \dfrac{\sinh\left[(a+c)x+b+d\right]}{2(a+c)} - \dfrac{\sinh\left[(a-c)x+b-d\right]}{2(a-c)} & \left[a^2 \neq c^2\right] \\[3mm] -\dfrac{x}{2}\cosh\left(b-d\right) + \dfrac{\sinh\left(2ax+b+d\right)}{4a} & \left[a = c\right] \end{cases}$$

$$(A.22)$$

$$\int \sinh\left(ax+b\right)\cosh\left(cx+d\right)dx = \begin{cases} \dfrac{\cosh\left[(a+c)x+b+d\right]}{2(a+c)} + \dfrac{\cosh\left[(a-c)x+b-d\right]}{2(a-c)} & \left[a^2 \neq c^2\right] \\[3mm] \dfrac{x}{2}\sinh\left(b-d\right) + \dfrac{\cosh\left(2ax+b+d\right)}{4a} & \left[a = c\right] \end{cases}$$

$$(A.23)$$

$$\int \cosh\left(ax+b\right)\cosh\left(cx+d\right)dx = \begin{cases} \dfrac{\sinh\left[(a+c)x+b+d\right]}{2(a+c)} + \dfrac{\sinh\left[(a-c)x+b-d\right]}{2(a-c)} & \left[a^2 \neq c^2\right] \\[3mm] \dfrac{x}{2}\cosh\left(b-d\right) + \dfrac{\sinh\left(2ax+b+d\right)}{4a} & \left[a = c\right] \end{cases}$$

$$(A.24)$$

$$\int_0^x x^2 e^{ax}\cos bx\,dx = \frac{e^{ax}}{(a^2+b^2)^3}\left(a_1\sin\left(bx\right) + a_2\cos\left(bx\right)\right) \tag{A.25}$$

where

$$a_1 = bx^2(a^2+b^2)^2 - 4abx(a^2-b^2) + 2b(3a^2-b^2) \tag{A.26}$$

$$a_2 = ax^2(a^2+b^2)^2 - 2x(a^4-b^4) + 2a(a^2-3b^2) \tag{A.27}$$

$$\int_0^x xe^{ax}\cos bx\,dx = \frac{e^{ax}}{(a^2+b^2)^2}\left(a_1\sin\left(bx\right) + a_2\cos\left(bx\right)\right) \tag{A.28}$$

where

$$a_1 = bx(a^2 + b^2) - 2ab \tag{A.29}$$
$$a_2 = ax(a^2 + b^2) - (a^2 - b^2) \tag{A.30}$$

$$\int_0^x x^2 e^{ax} \sin bx \, dx = \frac{e^{ax}}{(a^2 + b^2)^3}(a_1 \cos(bx) + a_2 \sin(bx)) \tag{A.31}$$

where

$$a_1 = -bx^2(a^2 + b^2)^2 + 4abx(a^2 - b^2) - 2b(3a^2 - b^2) \tag{A.32}$$
$$a_2 = ax^2(a^2 + b^2)^2 - 2x(a^4 - b^4) + 2a(a^2 - 3b^2) \tag{A.33}$$

$$\int_0^x x e^{ax} \sin bx \, dx = \frac{e^{ax}}{(a^2 + b^2)^2}(a_1 \cos(bx) + a_2 \sin(bx)) \tag{A.34}$$

where

$$a_1 = -bx(a^2 + b^2) + 2ab \tag{A.35}$$
$$a_2 = ax(a^2 + b^2) - (a^2 - b^2) \tag{A.36}$$

$$\int_0^x e^{ax} \cos bx \, dx = \frac{e^{ax}}{(a^2 + b^2)}(a \cos(bx) + b \sin(bx)) \tag{A.37}$$

$$\int_0^x e^{ax} \sin bx \, dx = \frac{e^{ax}}{(a^2 + b^2)}(a \sin(bx) - b \cos(bx)) \tag{A.38}$$

A.1.3 Linearization

Taylor series

A function $f(x)$, which is continuous and differentiable at a point x_0, can be expanded in an infinite polynomial series given below:

$$f(x) = f(x_0) + \left.\frac{df}{dx}\right|_{x_0} \Delta x + \left.\frac{d^2 f}{dx^2}\right|_{x_0} \frac{(\Delta x)^2}{2!} + \left.\frac{d^3 f}{dx^3}\right|_{x_0} \frac{(\Delta x)^3}{3!} + \cdots \tag{A.39}$$

where

$$\Delta x = x - x_0$$

and the derivatives as shown are evaluated at x_0 and are hence not functions of Δx.

Some functions expanded in Taylor series:

$$\sin(x_0 + \Delta x) = \sin x_0 + (\cos x_0)\Delta x - (\sin x_0)\frac{(\Delta x)^2}{2!} + \cdots \tag{A.40}$$

$$\cos(x_0 + \Delta x) = \cos x_0 - (\sin x_0)\Delta x - (\cos x_0)\frac{(\Delta x)^2}{2!} + \cdots \tag{A.41}$$

$$e^{(x_0 + \Delta x)} = e^{x_0}\left(1 + \Delta x + \frac{(\Delta x)^2}{2!} + \cdots\right) \tag{A.42}$$

Exercise A.1

Let $x_0 = 0$ and $\Delta x = 1$ or 2; then, take successively higher number of terms in the series, and tabulate the computed results and compare with the actual value of $\sin x$. How high can the value of Δx be if we want to truncate the series after the linear term, and still want the error to be less than 5%?

A.2 Ordinary Differential Equations

A.2.1 Linear Equations

A function $f(x)$ is said to be linear if it satisfies the following condition:

$$f(ax_1 + bx_2) = af(x_1) + bf(x_2) \tag{A.43}$$

where a and b are arbitrary constants.

A differential equation is said to be linear if each of its terms is a linear function of the dependent variable and its derivatives. The above condition is called the *principle of superposition* and is the litmus test for linearity of functions, operators, and equations.

Examples of linear functions of x include $ax + b$, $\frac{dx}{dt}$, and $\int x \, dt$, where t is another variable on which x depends. Some of the nonlinear functions are x^2, $\sin x$, and e^x.

Examples of linear differential equations are

$$\frac{dx}{dt} + 3x = \sin 2t \tag{A.44}$$

$$\frac{d^2x}{dt^2} + 5t\frac{dx}{dt} + 20t^2 = 0 \tag{A.45}$$

Note that the above equations are linear in spite of the appearance of nonlinear terms because they are nonlinear functions of the independent variable, t, and not of the dependent variable x or its derivatives.

Some nonlinear differential equations are

$$\frac{dx}{dt} + 3\sin x = t \tag{A.46}$$

$$\frac{d^2x}{dt^2} + x\frac{dx}{dt} = 0 \tag{A.47}$$

$$\left(\frac{dx}{dt}\right)^2 + x = \cos t \tag{A.48}$$

A.2.2 Constant and Variable Coefficient Equations

An equation is said to be a constant coefficient equation if the coefficients multiplying the dependent variable and its derivatives are not functions of the independent variable; the equation is a variable coefficient equation otherwise.

An example of a constant coefficient equation is

$$3\frac{d^2x}{dt^2} + 5x = 2t \tag{A.49}$$

and a variable coefficient equation would be

$$\frac{dx}{dt} + (\sin t)x = 0 \tag{A.50}$$

A.2.3 Solution of Linear Constant Coefficient Equations

The following treatment is limited to linear constant coefficient ordinary differential equations; i.e., we will only deal with equations of the form

$$C_n \frac{d^n x}{dt^n} + C_{n-1} \frac{d^{n-1} x}{dt^{n-1}} + \cdots + C_1 \frac{dx}{dt} + C_0 x = F(t) \tag{A.51}$$

where C_i are constants, and $F(t)$ is some function of the variable t.

Homogeneous solution

We need to solve an nth-order linear equation of the form

$$C_n \frac{d^n x}{dt^n} + C_{n-1} \frac{d^{n-1} x}{dt^{n-1}} + \cdots + C_1 \frac{dx}{dt} + C_0 x = 0 \tag{A.52}$$

The right-hand side has been set to zero because we are seeking a homogeneous solution. Given such an equation of any order, the standard procedure is to seek a solution in the form of an exponential function; i.e., we propose a *trial solution* in the form

$$x(t) = Ae^{st} \tag{A.53}$$

where A and s are presently unknown constants.

The above function is of course only a supposition and until it satisfies the differential equation will remain only a hypothesis. Hence, we substitute this assumed form into the differential equation to find out if it is indeed the correct form. Note that we could have chosen any form of the solution at this point, say $x(t) = At^2$, but have deliberately chosen the exponential form because we happen to know that it is the right form. The students are urged to try a "wrong" form for the solution to see what happens.

Substitution of the assumed form into the differential equation yields

$$\left(s^n C_n + s^{n-1} C_{n-1} + \cdots + sC_1 + C_0 \right) Ae^{st} = 0 \tag{A.54}$$

Ignoring the possibility that $A = 0$, which would imply a trivial solution $(x(t) \equiv 0)$, we get an nth-order *algebraic* equation in s:

$$s^n C_n + s^{n-1} C_{n-1} + \cdots + sC_1 + C_0 = 0 \tag{A.55}$$

The algebraic equation obtained above is called the *characteristic equation* for the given system and, when solved, will yield n roots, called the *characteristic roots*.

Our trial solution process for the given differential equation has now yielded n solutions, each in the exponential form assumed above (with the nth root substituted for s), and such that there is no restriction on the constant A. In other words, a function of the form $A_i e^{s_i t}$ will satisfy the given differential equation, *provided* s_i is a root of the characteristic equation given above and A_i is an arbitrary constant.

The final step is to note that if we have n solutions to a linear differential equation, then by the principle of superposition, the sum of the solutions is also a solution. The following expression is then the complete homogeneous solution to the differential equation:

$$x(t) = A_1 e^{s_1 t} + A_2 e^{s_2 t} + \cdots + A_n e^{s_n t} \tag{A.56}$$

Parenthetically, we note that the above expression is correct only if the characteristic roots are all distinct, which we have implicitly assumed.

Particular solution

The particular solution is the solution to the inhomogeneous differential equation

$$C_n \frac{d^n x}{dt^n} + C_{n-1} \frac{d^{n-1} x}{dt^{n-1}} + \cdots + C_1 \frac{dx}{dt} + C_0 x = F(t) \tag{A.57}$$

where C_i are constants, and $F(t)$ is some function of the variable t. There are several methods for obtaining the particular solution; in the following treatment, however, we will use the method of undetermined coefficients which amounts to making an intelligent guess. This guess depends upon what kind of function $F(t)$ is. Here, we will look at some forms of the function relevant to vibration and discuss the solution procedure.

Constant function   If the function $F(t) = K$, a constant, the solution is sought in the form of a constant; i.e.,

$$x_p(t) = B \tag{A.58}$$

The substitution of this assumed solution form into the differential equation yields

$$x_p(t) = B = \frac{K}{C_0} \tag{A.59}$$

Exponential function   If the function $F(t) = F_0 e^{kt}$, the solution is sought in the same form; i.e.,

$$x_p(t) = B e^{kt} \tag{A.60}$$

Substitution into the differential equation yields

$$x_p(t) = \frac{F_0}{\Delta} \tag{A.61}$$

where

$$\Delta = C_n k^n + C_{n-1} k^{n-1} + \cdots + C_1 k + C_0 \tag{A.62}$$

Note that the above solution is valid only if Δ is not equal to zero ($\Delta = 0$ implies that k is one of the characteristic roots).

Initial conditions

We have found above that an nth-order linear differential equation yields a solution that has n arbitrary constants associated with it. These constants are determined by the specification of initial conditions. Hence, in the case of an nth-order equation, we need n initial conditions. Substitution of the initial conditions into the solution form will yield n linear algebraic equations, which can be solved to determine the constants. It is important to bear in mind that the initial conditions are applied to the complete solution (including the particular solution).

A.3   Linear Algebra

A *matrix* is an ordered set of quantities placed in m rows and n columns. A *determinant* of a square matrix A (i.e., one in which the number of columns = the number of rows) is a single number and is denoted by det A or $|A|$.

Let the elements of a matrix A be denoted by a_{ij}, corresponding to the quantity located at the ith row and the jth column. Then the determinant of a matrix that has only one row and one column is defined to be that element. In other words, if

$$A = [a_{11}], \ \det A = |A| = a_{11}$$

If the matrix has more than one row, then the determinant is found by the following formula:

$$\det A = \sum_{j=1}^{n} a_{1j} C_{1j}$$

where the number C_{ij} is called the *cofactor* of a_{ij}. The cofactor is given by

$$C_{ij} = (-1)^{i+j} M_{ij}$$

where M_{ij}, called the *minor* of a_{ij}, is defined to be the determinant of the matrix formed by omitting the ith row and the jth column of the $n \times n$ matrix.

As an cxample, when the matrix A has two rows and columns,

$$A = \begin{pmatrix} a_{11} & a_{12} \\ a_{21} & a_{22} \end{pmatrix}$$

Then the determinant of A is given by

$$\det A = a_{11} C_{11} + a_{12} C_{12} = a_{11} \det [a_{22}] - a_{12} \det [a_{21}] = a_{11} a_{22} - a_{21} a_{12}$$

If the matrix A has three rows and columns,

$$A = \begin{pmatrix} a_{11} & a_{12} & a_{13} \\ a_{21} & a_{22} & a_{23} \\ a_{31} & a_{32} & a_{33} \end{pmatrix}$$

The determinant of A is given by

$$\det A = \begin{vmatrix} a_{11} & a_{12} & a_{13} \\ a_{21} & a_{22} & a_{23} \\ a_{31} & a_{32} & a_{33} \end{vmatrix}$$

$$= a_{11} C_{11} + a_{12} C_{12} + a_{13} C_{13}$$

$$= a_{11} \begin{vmatrix} a_{22} & a_{23} \\ a_{32} & a_{33} \end{vmatrix} - a_{12} \begin{vmatrix} a_{21} & a_{23} \\ a_{31} & a_{33} \end{vmatrix} + a_{13} \begin{vmatrix} a_{21} & a_{22} \\ a_{31} & a_{32} \end{vmatrix}$$

$$= a_{11} a_{22} a_{33} - a_{11} a_{32} a_{23} - a_{12} a_{21} a_{33} + a_{12} a_{31} a_{23} + a_{13} a_{21} a_{32} - a_{13} a_{31} a_{22}$$

A.3.1   Matrix Algebra

Theorem A.3.1

Matrix addition is associative and commutative.

$$(A + B) + C = A + (B + C)$$
$$A + B = B + A$$

Theorem A.3.2

Matrix multiplication has the following properties:

1. It is not commutative in general.

   $$AB \neq BA$$

2. It is distributive.

   $$A(B + C) = AB + AC$$
   $$(A + B)C = AC + BC$$

3. It is associative.

   $$(AB)C = A(BC)$$

4. The cancelation law is not true in general. $AB = 0$ does not imply that either $A = 0$ or $B = 0$, and $AB = AC$ does not imply that $B = C$.

Theorem A.3.3

The transpose of a matrix has the following properties:

1. $(A + B)^T = A^T + B^T$
2. $(A^T)^T = A$
3. $(AB)^T = B^T A^T$

Definition A.3.1

A square matrix A is symmetric if $A = A^T$; i.e., if its ith row and ith column are identical.

Theorem A.3.4

If A and B are symmetric,

1. $(AB)^T \neq AB$
2. $(B^T AB)^T = B^T AB$

Definition A.3.2

A skew symmetric matrix is defined to be a matrix such that $A^T = -A$.

Theorem A.3.5

Any matrix A can be written as a sum of symmetric and skew symmetric matrices:

$$A = \frac{1}{2}(A + A^T) + \frac{1}{2}(A - A^T)$$

Definition A.3.3

A square matrix that possesses an inverse such that $A^{-1}A = AA^{-1} = I$ is said to be *nonsingular*. A square matrix that does not possess an inverse is said to be *singular*.

Theorem A.3.6

If A and B are nonsingular, then

1. $(A^{-1})^{-1} = A$
2. $(AB)^{-1} = B^{-1}A^{-1}$

A.3.2   Linear Algebraic Equations

Theorem A.3.7

Let $Ax = b$ describe m equations in m unknowns. Then there exists precisely one solution x for every b if and only if $x = 0$ is the unique solution to $Ax = 0$. In other words, either $Ax = b$ has precisely one solution for every b or $Ax = 0$ has a nonzero solution x but not both can hold simultaneously.

Definition A.3.4

A *block diagonal* matrix is a square "partitioned" matrix in which the diagonal elements are square matrices and all other elements are null matrices.

Theorem A.3.8

A block diagonal matrix is nonsingular if and only if all the diagonal matrix elements are nonsingular.

Definition A.3.5

A set of vectors $v_1, v_2, \ldots, v_n$ is said to be *linearly dependent* if there exist numbers $a_1, a_2, \ldots, a_n$, not all zeros, such that

$$a_1 v_1 + a_2 v_2 + \cdots + a_n v_n = 0.$$

If the set is not linearly dependent, then it is said to be *linearly independent*.

A.3.3   The Eigenvalue Problem

Definition A.3.6

The polynomial $f(\lambda) = \det(A - \lambda I)$ is called the *characteristic polynomial* and the equation $f(\lambda) = 0$ is called the *characteristic equation* of the $n \times n$ matrix A. The *eigenvalues* of A are the scalars λ for which $Ax = \lambda x$ possesses nonzero solutions. The corresponding nonzero solutions x are the *eigenvectors* of A. The eigenvalues and eigenvectors together are called the *eigensystem* of A.

Theorem A.3.9

Let A be an $n \times n$ matrix.

1. There exists at least one eigenvector associated with each distinct value for an eigenvalue λ. If A and λ are real, then the eigenvector may be taken to be real.

2. If $\lambda_1, \lambda_2, \ldots, \lambda_n$ is a collection of distinct eigenvalues, and if $x_1, x_2, \ldots, x_n$ form a set of associated eigenvectors, then the set of eigenvectors is linearly independent.

Definition A.3.7

If there exists a nonsingular matrix P such that $P^{-1}AP = B$, then the matrix B is said to be *similar* to A, and we say that B is obtained from A by means of a *similarity transformation*.

Theorem A.3.10

1. Similar matrices have the same characteristic equation and the same eigenvalues.

2. If $P^{-1}AP = B$ and x is an eigenvector of A corresponding to the eigenvalue λ, then $P^{-1}x$ is an eigenvector of B corresponding to λ.

A good reference for linear algebra is *Applied Linear Algebra* [30].

A.4 Fourier Series

In nature as well as in engineered systems, we often experience periodic phenomena. It may be something very regular such as a simple sine wave, or something more complex as shown in Fig. A.2. In all of these cases, thanks to Fourier, we have a very useful theorem that goes as follows.

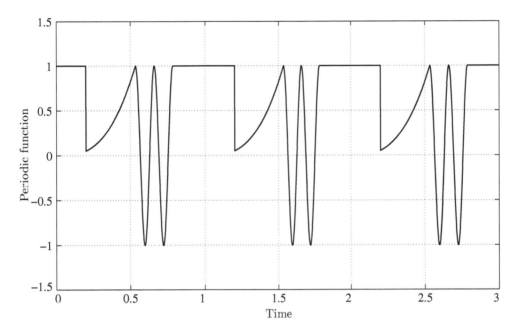

Figure A.2: *A periodic function with a period of 1 s*

Theorem A.4.1

Any periodic motion can be represented by a series of of sines and cosines that are harmonically related.

Although it sounds somewhat simplistic, this is an extremely useful theorem that comes up in many surprising places both in theory and practice, not only in vibration but also in diverse areas such as heat transfer, fluid mechanics, electrical systems, magnetism, economics, biological systems, etc. Mathematically, this theorem states the following and is valid for any function, $x(t)$, that is periodic with period T:

$$x(t) = \frac{a_0}{2} + \sum_j a_j \cos j\omega t + \sum_j b_j \sin j\omega t \tag{A.63}$$

where

$$\omega = \frac{2\pi}{T} \tag{A.64}$$

Note that you may see some other conventions in the literature, notably, with the constant coefficient; some authors use a_0 instead of $a_0/2$ that you see here. Of course, that is no problem as long as you stick to whichever definition you start with as the analysis proceeds. Note also that we say that the frequencies are harmonically related because the second term has a frequency that is twice the first, the frequency of the third is three times the first, and so on. In other words, they are related by integers.

Next, we need to derive expressions to determine the *Fourier coefficients*, a_j and b_j. To do this we use what is called the *orthogonality property* of the sine functions. We multiply both sides by $\cos k\omega t dt$, and integrate from 0 to T, where k is an integer.

$$\int_0^T x(t) \cos k\omega t \, dt = \int_0^T \frac{a_0}{2} \cos k\omega t \, dt$$

$$+ \int_0^T \sum_j a_j \cos j\omega t \cos k\omega t \, dt$$

$$+ \int_0^T \sum_j b_j \sin j\omega t \cos k\omega t \, dt \tag{A.65}$$

Now, using the formulae given elsewhere (Section A.2), we note that *all* the terms on the right-hand side go to zero except *one*: the one for which $k = j$. Then,

$$\int_0^T x(t) \cos k\omega t \, dt = \int_0^T \sum_j a_j \cos^2 k\omega t \, dt = \frac{a_j T}{2} \tag{A.66}$$

This leads to an expression to determine a_j (we changed k back to j, an arbitrary index):

$$a_j = \frac{2}{T} \int_0^T x(t) \cos j\omega t \, dt \tag{A.67}$$

Similarly, the other coefficients can be calculated:

$$a_0 = \frac{2}{T} \int_0^T x(t) \, dt \tag{A.68}$$

$$b_j = \frac{2}{T} \int_0^T x(t) \sin j\omega t \, dt \tag{A.69}$$

A.4.1 Even and Odd Functions

A function is said to be odd if

$$f(t) = -f(-t) \tag{A.70}$$

Or, an odd function is antisymmetric about the origin. Examples are t and $\sin \omega t$.

A function is said to be even if

$$f(t) = f(-t) \tag{A.71}$$

Or, an even function is symmetric about the origin. Examples are t^2 and $\cos \omega t$. Note that functions do not have to be one or the other; they can be neither odd nor even; for example, e^{at}.

Some basic mathematical doodling will convince you that the following is true:

$$\text{Odd function:} \quad \int_0^T f(t) \cos j\omega t \, dt = 0, \, \forall j \tag{A.72}$$

$$\text{Even function:} \quad \int_0^T f(t) \sin j\omega t \, dt = 0, \, \forall j \tag{A.73}$$

In other words, for an odd function, $a_j = 0$ for all j, and for an even function, $b_j = 0$.

A.5 Laplace Transforms

The Laplace transform of a function of time is defined by the following integral:

$$\mathcal{L} = \int_0^\infty f(t)\, e^{-st}\, dt \tag{A.74}$$

where s, called the Laplace variable, is complex. It is assumed that $f(t)$ is zero for $t < 0$.

A.5.1 Some Properties

Laplace transforms have some very useful properties. In the following, a, a_1, etc., are constants (not functions of time). Upper case denotes a Laplace transformed variable; for example,

$$F_1(s) \stackrel{\text{def}}{=} \mathcal{L}[f_1(t)] \tag{A.75}$$

- Scaling

$$\mathcal{L}[af(t)] = aF(s)$$

- Linearity

$$\mathcal{L}[a_1 f_1(t) + a_2 f_2(t)] = a_1 F_1(s) + a_2 F_2(s)$$

- Shift (in time)

$$\mathcal{L}[f(t-a)] = e^{-as} F(s)$$

- Shift (in Laplace variable)

$$\mathcal{L}[f(t)e^{-at}] = F(s+a)$$

- Initial value theorem

$$f(0+) = \lim_{s \to \infty} [sF(s)]$$

- Final value theorem

$$\lim_{t \to \infty} f(t) = \lim_{s \to 0} [sF(s)]$$

- Differentiation

$$\mathcal{L}\left[\frac{df}{dt}\right] = sF(s) - f(0)$$

$$\mathcal{L}\left[\frac{d^2 f}{dt^2}\right] = s^2 F(s) - \dot{f}(0) - sf(0)$$

- Integration

$$\mathcal{L}\left[\int_0^t f(\tau)d\tau\right] = \frac{F(s)}{s}$$

Table A.1 lists some typical functions of time and their Laplace transforms.

$f(t)$	$F(s)$
$\delta(t)$ (unit impulse)	1
1 (unit step)	$\dfrac{1}{s}$
t (unit ramp)	$\dfrac{1}{s^2}$
t^2 (parabolic)	$\dfrac{2}{s^3}$
t^n	$\dfrac{n!}{s^{n+1}}$
e^{-at}	$\dfrac{1}{s+a}$
$t^n e^{-at}$	$\dfrac{(n)!}{(s+a)^{n+1}}$
$\sin bt$	$\dfrac{b}{s^2+b^2}$
$\cos bt$	$\dfrac{s}{s^2+b^2}$
$e^{-at}\cos bt$	$\dfrac{s+a}{(s+a)^2+b^2}$
$e^{-at}\sin bt$	$\dfrac{b}{(s+a)^2+b^2}$
$e^{-at}f(t)$	$F(s+a)$
$\dfrac{df}{dt}$	$sF(s)-f(0)$
$\dfrac{d^2 f}{dt^2}$	$s^2 F(s)-sf(0)-\left.\dfrac{df}{dt}\right|_{t=0}$
$\dfrac{d^n f}{dt^n}$	$s^n F(s)-\displaystyle\sum_{k=0}^{n-1} s^{n-k-1}\left[\dfrac{d^k f(t)}{dt^k}\right]\Bigg|_{t=0}$
$\displaystyle\int_0^t f(t)\,dt$	$\dfrac{F(s)}{s}+\dfrac{1}{s}\left[\int f(t)\,dt\right]\big|_{t=0}$
$f(t)=f(t+T)$ (periodic)	$\dfrac{F(s)}{1-e^{-sT}}$, $f(t)$ written using step functions

Table A.1: *Laplace transform pairs*

Problems

A.1 A function $x(t)$ is of the form $C\sin(\omega t+\phi)$.

   (a) $C=5$, $\omega=10$ rad/s, $\phi=0$.

   (b) $C=5$, $\omega=1$ rad/s, $\phi=0$.

   (c) $C=3$, $\omega=10$ rad/s, $\phi=0$.

(d) $C = 5$, $\omega = 10$ rad/s, $\phi = \pi/2$.

(e) $C = 5$, $\omega = 10$ rad/s, $\phi = 30°$.

(f) $C = 5$, $\omega = 100$ rad/s, $\phi = -30°$.

For each of the sets of numerical values given above, use MATLAB and do the following.

(a) Plot $x(t)$ over at least three cycles.

(b) Determine the period from the graph and compare it to the analytical expression.

(c) Create one plot with (a) and (b) on the same scale.

(d) Create one plot with (a) and (e) on the same scale.

(e) Comment on the effect of the magnitudes of the amplitude, frequency and phase on the nature of the graphs.

A.2 Show that $A \sin(\omega t) + B \cos(\omega t) = C \sin(\omega t + \phi)$. Express C and ϕ in terms of A and B.

A.3 Show that $A \sin(\omega t) + B \cos(\omega t) = C \cos(\omega t + \phi)$. Express C and ϕ in terms of A and B.

A.4 Given $x(t) = 3 \sin 6t - 4 \cos 6t$, express it as $C_1 \sin(\omega t + \phi_1)$ and as $C_2 \cos(\omega t + \phi_2)$. Then, verify (using MATLAB) by sketching all the three expressions.

A.5 For each of the following cases, express as a sine wave and determine the frequency of oscillation. Sketch each of them using MATLAB:

(a) $x(t) = (2 + 5i)e^{i3t} + (2 - 5i)e^{-i3t}$

(b) $x(t) = (3 + 2i)e^{i6t} + (3 - 2i)e^{-i6t}$

(c) $x(t) = 5e^{2t} + 3e^{-2t}$

(d) $x(t) = (2 + 3i)e^{(-1+5i)t} + (2 - 3i)e^{(-1-5i)t}$

A.6 Expand $\sin(x)$ in a Taylor series (with two nonzero terms) about the following equilibrium points:

(a) $x_0 = 0$

(b) $x_0 = 30°$

(c) $x_0 = 90°$

A.7 Expand $\cos(x)$ in a Taylor series (with two nonzero terms) about the following equilibrium points.

(a) $x_0 = 0$

(b) $x_0 = 30°$

(c) $x_0 = 90°$

A.8 Linearize $\sin x \cos^2 x$ using Taylor series about $x_0 = 45°$. How much can x deviate from x_0 before the error becomes 10%?

A.9 Solve the following differential equations analytically. Compare and sketch the solutions; use initial conditions: $x(0) = 1$, $\dot{x}(0) = 0$:

(a) $\ddot{x} + 9x = 0$

(b) $\ddot{x} - 9x = 0$

A.10 Solve the following differential equations:

(a) $\ddot{x} + 4x = 0$

(b) $\ddot{x} + 4x = \sin t$

(c) $\ddot{x} + 4x = \sin 2t$

A.11 Solve the following differential equations:

(a) $\ddot{x} + 10\dot{x} + 100x = 0$

(b) $\ddot{x} + 10\dot{x} + 100x = \sin 3t$

(c) $\ddot{x} + 10\dot{x} + 100x = \sin 10t$

B An Introduction to MATLAB

If you don't know how to do something, you don't know how to do it with a computer.

—Anonymous

Some General Tips

Typing **help** will get you a list of available built-in and other functions. **help eig**, for example, will get you help on a specific function, in this case, **eig**, for solving eigenvalue problems. Highly recommended: if you are a new user, **demo** will take about 15 min and will be very useful. You can try several online tutorials at http://www.mathworks.com.

If you are in the middle of a session, and want to take a break, you can type **save filename**. This will save the names and values of every variable in a file by name "filename.mat"; the next time you get on to MATLAB, you can start where you left off by typing **load filename**. Type **help save** for more options.

The most powerful feature of MATLAB is in "m-files." These are command script files and typically contain just the usual MATLAB commands you would type in an interactive session. Note that these files have to be in ASCII, and hence you have to edit them using a text editor. The MATLAB editor is pretty good; type "edit" in MATLAB to invoke the editor.

Change to the directory where you have m-files by using "cd d:\test", for example. You can check the files in the current directory by typing "what". Variables in the workspace are checked by using "who". "lookfor" is a useful command: "lookfor plot" will list all commands that have anything to do with plot.

Some of the most common errors (along with annoying beeps) you will probably encounter are with respect to matrix operations. Matrices are easy to handle in MATLAB; however, you will need to know some essentials (the demo program can be quite useful here). When you need to multiply two matrices, "a" and "b," it is easy enough; you type $c = a * b$, and the output will be the matrix product. However, frequently, you do not need the matrix product; instead, you actually want to multiply every element of "a" by the corresponding element of "b." To accomplish that, you need to type **a .* b**. Similarly, if you want element by element division, type **a ./ b**.

If you get an error with a statement involving a vector that you cannot make sense of, as a possible quick-fix, try the same statement with its transpose; **a'** will get you the transpose of "a."

- MATLAB is case sensitive. So, "A" and "a" are different. All built-in commands have to be lower case.
- The symbol for power is ˆ; for example, the equivalent of x**3 of FORTRAN is **xˆ3** in MATLAB.
- The DO-loop structure in MATLAB is done using the **for** statement. There is an **if then else end** structure if you need to use it; in addition, there is also a **while ... end** statement. (Try **help for**.)
- No statement labels or statement numbers are necessary (and are not permitted).
- The number π is predefined; so are the imaginary number, i, and infinity, ∞.
- In m-files, you might want to put a semicolon at the end of your statements; such statements will not be echoed by MATLAB.
- Remember to comment your m-files liberally (they are after all programs); a "%" will cause MATLAB to treat the rest of that line as a comment, and will hence be ignored.
- A continuation line in an m-file is achieved by putting "..." at the end of the line.
- When viewing plots, it may be useful to use the **pause** command, which will freeze the screen until you hit a key. Also **hold on, hold off** are useful if you want to see one plot on top of another. "figure" will create a new figure window. If you create too many figure windows, they can all be closed by typing "close all" in MATLAB (rather than using the mouse).
- MATLAB has a subroutine structure: the syntax is **[y1, y2] = function(x1,x2,x3)**. Just as in FORTRAN and C, the variables within this function are local and do not interfere with the rest of the program. However, you can make variable values available for use within the **function** by using the **global** command in a main program (or m-file) that calls it; the **global** command works identical to the COMMON statement in FORTRAN.

Some Commands to Play With

```
help
help plot
3*2
sin(pi/2)
x=3;
y=8*x^2+5;
% predefined numbers
i, pi, 1/0
t=(0:.1:3*pi);
y=sin(t);
plot(t,y); grid;
xlabel('Time');
ylabel('x(t)');

figure;
plot(t,cos(t)); grid;
xlabel('Time'); ylabel('cos(t)');

clear
t=(0:.1:3*pi);
y=sin(t) .* t .* cos(t);
```

```
plot(t,y); grid;
xlabel('t');
ylabel('y(t)');
y= sin(t) ./ cos(t);
plot(t,y); grid;
xlabel('t'); ylabel('y(t)');

% matrices
a = [1 -1; 2 3]; b = [2 0;-3 5]; c = a * b; d = inv(a) * b;
determinant = det(a);

% loops
for i=1:1:5,
  x(i) = i;
end;

for i=-3:2:20,
  if (i<0)
    x(i) = - 2*i;
  else
    x(i) = i;
  end;
end;
```

Numerical Solution of a Nonlinear Algebraic Equation

Main program:

```
global m

m = 2;
initial_guess = .8;
soln = fzero('test',.5);


-------------------------------------
file: test.m
function y=test(x)
% supply the nonlinear equation

  global m

  y = x.^2 - m*sin(x) .* cos(x).^2;
-------------------------------------
```

Solution of a Single Differential Equation Using MATLAB

First prepare a file (in ASCII format) to define the equations to be integrated. This "subroutine" should take the independent variable (time) and the vector of dependent variables (x) as input and provide the vector of derivatives (xdot) as the output. Example file to solve the simple pendulum motion $(m\ell^2\ddot{\theta} + mg\ell\sin\theta = 0)$ is as follows.

```
function xdot=eqns(time,x)
% supply the EOM for the dynamics of a simple pendulum

global mass gravity length

xdot(1,1) = x(2);
xdot(2,1) = (-mass*gravity*length*sin(x(1)))/(mass*length^2);
```

Save the file as "name.m" (here, eqns.m) in the hard disk in the current directory. Get into MATLAB, and specify the initial conditions, and the range of the independent variable (tinit, tfinal). Also, the tolerance for integration can be specified (optionally).

```
global mass gravity length
mass=1.0;
gravity=9.81;
length=0.2;
tinit = 0.0;
tfinal = 3.0;
tspan=[tinit tfinal];
inicon = [0.1 0.0]'; %  ' denotes transpose
```

The integration is carried out by typing the following statement.

```
[time, soln] = ode23('eqns',tscan,inicon);
```

In this example, a plot of theta was obtained by the following command.

```
plot(time,soln(:,1));
```

Typing "help ode23" will get you specific help on using this routine. You could also use "ode45," which has a higher accuracy than "ode23." Also, type "help odeset" to get help on setting options (such as tolerance and stopping criteria). There also numerous examples you can browse such as "orbitode," "rigidode," and "vdpode."

Numerical Solution of Multiple ODEs

Suppose we need to solve the following system of differential equations:

$$M\ddot{x} + C\dot{x} + Kx = F(t) \tag{B.1}$$

We will use an example of a two-DOF system with an arbitrary force acting on the first degree of freedom. The matrices would be written up as shown in the following MATLAB code (Listing B.1).

Listing B.1: *Matrix definitions in MATLAB*

```
% matrices.m
  M = [1  0; ...
       0  3];
  K = [ 3  -2; ...
       -2   5]*1e+04;
  C = [ 1  -1; ...
       -1   3]*1e+01;
```

First, we need to arrange them in a standard form of first-order ODEs. In order to do that, we define a new coordinate vector

$$y = \begin{bmatrix} \dot{x} \\ x \end{bmatrix} \tag{B.2}$$

Then, the equations can be written in terms of this new vector as follows:

$$\begin{bmatrix} M & C \\ 0 & I \end{bmatrix} \begin{bmatrix} \ddot{x} \\ \dot{x} \end{bmatrix} + \begin{bmatrix} 0 & K \\ -I & 0 \end{bmatrix} \begin{bmatrix} \dot{x} \\ x \end{bmatrix} = \begin{bmatrix} F \\ 0 \end{bmatrix} \tag{B.3}$$

Note that the first equation is the same as the original equation and the second is simply an identity statement $(\dot{x} - \dot{x} = 0)$.

The equations are then in the form

$$A_1 \dot{y} + A_2 y = Q \tag{B.4}$$

Next, we get an explicit expression for $\dot{y}$:

$$\begin{aligned} \dot{y} &= -A_1^{-1} A_2 y + A_1^{-1} Q \\ &= Ay + BQ(t) \end{aligned} \tag{B.5}$$

Now, the equations are in a form suitable for computer solution. The MATLAB code to reorganize the matrices would be as shown in Listing B.2.

Listing B.2: *Reordering matrices in the first-order form*

```
% reorder .m
  A1=[M            C;  ...
         zeros(2)  eye(2)];
  A2=[ zeros(2)  K;
         −eye(2)     zeros(2)];
  B=inv(A1);
  A=−inv(A1)∗A2;
```

Suppose we wish to determine the response of the system to a force (acting only on the first degree of freedom) that is a function of time as shown in Fig. B.1.

The MATLAB description of the force in the form of a **function** is shown in Listing B.3.

Listing B.3: *Force defined in a MATLAB function*

```
% forcefcn .m
function [f]=forcefcn(t)

  tau1=0.5;
  tau2=1.5;
  tau3=2.0;
  freq=0.5∗pi/(tau1);

  if (t<tau1)
     f=3∗sin(freq∗t);
  elseif (t<tau2)
     f=3;
  elseif (t<tau3)
     f=3∗sin(freq∗(t−tau2)+pi/2);
  else
     f=0;
  end;
```

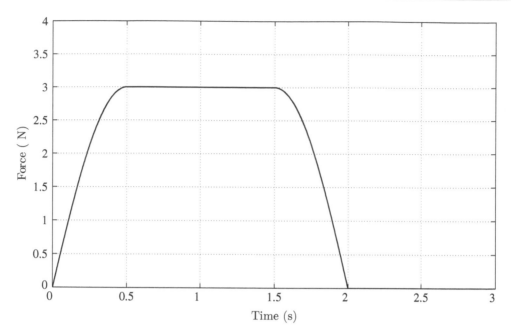

Figure B.1: *Excitation, F_1, as a function of time*

The equations of motion need to be specified in a **function** routine as shown in Listing B.4. Note that the excitation vector, Q, is all zeros except for the first element that corresponds to F_1 acting on the first degree of freedom. This routine is called by the MATLAB ODE solver and in turn calls the routine, **forcefcn** to evaluate the force at the current instant of time.

Listing B.4: *ODEs defined in a MATLAB function*

```
% equations.m
  function [ydot]=equations(t,y)
% supply the equations to be solved numerically

  global A B

% determine the excitation at the current time
  f=forcefcn(t);
% load it into the force vector
  F=[f; 0];
  Q=[F; zeros(2,1)];
% the ode's in first order form
  ydot = A*y + B*Q;
```

The main program that solves the equation would be as shown in Listing B.5.

Listing B.5: *Multiple DOF ODE solution*

```
% mdofode.m
% example program to solve multiple ode's by numerical integration

  global A B

% create matrices
  matrices;
```

```
% reorder into first order form
  reorder;
% call an ode solver
  initial_cond=zeros(4,1);
  tspan=[0 3];
  [t,y] = ode45('equations',tspan,initial_cond);
% plot
  forceplot;
  figure;
  plots;
```

Note that in the final solution, $y(1) = \dot{x}(1), y(2) = \dot{x}(2), y(3) = x(1)$, and $y(4) = x(2)$. A plot of the two degrees of freedom is shown in Fig. B.2.

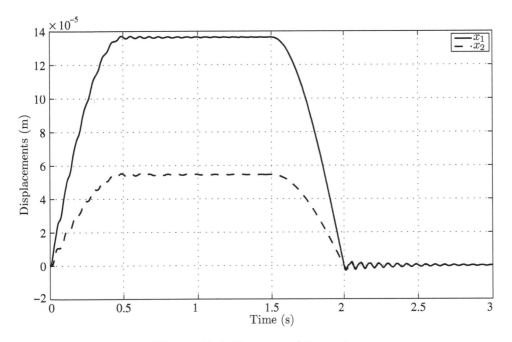

Figure B.2: *Response of the system*

The MATLAB code to produce this plot is as shown in Listing B.6.

Listing B.6: *MATLAB function to produce the plots*

```
% plots.m

  plot(t,y(:,3),t,y(:,4),'LineWidth',vibelinethickness)
  grid on;
  legend('x1','x2');
  xlabel('Time_(s)');
  ylabel('Displacements_(m)');
```

C The Greek Alphabet

ΔΟΣ ΜΟΙ ΠΟΕ ΣΤΩ ΚΑΙ ΚΙΝΩ ΤΗΝ ΓΗΝ*

—Archimedes (280-211 BC)

In this text as well as in many engineering and scientific contexts, we liberally use the Greek alphabet, which nicely sets apart the symbols of important quantities from the normal English text written in the Roman alphabet. The students, as well as a surprising number of professionals, seem to have trouble recognizing these symbols (often, we have seen omega (ω) being read as the English 'w'). Also, there is considerable unfamiliarity with the upper case characters. Hence, in the interest of making that passage easier, as well as with the secondary objective of advancing the knowledge of one of the great classical languages of the world, we present below the Greek alphabet along with a notational guide for commonly used quantities. Note that even if a common usage is not listed here, it can prove to be handy when you need to define your own symbols.

Upper case	Lower case	Name	Common usage	Upper case	Lower case	Name	Common usage
A	α	alpha	angle	Ξ	ξ	xi	
B	β	beta	angle	O	o	omicron	
Γ	γ	gamma		Π	π	pi	the famous ratio!
Δ	δ	delta					
E	ϵ	epsilon	small parameter	P	ρ	rho	density
Z	ζ	zeta	damping ratio	T	τ	tau	time constant
H	η	eta	loss factor	Υ	υ	upsilon	
Θ	θ	theta	angle	Φ	ϕ	phi	angle
I	ι	iota		X	χ	chi	a statistical distribution
K	κ	kappa					
λ	Λ	lambda	eigenvalue	Ψ	ψ	psi	angle
M	μ	mu	viscosity	Ω	ω	omega	frequency
N	ν	nu					

Table C.1: *The Greek Alphabet*

---

*Give me a lever long enough and a fulcrum on which to place it, and I shall move the world.

Glossary

Ah! Don't say that you agree with me. When people agree with me I always feel that I must be wrong.

—Oscar Wilde

- **Acceleration:** Second derivative of displacement.
- **Accelerometer:** A transducer that measures acceleration.
- **Amplitude:** Maximum value of a harmonic quantity; for example, displacement or force.
- **Base excitation:** The vibration isolation problem that arises from having to isolate systems from surrounding vibration.
- **Characteristic equation:** The algebraic equation that results from the eigenvalue problem (for example, Eqs. 7.14 and 9.18).
- **Characteristic roots:** Eigen-solutions, or solutions of the characteristic equation.
- **Conservative force:** A force whose work done does not depend on the path, but only on the end points; a non-dissipative force.
- **Conservation of energy:** The principle that all mechanical energy is conserved in a system if there are only conservative forces acting on it.
- **Continuous systems:** A model that includes the distributed property of systems which leads to partial differential governing equations in space and time. Distinct from *discrete systems*. Also called *distributed parameter systems*.
- **Convolution integral:** An integral that provides the response of a dynamic system to an arbitrary force (Eq. 6.24). Also called *Duhamel's integral*.
- **Coordinate:** A number used to describe the motion of a dynamic system; the number of independent coordinates should be equal to the number of degrees of freedom.
- **Coulomb friction:** A frictional mechanism that occurs when two dry surfaces slide against each other. Also called dry friction or sliding friction.
- **Critically damped system:** A system with a moderate amount of damping that results in a non-oscillatory response and is characterized by the nondimensional damping ratio ζ being exactly equal to 1.
- **Damping:** Also called friction; the mechanism by which energy of vibration is dissipated (usually).

- **Damping ratio:** A non-dimensional ratio, ζ, that characterizes strength of damping in the system; a critical value of $\zeta = 1$ separates dramatically different behavior in a damped system.
- **Degree of freedom (DOF):** The number of independent coordinates necessary to describe the motion completely.
- **Diagnostics:** The science of determining the state of a machine and to identify its potential faults from a measurement of its vibrational response.
- **Discrete systems:** An approximate model that assumes that the displacement variables are only functions of time. Also called *lumped parameter systems* as the assumption implies that properties such as stiffness and mass are spatially discrete. Distinct from *continuous or distributed parameter systems.*
- **Displacement:** Position of a particle or a point on a rigid body as a function of time.
- **Dry friction:** See Coulomb friction.
- **Dynamic magnification factor:** In harmonic excitation of a SDOF system, the ratio of the amplitude of dynamic response to an equivalent static displacement (Eq. 5.15).
- **Dynamic vibration absorber:** The concept of adding a tuned mass-spring oscillator to a vibrating system to reduce the vibration amplitudes due to harmonic excitation.
- **Dynamics:** Study of bodies in motion.
- **Eigenvalue problem:** An algebraic problem that arises in the analysis of dynamic systems. Solution yields eigenvalues and eigenvectors. The eigenvalues are related to the natural frequencies of vibration, and the eigenvectors are associated with the mode shapes.
- **Equation of motion (EOM):** The governing equation (often differential) whose solution would describe a motion variable.
- **Equilibrium position(s):** Static positions where a system could remain at rest; determined by setting all time derivatives and time functions to zero in an Equation of Motion.
- **Excitation:** Force or moment acting on a mechanical system (usually dynamic).
- **Excitation frequency:** Same as **forcing frequency**.
- **Exciter:** An experimental device that is used to deliver a force of known characteristics to a dynamic system in order to characterize the system.
- **Expansion theorem:** This theorem states that any possible motion of a linear vibrating system can be expressed as a linear combination of eigenvectors (Eq. 7.127). This result follows from the property of linear independence of eigenvectors.
- **Final value theorem:** A theorem that relates the final value of a function to a limit in the Laplace domain (Section 6.3.1). Useful to find steady state values of response variables.
- **Forced response:** The response of a dynamic system to an external force, determined by obtaining the particular solution to the governing differential equation. Distinct from free response.
- **Forcing frequency:** Frequency of a harmonic force or moment (excitation).
- **Fourier series:** An infinite series of sines and cosines that can be considered to be mathematically equivalent to a periodic function; follows from Fourier theorem.
- **Free body diagram (FBD):** A depiction of a system separated from its environment showing all the forces acting on it.
- **Free response:** The dynamic response of a system without any external forces. Computed using the homogeneous solution. Distinct from forced response.
- **Frequency:** Number of times per unit time a periodic quantity repeats itself.
- **Frequency ratio:** Often, the ratio of forcing frequency to a natural frequency (ω/ω_n).

- **Frequency response function:** The response of a dynamic system to a harmonic force plotted as the amplitude of the displacement versus the forcing frequency.
- **Galilean reference frame:** Also called Newtonian reference frame; a frame of reference fixed with respect to the center of the universe. In practical terms, this is often a fixed point in the lab.
- **Harmonic:** Any function that is expressible as $A \sin (\omega t + \phi)$; A - amplitude, ω - frequency, ϕ - phase. A harmonic function is always **periodic** (but the converse is not true).
- **Homogeneous solution:** The solution of an ordinary differential equation after removing all forces are constant or depend upon time. The result is also called complementary solution and represents the free response of a dynamic system.
- **Hysteresis:** Characterized by different behavior in loading and unloading of a material as evidenced by a loop in the stress-strain curve (as opposed to a single line). Called hysteretic damping and leads to loss of energy given by the area enclosed by the hysteresis loop.
- **Impulse:** In dynamics, it is the integral of force over a specified period of time. An impulse function in general can be defined to be a quantity that lasts for an infinitesimal amount of time, but whose impulse magnitude is finite.
- **Impulse response function:** Response of a dynamic system to a unit impulse function.
- **Impulsive force:** An impulsive force is a large force that acts over a short amount of time, and is typically encountered in collisions.
- **Kinematics:** Study of the geometry of motion (i.e., displacement, velocity, and acceleration).
- **Kinetic energy, law of:** Work done by a force from point 1 to point 2 is equal to the change in kinetic energy (Eq. 2.61).
- **Kinetics:** Study of forces that cause motion.
- **Laplace transform:** An integral transform that converts a function of time to that of another variable (s) called the Laplace variable (Eq. 6.74). Useful technique to solve linear differential equations by converting them to algebraic equations.
- **Linearization:** The process of converting a nonlinear equation to an approximate linear one by making approximations of small motions about equilibrium positions. Often carried out by using Taylor series.
- **Logarithmic decrement:** Logarithm of the successive amplitudes of damped motion; used to estimate damping from experiment.
- **Mass moment of inertia:** A mass distribution property defined by integrating the product of mass and the square of the distance from a coordinate axis.
- **Material damping:** damping that occurs within the structure of a vibrating material. Also called internal friction or hysteretic damping.
- **Modal analysis:** A procedure by which general free and forced response of an MDOF system can be obtained using a linear combination of eigenvectors (Section 7.4.4).
- **Modal damping:** The concept of an equivalent damping ratio for each mode in an MDOF system analogous to the damping ratio in a SDOF system.
- **Mode shape:** A plot of the eigenvector displaying the relative magnitudes of the different coordinates oscillating at one of the natural frequencies in a particular mode.
- **Natural frequencies:** Frequencies of free (or unforced) vibration.
- **Nonlinear system:** A system that does not obey the principle of superposition.
- **Normalization:** The process of specifying a multiplier for an eigenvector such that its weighted inner product is equal to 1 (Eq. 7.106). A normalization is necessary since all eigenvectors are arbitrary to a multiplicative constant.

- **Numerical simulation:** Use of a computer to solve a system of equations as opposed to an *analytical* solution.
- **Orthogonality, principle of:** A special property of eigenvectors (Eq. 7.123 and Eq. 7.124) and eigenfunctions that enables one to decouple the differential equations of motion.
- **Overdamped system:** A system with a large amount of damping that results in a non-oscillatory response which is exponentially damped. Characterized by the nondimensional damping ratio ζ being larger than 1.
- **Particle:** A point mass; an idealization that is assumed to have mass but not to occupy any space.
- **Particular solution:** The solution of a nonhomogeneous differential equation; leads to the forced response in a dynamic system.
- **Percent overshoot:** A measure of the underdamped response of a system that quantifies the peak value in transient response (Eq,. 6.143).
- **Period:** The amount of time it takes to complete one cycle when a quantity is repeating in a regular fashion.
- **Periodic:** Any function that repeats itself with time.
- **Phase lag:** Often, this refers to the phase difference between the vibrational displacement and the force in harmonic excitation.
- **Phasor:** A representation of a harmonic function (such as displacement, velocity or force) in the form of a rotating vector on a complex plane.
- **Polynomial eigenvalue problem:** A general form of the eigenvalue problem that results from having a damped system (Eq. 7.184).
- **Potential energy:** Defined for a conservative force as the potential of that force to do work; defined by Eq. 2.63.
- **Rayleigh damping:** A kind of damping mechanism which results in a damping matrix that is proportional to a combination of stiffness and mass matrices (Eq. 7.174). Also called proportional damping.
- **Resonance:** The phenomenon of high vibrational amplitudes in harmonic excitation when the forcing frequency is close to the natural frequency.
- **Response:** Displacement as a function of time.
- **Response spectrum:** In periodic vibrational analysis, plot of the amplitudes of the harmonic responses versus the harmonic.
- **Rigid body:** A body that is rigid (not flexible); an idealization that has mass and can translate as well as rotate.
- **Root mean square (RMS):** The square root of the average value of a square of a function over a period of time. Defined in order to gauge the magnitude of an oscillating function whose average value itself would be zero, but whose RMS value would be nonzero and would hence give us a good estimate.
- **Rotating unbalance:** A pervasive problem of vibration problems caused in rotating systems due to a mass imbalance that becomes especially acute at large speeds.
- **Rotor dynamics:** The discipline of dynamic analysis and design of rotating systems such as turboshafts, engines, etc.
- **Separation of variables:** A method of solving partial differential equations by separating the functions of spatial coordinate (x) and time (t), and setting them equal to a separation constant.

- **Settling time:** A measure of the transient response of a dynamic system to determine how long it takes for a dynamic system to come within 2% of its steady state value (Eq. 6.144).
- **Stabilization:** Converting an unstable system to a stable one by modifying it in some manner.
- **Stability:** The property of a system that determines if the system returns to an equilibrium position after being disturbed.
- **Steady state response:** The response of a dynamic system after an initial transient period; steady state response can be constant or time-varying; in the latter case, it would need to be periodic.
- **Step response:** Response of a dynamic system to a step function.
- **Superposition, principle of:** For a dynamical system, it is the principle that adding two responses from two individual forces would be the same as the result when both forces act simultaneously.
- **Transducer:** A device that converts signals from one form to another; here, it is often a sensor that measures and converts displacement, velocity, acceleration or force into an electrical quantity such as voltage or current.
- **Transfer function:** The ratio of Laplace transform of an output to that of an input in a dynamic system; often, the output is displacement or velocity, and the input is a force (Eq. 6.86).
- **Transient response:** The response of a dynamic system in the short term; distinct from steady state response.
- **Transmissibility ratio:** In vibration isolation, a nondimensional ratio defined by the ratio of amplitude of force transmitted to that of the disturbing force.
- **Underdamped system:** A system with a small amount of damping that results in an oscillatory response which is exponentially damped. Characterized by the nondimensional damping ration ζ being less than 1.
- **Velocity:** Time derivative of displacement.
- **Vibrometer:** A sensor that measures vibrational quantity, often displacement.
- **Work:** A scalar quantity defined by Eq. 2.58.

Bibliography

We are like dwarfs on the shoulders of giants, so that we can see more than they, and things at a greater distance, not by virtue of any sharpness of sight on our part, but because we are carried high and raised up by their giant size.

—Bernard of Chartres (12th century)

[1] Dugas, R., 1988. *A History of Mechanics*. Dover Publications, New York, NY.

[2] Burton, R., 1958. *Vibration and Impact*. Addison-Wesley Publishing House, Boston, MA.

[3] Dimarogonas, A. D., and Haddad, S., 1992. *Vibration for Engineers*. Prentice Hall, Englewood Cliffs, NJ.

[4] Rayleigh, L., 1945. *Theory of Sound*. Dover Publications, New York, NY.

[5] Routh, E. J., 1905. *Dynamics of a System of Rigid Bodies: Elementary Part*. Dover Publications, New York, NY.

[6] Rao, J. S., 2011. *History of Rotor Dynamics*. Springer Verlag, New York, NY.

[7] MIL-STD-167-1, M. S. *Mechanical Vibrations of Shipboard Equipment*. Naval Ship Systems Command, US Department of Defense, Washington, DC.

[8] Nataraj, C., and Pietrusko, R. G., 2005. "Dynamic response of rigid rotors supported on rolling element bearings with an outer raceway defect". In Proceedings of the ASME International Design Engineering Technical Conferences, September 24–28.

[9] Greenwood, D. T., 1988. *Principles of Dynamics*, second ed. Prentice Hall, Englewood Cliffs, NJ.

[10] Strogartz, S. H., 1994. *Nonlinear Dynamics and Chaos*. Westview Press, Boulder, CO.

[11] Nataraj, C., 2010. *Harris' Shock and Vibration Handbook*. McGraw Hill, New York, NY, ch. 4, pp. 4.1–4.45.

[12] Birkhoff, G., and Rota, G.-C., 1978. *Ordinary Differential Equations*. John Wiley & Sons, New York, NY.

[13] Heath, T. L., ed., 1953. *The Works of Archimedes with the Method of Archimedes*. Dover Publications, New York, NY.

[14] Mechanics Magazine, 1824. *Cover*, Vol. 2. Knight & Lacey, London.

[15] Wu, S.-T., 2009. "Active pendulum vibration absorbers with a spinning support". *Journal of Sound and Vibration,* **323**, pp. 1–16.

[16] Timoshenko, S., Young, D. H., and W. Weaver, J., 1974. *Vibration Problems in Engineering*, fourth ed. John Wiley & Sons, New York, NY.

[17] Mahdavi, M. H., Farshidianfar, A., Tahani, M., Mahdavi, S., and Dalir, H., 2007. "A more comprehensive modeling of atomic force microscope cantilever". *Ultramicroscopy*, **109**, pp. 54–60.

[18] Coermann, R. R., 1962. "The mechanical impedance of the human body in sitting and standing positions at low frequencies". *Human Factors*, **4**, pp. 227–253.

[19] Kappaganthu, K., and Nataraj, C., 2011. "Optimal biped design using a moving torso: Theory and experiments". In *Biped Robots*, A. C. P. Filho, ed. InTech Publishers, Rijeka, Croatia, ch. 3, pp. 35–54.

[20] Wagner, L. F., and Griffin, J. H., 1993. "A continuous analog model for grouped-blade vibration". *Journal of Sound and Vibration*, **165**(3), pp. 421–438.

[21] Shah, B. M., Pillet, D., Bai, X.-M., Keer, L. M., Wang, Q. J., and Snurr, R. Q., 2009. "Construction and characterization of a particle based thrust damping system". *Journal of Sound and Vibration*, **326**, pp. 489–502.

[22] Cheng, S., Darivandi, N., and Ghrib, F., 2010. "The design of an optimal viscous damper for a bridge stay cable using energy-based approach". *Journal of Sound and Vibration*, **329**, pp. 4689–4704.

[23] Wu, J. Z., Dong, R. G., Welcome, D. E., and Xu, X. S., 2010. "A method for analyzing vibration power absorption density in human finger tip". *Journal of Sound and Vibration*, **329**, pp. 5600–5614.

[24] Lazan, B. J., 1968. *Damping of Materials and Members in Structural Mechanics*. Pergamon Press, Oxford.

[25] Nashif, A. D., Jones, D. I. G., and Henderson, J. P., 1985. *Vibration Damping*. John Wiley & Sons, New York, NY.

[26] Den Hartog, J. P., 1931. "Forced vibrations with combined Coulomb and viscous damping". *Transactions of ASME*, **53**, pp. APM–107.

[27] Hartog, J. P. D., 1940. *Mechanical Vibrations*. McGraw Hill, New York.

[28] Obert, E. F., 1973. *Internal Combusion Engines and Air Pollution*. Harper & Row, New York, NY.

[29] Wei, L., and Griffin, J., 1998. "The prediction of seat transmissibility from measures of seat impedance". *Journal of Sound and Vibration*, **214**(1), pp. 121–137.

[30] Noble, B., and Daniel, J. W., 1988. *Applied Linear Algebra*, third ed. Prentice-Hall, Englewood Cliffs, NJ.

[31] Wikipedia, 2011. Tower of Hanoi.

[32] Thomson, W. T., 1981. *Theory of Vibration with Applications*, second ed. Prentice-Hall, Inc., Englewood Cliffs, NJ.

[33] Allen, G., 1978. "A critical look at biomechanical modeling in relation to specifications of human tolerance of vibration and shock". In AGARD Conference Proceedings, no. 253, Paper A25-5, pp. 6–10.

[34] Suggs, C. W., Abrams, C. F., and Stikeleather, L. F., 1969. "Application of a damped spring-mass human vibration simulator in vibration testing of vehicle seats". *Ergonomics*, **12**, pp. 79–90.

[35] Khazanov, Y., 2007. "Dynamic vibration absorbers – application with variable speed machines". *Pumps and Systems*, August, pp. 114–119.

[36] Brennan, M. J., Day, M. J., and Randall, R. J., 1998. "An experimental investigation into the semi-active and active control of longitudinal vibrations in a large tie-rod structure". *Journal of Vibrations and Acoustics*, **120**, Issue 1, pp. 1–12.

[37] Hochlenert, D., Spelsberg-Korspeter, G., and Hagedorn, P., 2010. "A note on safety-relevant vibrations induced by brake squeal". *Journal of Sound and Vibration*, **329**, pp. 3867–3872.

[38] Zhao, Y., Weichao Sun, and Gao, H., 2010. "Robust control synthesis for seat suspension systems with actuator saturation and time varying input delay". *Journal of Sound and Vibration*, **329**, pp. 4335–4353.

[39] Taylor, C. M., Turner, S., and Sims, N. D., 2010. "Chatter, process damping, and chip segmentation in turning: A signal processing approach". *Journal of Sound and Vibration*, **329**, pp. 4922–4935.

[40] Prabakar, R. S., Sujatha, C., and Narayanan, S., 2009. "Optimal semi-active preview control response of a half car vehicle model with magnetorheological damper". *Journal of Sound and Vibration*, **326**, pp. 400–420.

[41] Dong, B., Lin, M. M., and T. Chu, M., 2009. "Parameter reconstruction of vibration systems from partial eigeninformation". *Journal of Sound and Vibration*, **327**, pp. 391–401.

[42] Guasch, O., 2009. "Direct transfer functions and path blocking in a discrete mechanical system". *Journal of Sound and Vibration*, **321**, pp. 854–874.

[43] Galvin, P., Romero, A., and Dominguez, J., 2010. "Fully three-dimensional analysis of high speed train track soil structure dynamic interaction". *Journal of Sound and Vibration*, **329**, pp. 5147–5163.

[44] Liang, C.-C., and Feng Chiang, C., 2006. "A study on biodynamic models of seated human subjects exposed to vertical vibration". *International Journal of Industrial Ergonomics*, **36**(10), pp. 869–890.

[45] Wan, Y., and Schimmels, J. M., 1995. "A simple model that catures the essential dynamics of a seated human exposed to whole body vibration". *Advances in Bioengineering*, **31**, pp. 333–334.

[46] Boileau, P. E., and Rakheja, S., 1998. "Whole-body vertical biodynamic response characteristics of the seated vehicle driver: Measurement and model development". *International Journal of Industrial Ergonomics*, **22**, pp. 449–472.

[47] Qassem, W., Othman, M. O., and Abdul-Majeed, S., 1994. "The effects of vertical and horizontal vibrations on the human body". *Medical Engineering Physics*, **16**, pp. 151–161.

[48] Qassem, W., and Othman, M. O., 1996. "Vibration effects on sitting pregnant women—subjects of various masses". *Journal of Biomechanics*, **29**(4), pp. 493–501.

[49] Wikimedia, 2006, http://commons.wikimedia.org/wiki/File:Stockbridge_Damper.jpg.

[50] Wikimedia, 2008, http://commons.wikimedia.org/wiki/File:Stockbridge_damper-closeup.jpg.

[51] Rens, J., Clark, R. E., and Howe, D., 2001. "Vibration analysis and control of reciprocating air-compressors". *International Journal of Applied Electromagnetics and Mechanics*, **15**, pp. 155–162.

[52] Rivaz, H., and Rohling, R., 2007. "An active dynamic vibration absorber for a hand-held vibro-elastography probe". *Trans. ASME, Journal of Vibration and Acoustics*, **129**, February, pp. 101–112.

[53] Chatterjee, S., 2010. "Optimal active absorber with internal state feedback for controlling resonant and transient vibration". *Journal of Sound and Vibration,* **329**, pp. 5397–5414.

[54] Yoon, J.-Y., and Singh, R., 2010. "Indirect measurement of dynamic force transmitted by a nonlinear hydraulic mount under sinusoidal excitation with focus on super-harmonics". *Journal of Sound and Vibration,* **329**, pp. 5249–5272.

[55] Megahed, S. M., and El-Razik, A. K. A., 2010. "Vibration control of two degrees of freedom system using variable inertia vibration absorbers: Modeling and simulation". *Journal of Sound and Vibration,* **329**, pp. 4841–4865.

[56] Park, J., Wang, S., and Malcolm J. Crocker, 2009. "Mass loaded resonance of a single unit impact damper caused by impacts and the resulting kinetic energy influx". *Journal of Sound and Vibration,* **323**, pp. 877–895.

[57] Cheng, C. C., Kuo, C. P., Wang, F. C., and Cheng, W. N., 2009. "Moving follower rest design using vibration absorbers for ball screw grinding". *Journal of Sound and Vibration,* **326**, pp. 123–136.

[58] Lee, J. S., Kwon, S.-D., Kim, M.-Y., and Yeo, I. H., 2009. "A parametric study on the dynamics of urban transit maglev vehicle running on flexible guideway bridges". *Journal of Sound and Vibration,* **328**, pp. 301–317.

[59] Ding, L., Hao, H., and Zhu, X., 2009. "Evaluation of dynamic vehicle axle loads on bridges with different surface conditions". *Journal of Sound and Vibration,* **323**, pp. 826–848.

[60] Rao, L. V. V. G., and Narayanan, S., 2009. "Sky-hook control of nonlinear quarter car model traversing rough road matching performance of LQR control". *Journal of Sound and Vibration,* **323**, pp. 515–529.

[61] Ok, S.-Y., Song, J., and Park, K.-S., 2009. "Development of optimal design formula for bi-tuned mass dampers using multi-objective optimization". *Journal of Sound and Vibration,* **322**, pp. 60–77.

[62] Raman, C. V., 1909. "The small motion at the nodes of a vibrating string". *Nature,* **82**, November 4, p. 9.

[63] Meirovitch, L., 1980. *Computational Methods in Structural Dynamics.* Springer Verlag, New York, NY.

[64] Abramowitz, M., and Stegun, I. A., eds., 1972. *Handbook of Mathematical Functions.* Dover Publications, New York, NY.

[65] Warwick, G., 2008. NASA tackles vibration risk in Ares I launcher, April 4, http://www.flightglobal.com/articles/2008/04/04/222778/nasa-tackles-vibration-risk-in-ares-i-launcher.html.

[66] Andrén, L., Håkansson, L., Brandt, A., and Claesson, I., 2004. "Identification of dynamic properties of boring bar vibrations in a continuous boring operation". *Mechanical Systems and Signal Processing,* **18**, pp. 869–901.

[67] Piersol, A. G., and Paez, T. L., eds., 2010. *Harris' Shock and Vibration Handbook,* sixth ed. McGraw-Hill, New York, NY.

[68] Broch, J. T., 1984. *Mechanical Vibration and Shock Measurements.* Brüel & Kjaer, Denmark.

[69] Smith, D., 1974. A case study and analysis of the Tacoma narrows bridge failure. Tech. Rep. 99.497 Engineering Project, Carleton University, Ottawa, Canada, March 29.

[70] Au, F. T. K., Wang, J. J., and Cheung, Y. K., 2001. "Impact study of cable-stayed bridge under railway traffic using various models". *Journal of Sound and Vibration,* **240**(3), pp. 447–465.

[71] Berlioz, A., Hagopian, J. D., Dufour, R., and Draoui, E., 1996. "Dynamic behavior of a drill-string: Experimental investigation of lateral instabilities". *Journal of Vibration and Acoustics*, **118**(3), pp. 292–298.

[72] Chatterjee, S., and Mahata, P., 2009. "Controlling friction induced instability by recursive time delayed acceleration feedback". *Journal of Sound and Vibration*, **328**, pp. 9–28.

[73] Chatterjee, S., and Mahata, P., 2009. "Time delayed absorber for controlling friction driven vibration". *Journal of Sound and Vibration*, **322**, pp. 39–59.

[74] Chen, C. H., Wang, K. W., and Shin, Y. C., 1994. "An integrated approach toward the dynamic analysis of high-speed spindles: Part I—system model". *Journal of Vibration and Acoustics*, **116**(4), pp. 506–513.

[75] Cheng, C. C., and Shiu, J. S., 2001. "Transient vibration analysis of a high-speed feed drive system". *Journal of Sound and Vibration*, **239**(3), pp. 489–504.

[76] Chonan, S., Jiang, Z. W., and Shyu, Y. J., 1992. "Stability analysis of a 2" floppy disk drive system and the optimum design of the disk stabilizer". *Journal of Vibration and Acoustics*, **114**(2), pp. 283–286.

[77] Colding-Jorgensen, J., 1993. "Rotor whirl measurements on a long rotating cylinder partially filled with liquid". *Journal of Vibration and Acoustics*, **115**(2), pp. 141–144.

[78] Ehrich, F. F., 1992. "Spontaneous sidebanding in high speed rotordynamics". *Journal of Vibration and Acoustics*, **114**(4), pp. 498–505.

[79] Ehrich, F., 1993. "Rotor whirl forces induced by the tip clearance effect in axial flow compressors". *Journal of Vibration and Acoustics*, **115**(4), pp. 509–515.

[80] Ehrich, F., 1995. "Nonlinear phenomena in dynamic response of rotors in anisotropic mounting systems". *Journal of Vibration and Acoustics*, **117**(B), pp. 154–161.

[81] Ferri, A. A., 1995. "Friction damping and isolation systems". *Journal of Vibration and Acoustics*, **117**(B), pp. 196–206.

[82] Germay, C., Denol, V., and Detournay, E., 2009. "Multiple mode analysis of the self-excited vibrations of rotary drilling systems". *Journal of Sound and Vibration*, **325**(1-2), pp. 362–381.

[83] Hu, H. Y., and Jin, D. P., 2001. "Non-linear dynamics of a suspended travelling cable subject to transverse fluid excitation". *Journal of Sound and Vibration*, **239**(3), pp. 515–529.

[84] Jen, M. U., and Magrab, E. B., 1993. "Natural frequencies and mode shapes of beams carrying a two degree-of-freedom spring-mass system". *Journal of Vibration and Acoustics*, **115**(2), pp. 202–209.

[85] Ju, S.-H., 2009. "Finite element investigation of traffic induced vibrations". *Journal of Sound and Vibration*, **321**, pp. 837–853.

[86] Kakizaki, T., Mizukami, M., and Kogure, K., 1991. "Dynamic analysis and control of a double-carriage rotary actuator for accurate magnetic disk head positioning". *Journal of Vibration and Acoustics*, **113**(4), pp. 434–440.

[87] Kappaganthu, K., and Nataraj, C., 2009. "A biped with a moving torso". *International Journal of Humanoid Robotics*, December.

[88] Karnopp, D., 1995. "Active and semi-active vibration isolation". *Journal of Vibration and Acoustics*, **117**(B), pp. 177–185.

[89] Kato, M., Kurohashi, M., and Aoshima, M., 1993. "Dynamic behavior of valves with pneumatic chambers for reciprocating compressors". *Journal of Vibration and Acoustics*, **115**(4), pp. 371–376.

[90] Kim, S., and Singh, R., 2001. "Vibration transmission through an isolator modelled by cotinuous system theory". *Journal of Sound and Vibration,* **248**(5), pp. 925–953.

[91] Lee, B.-H., and Lee, C.-W., 2009. "Model based feed forward control of electromagnetic type active control engine-mount system". *Journal of Sound and Vibration,* **323**, pp. 574–593.

[92] Lieh, J., 1993. "Semiactive damping control of vibrations in automobiles". *Journal of Vibration and Acoustics,* **115**(3), pp. 340–343.

[93] Lin, J., Lin, C. C., and Lo, H. S., 2009. "Pseudo inverse Jacobian control with grey relational analysis for robot manipulators mounted on oscillatory bases". *Journal of Sound and Vibration,* **326**, pp. 421–437.

[94] Matsumoto, Y., and Griffin, M. J., 2000. "Comparison of biodynamic responses in standing and seated human bodies". *Journal of Sound and Vibration,* **238**(4), pp. 691–704.

[95] Mitchiner, R. G., and Leonard, R. G., 1991. "Centrifugal pendulum vibration absorbers— theory and practice". *Journal of Vibration and Acoustics,* **113**(4), pp. 503–507.

[96] Moustafa, K. A. F., and El-Gebeily, M., 1991. "Free vibration characteristics of structures with nonlinear underlying soil". *Journal of Vibration and Acoustics,* **113**(1), pp. 95–99.

[97] Pinkaew, T., and Fujino, Y., 2001. "Effectiveness of semi-active tuned mass dampers under harmonic excitation". *Journal of Engineering Structures,* **23**, pp. 850–856.

[98] Shin, K., 2004. "Analysis of friction induced disc brake noise using simple mathematical models". *Noise and Vibration,* **35**, pp. 22–26.

[99] Sinha, J. K., Singh, S., and Rao, A. R., 2001. "Finite element simulation of dynamic behaviour of open-ended cantilever pipe conveying fluid". *Journal of Sound and Vibration,* **240**(1), pp. 189–194.

[100] Smith, R., 2001. "Changing the effective mass to control resonance problems". *Sound and Vibration,* May, pp. 14–17.

[101] Stolte, J., and Benson, R. C., 1992. "Dynamic deflection of paper emerging from a channel". *Journal of Vibration and Acoustics,* **114**(2), pp. 187–193.

[102] Sun, J. Q., Jolly, M. R., and Norris, M. A., 1995. "Passive, adaptive and active tuned vibration absorbers—a survey". *Journal of Vibration and Acoustics,* **117**(B), pp. 234–242.

[103] Venkatasubramanian, S. H., and White, W. N., 1993. "The influence of axial-torsional coupling on the natural frequencies of an aerial cable". *Journal of Vibration and Acoustics,* **115**(3), pp. 271–276.

[104] Wang, W. R., and Chang, C. N., 1994. "Dynamic analysis and design of a machine tool spindle-bearing system". *Journal of Vibration and Acoustics,* **116**(3), pp. 280–285.

[105] Yoshimura, T., Kume, A., Kurimoto, M., and Hino, J., 2001. "Construction of an active suspension system of a quarter car model using the concept of sliding mode control". *Journal of Sound and Vibration,* **239**(2), pp. 187–199.

[106] Yue, Y., Xie, J. H., and Xu, H. D., 2009. "Symmetry of the Poincaré map and its influence on bifurcations in a vibro impact system". *Journal of Sound and Vibration,* **323**, pp. 292–312.

[107] Zuo, L., and Nayfeh, S. A., 2006. "The two-degree-of-freedom tuned-mass damper for suppression of single-mode vibration under random and harmonic excitation". *Journal of Vibration and Acoustics,* **128**, pp. 56–65.

[108] Bisplinghoff, R. L., and Ashley, H., 1975. *Principles of Aeroelasticity.* Dover Publications, New York, NY.

[109] Crede, C. E., 1965. *Shock and Vibration Concepts in Engineering Design*. Prentice-Hall, Englewood Cliffs, NJ.

[110] Mansfield, N. J., 2005. *Human Response to Vibration*. CRC Press, Boca Raton, FL.

[111] Meirovitch, L., 1967. *Analytical Methods in Vibrations*. Prentice Hall, Englewood Cliffs, NJ.

[112] Ruzicka, J. E., ed., 1959. *Structural Damping*. ASME, New York, NY.

[113] Khazanov, Y., 2007. "Dynamic vibration absorbers—application with variable speed machines". *Pumps & Systems*, pp. 114–119.

[114] Lazan, B. J., 1959. "Energy dissipation mechanisms in structures, with particular reference to material damping". In *Structural Damping*, J. E. Ruzicka, ed. ASME, New York, NY, pp. 1–34.

[115] Ting-Kong, C., 1999. "Design of an adaptive dynamic vibration absorber". Master's thesis, The University of Adelaide, South Australia, April.

[116] Antonides, G. P., 1966. Longitudinal vibration of propulsion system on USS Simon Lake (AS-33). Tech. Rep., David Taylor Model Basin, Washington DC Acoustics and Vibration Lab, January.

Index

Originality is the fine art of remembering what you hear but forgetting where you heard it.

—Laurence J. Peter

9 781408 072653